T0259236

Mathematical Concepts and Methods in Modern Biology

Using Modern Discrete Models

ELSEVIER *science &* *technology books*

 Companion Web Site:

http://booksite.elsevier.com/9780124157804

Mathematical Concepts and Methods in Modern Biology: Using Modern Discrete Models
Raina Robeva and Terrell Hodge, *Editors*

Resources for Professors:

- All figures from the book available as both Power Point slides and .jpeg files
- Links to web sites carefully chosen to supplement the content of the textbook
- The web site contains computer code files, data, and additional material and projects
  in the form of appendices mentioned in each chapter as well as solutions to all chapter
  and project exercises. Please access the web site for this extra material
- Contact the editors with questions and/or suggestions

TOOLS FOR ALL YOUR TEACHING NEEDS
textbooks.elsevier.com

ACADEMIC PRESS

Mathematical Concepts and Methods in Modern Biology

Using Modern Discrete Models

Edited by

Raina Robeva

Sweet Briar College, Sweet Briar, VA, USA

Terrell L. Hodge

Western Michigan University,
Kalamazoo, MI, USA

AMSTERDAM • BOSTON • HEIDELBERG • LONDON
NEW YORK • OXFORD • PARIS • SAN DIEGO
SAN FRANCISCO • SINGAPORE • SYDNEY • TOKYO

Academic Press is an Imprint of Elsevier

Academic Press is an imprint of Elsevier
32 Jamestown Road, London NW1 7BY, UK
225 Wyman Street, Waltham, MA 02451, USA
525 B Street, Suite 1800, San Diego, CA 92101-4495, USA

Notice
No responsibility is assumed by the publisher for any injury and/or damage to persons or property as a matter of products liability, negligence or otherwise, or from any use or operation of any methods, products, instructions or ideas contained in the material herein. Because of rapid advances in the medical sciences, in particular, independent verification of diagnoses and drug dosages should be made.

British Library Cataloguing-in-Publication Data
A catalogue record for this book is available from the British Library

Library of Congress Cataloging-in-Publication Data
A catalog record for this book is available from the Library of Congress

ISBN: 978-0-12-415780-4

For information on all Academic Press publications
visit our website at www.store.elsevier.com

Contents

CHAPTER 3   **Inferring the Topology of Gene Regulatory Networks: An Algebraic Approach to Reverse Engineering 75**

Brandilyn Stigler and Elena Dimitrova

CHAPTER 4   **Global Dynamics Emerging from Local Interactions: Agent-Based Modeling for the Life Sciences 105**

David Gammack, Elsa Schaefer and Holly Gaff

Contributors

Sungwoo Ahn
Department of Mathematical Sciences, Indiana University-Purdue University Indianapolis, IN, USA

Robin Davies
Department of Biology, Sweet Briar College, Sweet Briar, VA, USA

Elena Dimitrova
Department of Mathematical Sciences, Clemson University, Clemson, SC, USA

Holly Gaff
Department of Biological Sciences, Old Dominion University, Norfolk, VA, USA

David Gammack
Department of Mathematics, Marymount University, Arlington, VA, USA

Aaron Garrett
Department of Mathematical, Computing, and Information Sciences, Jacksonville State University, Jacksonville, AL, USA

David Haws
IBM T. J. Watson Research Center, Yorktown Heights, NY, USA

Franziska Hinkelmann
Mathematical Biosciences Institute, The Ohio State University, Columbus, OH, USA

Terrell L. Hodge
Department of Mathematics, Western Michigan University, Kalamazoo, MI, USA

Winfried Just
Department of Mathematics, Ohio University, Athens, OH, USA

Bessie Kirkwood
Department of Mathematical Sciences, Sweet Briar College, Sweet Briar, VA, USA

James Kirkwood
Department of Mathematical Sciences, Sweet Briar College, Sweet Briar, VA, USA

Reinhard Laubenbacher,
Virginia Bioinformatics Institute, Virginia Tech, Blacksburg, VA, USA

Matt Oremland
Virginia Bioinformatics Institute, Virginia Tech, Blacksburg, VA, USA

Raina Robeva
Department of Mathematical Sciences, Sweet Briar College, Sweet Briar, VA, USA

Elsa Schaefer
Department of Mathematics, Marymount University, Arlington, VA, USA

Janet Steven
Department of Biology, Sweet Briar College, Sweet Briar, VA, USA

Brandilyn Stigler
Department of Mathematics, Southern Methodist University, Dallas, TX, USA

David Terman
Department of Mathematics, Ohio State University, Columbus, OH, USA

Necmettin Yildirim
Division of Natural Sciences, New College of Florida, Sarasota, FL, USA

Ruriko Yoshida
Department of Statistics, University of Kentucky, Lexington, KY, USA

Preface

In its report, *A New Biology for the 21st Century,*[1] the National Research Council defines the essence of the New Biology as "...re-integration of the many sub-disciplines of biology, and the integration into biology of physicists, chemists, computer scientists, engineers, and mathematicians to create a research community with the capacity to tackle a broad range of scientific and societal problems." The report stipulates that "...the emergence of the New Biology signals the need for changes in how scientists are educated and trained" and calls for substantive changes in interdisciplinary education at the junction of mathematics and biology at both the undergraduate and graduate levels. This report echoes many of the recommendations of an earlier influential report, *Bio 2010,*[2] that "...each institution of higher education reexamine its current curricula..." and concludes that "...College and university administrators, as well as funding agencies, should support mathematics and science faculty in the development or adaptation of techniques that improve interdisciplinary education for biologists."

Due to the high profiles of these reports, it is now widely accepted that a main push in biology during the coming decades will be toward an increasingly quantitative understanding of biological functions, and that the new generation of biologists will routinely use mathematical models and computational approaches to frame hypotheses, design experiments, and analyze results. A 2010 Society for Industrial and Applied Mathematics (SIAM) white paper, *Mathematics: An Enabling Technology for the New Biology*, further underscores the critical role that mathematicians and statisticians are asked to play toward accomplishing the New Biology's aims. This white paper also recommends increased federal support to ensure a pipeline of such adequately trained professionals, starting at the undergraduate level.

It is thus critically important that the training of the new biologists and their collaborators, whether coming through biology or through other areas of the natural and mathematical sciences, facilitates access to a rich toolbox of diverse mathematical approaches. New educational guidelines and recommendations linked with the reports above and with others,[3] have catalyzed various educational discussions and curricular changes. In particular, in the past few years the number of undergraduate and graduate programs in mathematical and computational biology has increased, institutions have added new courses in mathematical biology linked with ongoing research in biology, and the American Mathematical Society, the Mathematical Association of America (MAA), The National Science Foundation (NSF), the National Institutes of Health, and the NSF Mathematical Sciences Institutes are funding faculty

---

[1] A New Biology for the 21st Century (2009). The National Academies Press, Washington, DC.
[2] BIO2010: Transforming Undergraduate Education for Future Research Biologists (2003). The National Academies Press, Washington, DC.
[3] See, e.g., Vision and Change in Undergraduate Biology Education: A Call to Action (2009). AAAS, Washington, DC.

development workshops, research-related experiences, and specialized research conferences in mathematical biology for undergraduates.

However, while traditional mathematical biology topics using difference equations, differential equations, and continuous dynamical systems have to a large extent worked their way into the classroom and have become standard curriculum, mathematical techniques from modern discrete mathematics (encompassing traditional discrete mathematics with combinatorics and graph theory, as well as linear algebra, algebraic geometry, and modern abstract algebra) have remained relatively invisible in these curricular changes. The 2010 SIAM white paper cited above calls for increased support in a number of mathematical subfields with strong ties to modern discrete mathematics, as there is mounting evidence that novel algebraic methods are being used with great success in current mathematical biology research. These include Boolean networks, finite/polynomial dynamical systems (including many agent-based models), elements of graph theory, Petri nets, and Gröbner (Groebner) bases and other elements from algebraic geometry and modern algebra. In spite of their accessibility to undergraduates, these topics are almost entirely absent from the undergraduate mathematical biology training landscape. Thus, while novel applications of theories from modern discrete mathematics are finding increasing use in the rapidly evolving field of mathematical biology, the already existing gap between research and education is growing wider, particularly in the area of undergraduate education. While students interested in mathematical biology have relatively easy access to courses that utilize analytic methods, and generally have an adequate exposure to such methods before deciding upon a graduate program, students interested in learning about modern discrete mathematical approaches to mathematical biology topics have fewer doors visibly open to them, and indeed may not even know such approaches exist. Faculty who want to teach courses utilizing differential equations models now have ready access to a fair number of texts and textbook resources (including the textbook *An Invitation to Biomathematics* by Robeva *et al.* published by Elsevier in 2008) focusing primarily on the use of analytic mathematical methods in biology. In contrast, materials applying modern discrete mathematical methods in biology are generally widely scattered, and, outside of a select set of topics,[4] there are practically no educational resources reflecting the importance of algebraic methods in many of the fast-growing areas of mathematical biology. In the cases when sources for the latter are available (the 2005 text *Algebraic Statistics for Computational Biology* by Pachter and Sturmfels, published by Cambridge University Press, is an important example), the level of presentation is not necessarily aimed at the true beginner and may be more appropriate for graduate level training.

---

[4]E.g., some aspects of phylogenetics as they appear in portions of *Mathematical Biology* by Allman and Rhodes or more specialized texts like Semple and Steel's book *Phylogenetics* (Oxford University Press, 2003), Felsenstein's *Inferring Phylogenies* (Sinauer Associates, 2003), as well as combinatorial mathematics in Waterman's *Introduction to Computational Biology* (Chapman and Hall/CRC, 1995; second edition coming out in 2012).

We hope that our volume *Mathematical Concepts and Methods in Modern Biology: Using Modern Discrete Methods* will bring undergraduate students (and faculty interested in teaching them) face-to-face with more applications of modern discrete mathematics to biology. In its choice of topics and style of approach, this volume is not intended to be a comprehensive treatment of all current uses of modern discrete methods in biology, but to provide passageways to a diverse and expansive landscape. Consequently, the collection of chapters comprising the volume are designed to be largely independent from one another and can be viewed as "modules" for classroom use, as independent studies, as starting points for undergraduate research projects, or even as gentle entryways for more mathematically oriented readers. Each chapter begins with a question from modern biology, followed by the description of certain mathematical methods and theory appropriate in the search for answers. As such, the chapters can be viewed as fast-track pathways through the problem that begin by laying out the biological foundation, proceed by covering the relevant mathematical theory, and end by highlighting connections with ongoing research and current publication.

Multiple exercises and projects are embedded within the chapters, giving instructors the flexibility to cover material only up to a certain point and ignore later sections that may require higher mathematical sophistication. Embedding the exercises ensures that only material which has already been covered is needed for their execution. Many of the projects and exercises utilize specialized software, exemplifying the notion that familiarity and experience with computing applications which implement the mathematical theory are critical elements of the "modern biology" skills set. We have been particularly mindful of designing the exercises in a way that requires only the use of freely-available applications or mainstream proprietary software that is commonly available on college and university campuses (e.g., MATLAB).

Even though the chapters are to a large extent independent and self-contained, they are grouped, wherever appropriate, by common biological or mathematical threads. They are not organized by level of mathematical difficulty. A chapter appearing later in the volume should not be assumed to require a higher level of mathematical prerequisites. However, when the chapters consider similar biological questions or make use of the same mathematical theory, earlier chapters will usually contain more introductory details. In this sense, it would be beneficial to cover Chapters 1–3 in this order, as Chapters 2 and 3 expand upon the mathematical foundation presented in Chapter 1. We recommend the same for the following clusters: Chapters 4 and 5; Chapters 7 and 8, and (perhaps to a lesser degree) Chapters 9 and 10. Chapter 6 is self-contained. The highest level of mathematical proficiency reached in each chapter may vary significantly from topic to topic. The list below presents a brief summary of the chapters' topics, highlights the assumed mathematical background for each chapter, and provides information regarding possible course adoptions and use of specialized software.

Chapter 1. Mechanisms of Gene Regulation: Boolean Network Models of the Lactose Operon in *Escherichia coli,* by Raina Robeva, Bessie Kirkwood, and Robin Davies.

The transcription of genes (mRNA synthesis) and translation of mRNA (protein synthesis) are energetically expensive processes and cells have the ability to make certain proteins only when the environmental conditions warrant. Otherwise, if a cell had to make all of its proteins all of the time, it would be expending a lot of cellular energy in the making of proteins for which it has no use. Understanding the relevant mechanisms of gene expression, controlled via so-called gene regulatory networks, is thus critically important to understanding the regulation of cellular behavior. The lactose (*lac*) operon is a relatively simple but important example of a gene regulatory network for the metabolism of lactose in the bacterium *E. coli*. Since its discovery in the late 1950s, the *lac* operon has served as a model system for understanding many aspects of gene regulation.

The chapter is an introduction to mathematical modeling with Boolean networks in the context of gene regulatory networks, using the *lac* operon as a main example. Students who are prepared mathematically to enroll in a discrete mathematics course can read this chapter and work through all exercises. No specific mathematical background is required, as the chapter includes a primer on Boolean arithmetic. All substantive computations beyond the initial introductory exercises are done using the web-based suite *DVD*, which is freely available. Even though no prior knowledge of modern algebra is required, students enrolled in an undergraduate modern algebra course that covers algebraic rings and ideals can use elements of the chapter to introduce and motivate the question of solving polynomial systems of equations and the connections with Groebner bases of polynomial ideals. The chapter provides an online appendix on using Groebner bases for solving systems of polynomial equations.

Chapter 2. Bistability in the Lactose Operon of *Escherichia coli*: A Comparison of Differential Equation and Boolean Network Models, by Raina Robeva and Necmettin Yildirim.

Bistability is the ability of a system to achieve two different steady states under the same external conditions. The *lac* operon of *E. coli* is a bistable system: under certain external conditions, the *lac* operon may be turned on or turned off depending on the history of the cell (determined by the environmental conditions under which it has been grown). The chapter introduces several ordinary differential equation (ODE) models of the *lac* operon and their Boolean network analogs and compares these two types of models with regard to their ability to capture the bistable behavior of the *lac* operon system. The ODE and Boolean parts of the chapter could be considered independent if the reader would be willing to accept the ODE models without justification. Some of the exercises related to the ODE models require MATLAB, while the Boolean networks are analyzed using DVD, as in Chapter 1. The first part of the chapter is appropriate as an introduction to the modeling of biochemical reactions in differential equations courses, while the second part is appropriate for courses in discrete mathematics. The entire chapter can be used in a mathematical biology course, or as a student research project to highlight connections between abstract algebra and differential equations in the context of gene regulation.

Chapter 3. Inferring the Topology of Gene Regulatory Networks: An Algebraic Approach to Reverse Engineering, by Brandilyn Stigler and Elena Dimitrova.

Key features of gene regulatory networks can be represented diagrammatically through graphs whose nodes are genes or gene-related products, and whose interactions are, at least partially, captured through certain types of edges. The topology of a gene regulatory network is the essential shape of this graph. It is a very important and difficult biological task to try, from knowing only partial information (generally observed only during snapshots in time) about the expressions of genes or gene products, to infer this topology, and hence discover the relationships (edges) among the nodes.

This chapter uses aspects of the algebra of polynomials to recreate such networks from time series data when the levels of gene or gene product expression can be captured by a finite number of states. Such systems generalize the Boolean models treated previously in Chapters 1 and 2. At their most elementary level, they can be approached through elementary multivariable polynomials by a reader familiar with modular arithmetic (an approach also taken initially in Chapter 5). The presentation in Chapter 3 generally assumes familiarity with elementary modern algebra at the level of rings and ideals and is appropriate for an undergraduate modern algebra course. Some advanced topics such as the Chinese remainder theorem for rings, the ideal-variety correspondence of algebraic geometry, primary decomposition of ideals, and Jacobson radical of an ideal make an appearance, but one need not be familiar with these more advanced concepts in order to work through the entire chapter. Some exercises do require the reader to compute the intersection of ideals in a polynomial ring, the primary decomposition of an ideal, and the Jacobson radical of an ideal, but readers with only an elementary background in modern algebra (and, just as well, those with more experience!) can perform these computations quite easily using the freely available computational algebra system *Macaulay 2*.

Chapter 4. Global Dynamics Emerging from Local Interactions: Agent-Based Modeling for the Life Sciences, by David Gammack, Elsa Schaefer and Holly Gaff.

Biological research into areas as widely varied as the population dynamics of prairie dog colonies and tick populations, bird flocking and evolutionary patterns, impact of individual behavioral choices on important societal problems, disease spreading, and blood vessel growth and leukocyte rolling, has been pursued through the use of scientific models that are agent-based. This chapter is an introduction to agent-based (also called individual-based) modeling through *Netlogo* (available for free download). It does not require any mathematical background except for some very elementary probability and provides the reader, through a rich set of hands-on exercises, with the opportunity to observe how the global behavior of a complex system of interacting "agents" arises from the local rules established for their interactions. The examples and projects presented in the chapter cover a wide range of models and topics, from basic classroom illustrations to models being used in ongoing research, including the following agent-based models that are examined and analyzed in detail: a model of axon guidance, a model for the spread of cholera, and two models describing the dynamics of tick-borne diseases. The chapter would be useful for mathematical modeling classes and in introductory programming classes.

Chapter 5. Agent-based Models and Optimal Control in Biology: A Discrete Approach, by Reinhard Laubenbacher, Franziska Hinkelmann, and Matt Oremland.

In this chapter, a wide class of agent-based models is investigated through several concrete examples and captured mathematically as polynomial dynamical systems over finite fields. This approach uses multivariable polynomials to represent the transitions between agents' states in time and polynomial functions to encode the dynamics of the entire system. It provides a broad mathematical framework for analyzing agent-based models, finding the long-term dynamic behavior of the systems, and implementing optimal control strategies.

The first seven sections of the chapter require very little mathematical background, although the first section would be best understood by a reader with some background in elementary differential equations. Section 8 can serve as a brief introduction to finite fields and to polynomial dynamical systems over finite fields (introduced in Chapter 3 as well). No modern algebra is required as a prerequisite here, though it would certainly be helpful. This section could also be used as motivation to learn more about polynomial rings and ideals over finite fields. The last section is more advanced and would be most appropriate for use in modern algebra courses or with students who have had a proof-based course in discrete mathematics and/or are engaged in student research.

Many of the chapter examples and one of the chapter exercises require the use of *Netlogo*. The web-based and freely-available application suite ADAM is used for obtaining and visualizing the characteristics of polynomial dynamical systems.

Chapter 6. Neuronal Networks: A Discrete Model, by Winfried Just, Sungwoo Ahn, and David Terman.

It is commonly believed that everything the brain does is the result of the collective electrical activity of neurons. Neurons communicate with other neurons by synaptic connections forming complex neuronal networks. Simple discrete dynamical system models of neuronal dynamics can be constructed by assuming that at any given time step each neuron can either fire or be at rest, that after it has fired each neuron needs to be at rest for a specified refractory period, and that the firing of a neuron is induced by firing of a sufficient number of other neurons with synaptic connections to it.

This chapter explores the relationship between the network connectivity and important features of the network dynamics such as the number and lengths of attractors, lengths of transients, and sizes of the basin of attraction. A variety of mathematical tools, ranging from combinatorics to probability theory, are used. The chapter also discusses some issues involved in choosing the appropriate model for a given biological system, including a result on the relation between the discrete dynamical systems models introduced in the chapter and certain more detailed ODE models. For the first four sections, students should have some experience with elementary notions of discrete mathematics such as the greatest common divisor, modular arithmetic, and the floor function at the level of writing proofs. Familiarity with graph theory would be beneficial, but is not required. Sections 5 and 6 require basic background in discrete probability. The material would be most appropriate for courses that assume proof-based discrete mathematics as a prerequisite. Some basic knowledge of ordinary differential equations is assumed in section 7. Online supplemental material containing extensions of the mathematical theory and providing a number of additional projects and exercises is also included. Use of MATLAB is suggested for some exercises and projects, and specialized MATLAB code is made available as part of the online supplement.

Chapter 7. Predicting Population Growth: Modeling with Projection Matrices, by Janet Steven and James Kirkwood.

In many models of population growth, life stages are defined based on morphological changes during growth, or changes in size. In some organisms, development leads to natural categories; seeds, seedlings, and reproductive plants, for example, or egg, larva, pupa, and adult in butterflies. In other organisms, sometimes it makes more sense to categorize individuals on the basis of age. Matrix algebra is often used to build models that incorporate the different stages an organism goes through during its life. The model can then be used to predict both the overall growth of the population and the distribution of individuals across these life stages.

The first several sections of the chapter provide an introduction to the modeling of population dynamics with projection matrices, through segmentation into various life stages. For these sections, only the very basics of matrix algebra are required (e.g., matrix notation, matrix multiplication, vectors), and concrete applications to a ginseng population are explored. Section 8 and beyond use linear algebra (eigenvalues and eigenvectors) to determine the steady-state stage distribution of a population. Familiarity with elementary linear algebra is a necessary prerequisite for these later sections. The chapter provides MATLAB and R commands for performing the necessary matrix operations, but *GNU Octave* can be used as a free alternative. Graphing calculators (e.g., the *TI-89*) may also be used to perform the calculations. Early material would be appropriate for any course introducing basic matrix theory, while the later material would be appropriate for linear algebra courses and could be used to demonstrate an important application of eigenvectors.

Chapter 8. Metabolic Pathways Analysis: A Linear Algebraic Approach, by Terrell L. Hodge.

At the cellular level, metabolic processes are biochemical reaction systems that enable a cell to extract energy and other necessities for life from nutrients, and to build new structures it needs to live and reproduce. The chains of biochemical reactions involved are called metabolic pathways, and the manipulation of them, and the complex networks into which they fit, is the domain of metabolic engineering. In this chapter, the underlying pathways and networks of metabolism are modeled mathematically through the use of matrix analysis and linear algebra associated to these systems of biochemical reaction equations. The initial material can be used to motivate the basics of matrix representations of linear equations, and the remainder fits well into a course covering the fundamentals of linear algebra, including analyzing null spaces, interpreting linear independence, bases, and more. Graphing calculators or standard mathematics software may be used to carry out calculations. A tutorial for a freely downloadable package *ExPA* appears in the supplementary materials.

Chapter 9. Identifying CpG Islands: Sliding Window and Hidden Markov Model Approaches, by Raina Robeva, Aaron Garrett, James Kirkwood, and Robin Davies.

In the strings of adenine (A), cytosine (C), guanine (G), and thymine (T) out of which DNA is formed, the dinucleotide CG appears with a probability that differs notably from what naïve randomness would predict. Regions with relatively low frequencies of the CG nucleotide contain clusters, known as "CpG islands," within which the CG content is much higher. CpG islands are often associated with the promoter regions of genes. Methylation of these promoter islands is associated with the transcriptional silence of the gene while promoter-associated CpG islands in constitutively-expressed housekeeping genes are unmethylated. Inappropriate methylation of the CpG islands in tumor suppressor promoters has been associated with the development of numerous human cancers. Thus, identifying the locations of CpG islands in DNA sequences is an important task.

In this chapter, a heuristic model for locating CpG islands using sliding windows is briefly introduced, followed by mathematical methods based on hidden Markov models. Familiarity with discrete probability (e.g., conditional probability, independence, geometric distribution) and finite Markov chains is assumed for the whole chapter, although a brief refresher on Markov chains is included. Many introductory and intuitive examples are included in order to illustrate the nature of hidden Markov models and their application as modeling tools for locating CpG islands in the genome. The natural place for the material would be in a discrete probability course, but the chapter can also be used in computer science courses since it covers decoding and training algorithms. The companion suite of freely-available web-based software applications *CpG Educate* is utilized for many of the chapter projects and exercises. The chapter includes an online project "Investigating Predicted Genes" appropriate for biology courses with no mathematics prerequisites.

Chapter 10. Phylogenetic Tree Reconstruction: Geometric Approaches, by David Haws, Terrell Hodge, and Rudy Yoshida.

Comparing the DNA sequences of individual species or groups of related species can often provide essential insights into evolutionary biology. This chapter's topic is the recovery of the evolutionary history of gene families, species, or other levels of biological organisms by means of phylogenetic trees, easily pictured as the equivalent of "family trees" but created only from DNA sequence data of the "family" members alive today, with no prior knowledge of their ancestors and their relationships. Reconstructing the evolutionary history of genes or organisms, based on molecular and genetic data, has a multiplicity of modern applications. The most obvious and historically revolutionary application is the classification of species and organisms not by their outward looks (classical taxonomy via morphology), but by their genetic similarities. Tracing the evolutionary history of genetic data has also informed our understanding of human and animal population movements across the globe over generations and millennia. In addition, phylogenetic tree reconstruction makes it possible to track, prepare for, and try to attack outbreaks of disease, such as HIV or the flu. As another important outcome, knowledge of phylogenetic trees has made it possible to reconstruct, e.g., biochemically recreate, potential ancestors of genes and to then use these ancestors to test hypotheses about their roles in the evolution of traits.

Through the study of a subset of tree reconstruction methods, the "distance-based" methods, such phylogenetic trees are represented as points in a high-dimensional real vector space, and the process of finding of a good tree that fits the real-world sequence data is treated as a geometric projection in this space. Freely accessible on-line programs are used to illustrate phylogenetic trees and implement some tree reconstruction methods. The first section can be used early in an elementary discrete mathematics or linear algebra course to introduce elementary matrix notation, basics on trees (as graphs), and high-dimensional spaces, in a biologically relevant context. Later sections explore two key distance-based tree reconstruction methods, and the relationship between them, through geometric structures in the aforementioned space, including cones and the use of linear optimization over a convex polytope whose vertices correspond to certain phylogenetic trees.

For the book as a whole, supplemental materials, including online projects, software and data files, appendices and extensions of the chapter materials, are gathered together and are available from the volume's web site (http://booksite.elsevier.com/9780124157804). Complete solutions to the chapter exercises and guidelines for the projects are also available from the website.

The materials authored or co-authored by Robeva and Hodge grew out of a set of educational modules based upon work supported by the NSF under grant DUE-0737467.[5] This material has been tested in multiple classroom settings at Sweet Briar College and at Western Michigan University. Those materials were also used in the faculty professional development PREP workshops "Mathematical Biology: Beyond Calculus" sponsored by the MAA (under NSF grant DUE-0817071) and offered in 2010 and 2011 at Sweet Briar College. We greatly appreciate these organizations' support.

We express our sincere gratitude to all of the authors who contributed their excellent work to this volume. We thank our wonderful editorial team at Elsevier and specifically our project managers, Catherine (Cassie) Mullane and Julia Haynes, and editor, Christine Minihane. We are particularly indebted to our former Elsevier editor, Patricia Osborn, who embraced this project early on, encouraged us to pursue it, and invested many hours of her time at the planning stages to ensure the publication of this collection. Robert Kipka at Western Michigan University was indispensable during the book's initiation and editing stages and we thank him warmly for his time and dedication. Finally, we thank our husbands, Boris Kovatchev and Robert McNutt, for their patience and support throughout this process.

Raina S. Robeva
Terrell L. Hodge
August 20, 2012

---

[5]Any opinions, findings, and conclusions or recommendations expressed in this material are those of the authors and do not necessarily reflect the views of the National Science Foundation.

Mechanisms of Gene Regulation: Boolean Network Models of the Lactose Operon in *Escherichia coli*

Raina Robeva*, Bessie Kirkwood* and Robin Davies†

*Department of Mathematical Sciences, Sweet Briar College, Sweet Briar, VA, USA
†Department of Biology, Sweet Briar College, Sweet Briar, VA, USA

1.1 INTRODUCTION

Understanding the mechanisms of gene expression is critically important for understanding the regulation of cellular behavior. Transcription of genes (messenger RNA (mRNA) synthesis), translation of mRNA (protein synthesis), degradation of mRNA and proteins, and protein–protein interactions are all involved in the control of gene expression where proteins may bind with DNA, with mRNA, and with other proteins, leading to complex networks of interactions. Cells have many more genes than they need to express under any given set of environmental conditions. Transcription and translation are energetically expensive processes, so cells should only express the genes required for the environmental circumstances in which they find themselves.

Certain genes, often termed *housekeeping genes*, are required to support basic life processes and are expressed continuously. The expression of many other genes, though, is contingent upon environmental or physiological factors. The expression of these *regulated genes* is controlled by the cell to ensure efficient use of its energy and materials. In this chapter we will focus on a set of regulated genes in *Escherichia coli* (*E. coli*), which are expressed only when lactose is the sole sugar available.

The most efficient point for controlling gene expression is at the level of transcription where the cell can control whether or not the gene is transcribed, at what rate it is transcribed, and under what conditions transcription occurs. Bacteria control transcription through the binding of specific proteins to their DNA. Some DNA binding proteins block transcription, while others cause the DNA to bend in a manner that facilitates the action of RNA polymerase. Still other proteins, the polymerase-associated sigma factors, confer sequence-specific binding ability on the RNA polymerase, allowing it to transcribe genes accurately. In the more complex eukaryotic cells of higher organisms, multiple proteins may bind to multiple sites in the DNA, and protein–protein interactions are also involved in the control of transcription.

DNA sequences controlling transcription may be found at considerable distances from the gene to be controlled and may be brought to the vicinity of the RNA polymerase binding site (the promoter) by the bending of the DNA. This highly complex structure presents significant experimental challenges in the process of understanding and describing cellular behavior.

Mathematics provides a formal framework for organizing the overwhelming amounts of disparate experimental data and for developing models that reflect the dependencies between the system's components. Different types of mathematical models have been developed in an attempt to capture gene regulatory mechanisms and dynamics.

Various broad classifications are used in reference to such models. *Deterministic* models generate the exact same outcomes under a given set of initial conditions while in *stochastic* models the outcomes will differ due to inherent randomness. *Dynamic* models focus on the time-evolution of a system while *static* models do not consider time as part of the modeling framework. Among the dynamic models, *time-continuous* models utilize time as a continuous variable, while in *time-discrete* models time can only assume integer values. *Space-continuous* models refer to situations where the model variables can assume a continuum of values while in *space-discrete* models those variables can only assume values from a finite set. Space-continuous models of gene regulation are often constructed in the form of differential equations (in the case of continuous time) or difference equations (in the case of discrete time) and focus on the fine kinetics of biochemical reactions. We will refer to such models as *DE models*. Discrete-time models built from functions of finite-state variables are referred to as *algebraic models*.

In a DE model, all variables assume values from within biologically feasible ranges. Modelers usually need comprehensive knowledge of the interactions between variables, which may include detailed information of recognized control mechanisms, rates of production and degradation, minimal and maximal biologically relevant concentrations, and so on. In an algebraic model only values from a finite set are allowed. The special case of a *Boolean network* allows only two states, e.g., 0 and 1, generally representing the absence or presence of gene products in a model of gene regulation. In contrast to DE models, the information necessary to construct a Boolean model requires only a conceptual understanding of the causal links of dependency. Thus, in general, DE models are quantitative while Boolean models are qualitative in nature.

Historically, DE models have been the preferred type of mathematical models used in biology. This type of dynamical modeling has proved to be essential for problems in ecology, epidemiology, physiology, and endocrinology, among many others. Boolean models were first introduced to biology in 1969 to study the dynamic properties of gene regulatory networks [1]. They are appropriate in cases where network dynamics are determined by the logic of interactions rather than finely tuned kinetics, which may often be unknown.

In this chapter we present some of the fundamentals of creating Boolean network models for one of the simplest and best understood mechanisms of gene regulation: the *lactose* (*lac*) *operon* that controls the transport and metabolism of lactose in *E. coli*. Since the seminal work by Jacob and Monod [2,3], the *lac* operon has become one of the most widely studied and best understood mechanisms of gene regulation. It has also been used as a test system for virtually every mathematical method of modeling gene regulation (see, e.g., [4–10]).

The rest of the chapter is organized as follows: In Section 1.2 we outline the basic structure of the *lac* operon. This section is meant only as a quick introduction and is not comprehensive in any way (see [11] for a more thorough introduction). Section 1.3 focuses on the construction and initial testing of a mathematical model with an emphasis on Boolean networks. A primer on Boolean algebra is included to make the chapter self-contained. We consider several Boolean models of the *lac* operon, then introduce and utilize the web-based suite of applications *Discrete Visualizer of Dynamics* [12] to perform initial testing and validation of the models. In Section 1.4 we turn to the question of determining the steady states (fixed points) of Boolean networks, casting the question in the broader context of polynomial dynamical systems and the use of Groebner bases for solving systems of polynomial equations. In Section 1.5 we point out directions for extending and generalizing the models and provide some concluding remarks regarding the possible use of this material in the undergraduate mathematics and biology curricula.

1.2   *E. COLI* AND THE *LAC* OPERON

E. coli is a short rod-shaped bacterium which is a common intestinal resident of mammals and birds. It has been the object of extensive study for decades. DNA replication, transcription, and translation were all elucidated in *E. coli* before they were studied in eukaryotic cells. Its physiology is well-understood and its entire gene sequence is known. For an overview of its importance to the study of genetics, see, for example, [13].

Since it lives in the intestines, any given *E. coli* bacterium's nutrition depends upon the diet of the animal whose digestive tract it inhabits. Digestion of the complex biomolecules in the foods consumed by the animal generally provides the bacterium with all of the simple biomolecules it needs. Digestion of starches provides the monosaccharide (simple sugar) glucose, digestion of proteins provides all of the amino acids, and, whenever milk is consumed by the host, *E. coli* is also exposed to lactose (milk sugar). Lactose is a disaccharide consisting of one glucose sugar linked to one galactose sugar. Galactose is a six carbon simple sugar which is an isomer of glucose. It has the same chemical formula as glucose but differs in the position of one hydroxyl group. Like glucose, galactose can be used as an energy source, although some additional enzymatic manipulation will be required.

In order to import sugars across their plasma membranes, cells must produce specific sugar *transport proteins* (e.g., glucose requires a glucose transporter; lactose requires a lactose transporter, and so on). Once the sugar has been imported into the cell, specific enzymes will act on the sugar, either to use it to make a required cellular molecule (in the constructive processes collectively called *anabolism*) or to break it down in order to harvest its chemical energy in the form of ATP (in the destructive processes collectively called *catabolism*). Glucose is the preferred energy source for all cells, as it is used in *glycolysis*, the first and most fundamental energy-providing pathway of both anaerobic and aerobic cellular respiration.

The interactions between sugars and their transport proteins and the enzymes involved in their utilization within the cell are very specialized and cells must have the capacity to make specific proteins for every sugar they take up and use. However, it would not make sense for the *E. coli* to make the lactose transport protein if no lactose is present in its environment. It would also be inefficient to make the enzyme necessary to break lactose into glucose and galactose if there were no lactose inside the cell. If a cell had to make all of its proteins all of the time, it would be expending a lot of cellular energy in the making of proteins for which it has no use. Instead, cells have the ability to make certain of their proteins only when the environmental conditions warrant. Such proteins are called *inducible proteins*, because their synthesis is induced as a consequence of a certain cellular condition. The proteins needed for lactose uptake and utilization are inducible proteins, as they are only made if lactose is present and glucose, the preferred energy source, is not. Their concentrations increase 1000-fold under these conditions. Inducible genes belong to the group of regulated genes, in that they are only transcribed under certain specific conditions.

In *E. coli*, when lactose is the only sugar present, the transporter protein *lactose (lac) permease* and the enzyme *β-galactosidase* are both required in order to utilize it. *Lac* permease is a transmembrane protein which binds the disaccharide and brings it across the plasma membrane into the cytoplasm of the cell. The *lac* permease mediated transport of lactose through the cellular membrane is reversible, meaning that high levels of lactose inside the cell result in reverse transport of lactose to the outside of the cell. *β*-galactosidase catalyzes the hydrolysis of lactose into glucose and galactose. It also catalyzes the rearrangement of lactose into allolactose. These two proteins, lactose permease and *β*-galactosidase are produced by a tightly coordinated mechanism described by Francois Jacob and Jacques Monod and termed the *lac operon*. In the *lac* operon, due to the arrangement of its genes *LacZ*, *LacY*, and *LacA*, a single messenger RNA encodes both *β*-galactosidase and lactose permease, as well as a third enzyme, *transacetylase*, not involved in the metabolism of lactose (see Figure 1.1). Thus, when the *lac* promoter is active, all three proteins are produced.

Adjacent to the *lac* genes *lacZ*, *lacY*, and *lacA*, is the gene *LacI* that encodes a regulatory protein, the *lac repressor* (see Figure 1.2A). When there is no lactose present in the cell's environment, the *lac* repressor protein will bind to the operator, preventing the RNA polymerase from producing the *lac* mRNA. When lactose is present in the medium outside of the cell, a small amount of it will be transported into the cell by the few lactose permease molecules found in the plasma membrane. Once the lactose is inside the cell, it is converted to allolactose by the few *β*-galactosidase molecules present. Allolactose binds the *lac* repressor and causes it to undergo a conformational change, such that it can no longer bind to the operator (see Figure 1.2B). As a result, the RNA polymerase is able to read right through, produce the mRNA, and the three proteins are produced. This causes more lactose to be brought

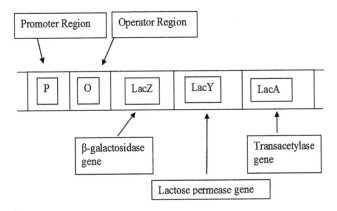

FIGURE 1.1

The structure of the *lac* operon. The RNA polymerase binds at the promoter and then proceeds along the *LacZ*, *LacY*, and *LacA* genes, copying them into a single mRNA which encodes the three proteins β-galactosidase, lactose permease, and transacetylase. The operator is a controlling region. See the text for more details.

into the cell by lactose permease and then broken down by β-galactosidase. We say that the operon is on. When all of the lactose is used up, there will be no allolactose, so the *lac* repressor protein will bind the operator and the synthesis of mRNA will stop. Levels of lactose permease and β-galactosidase will fall. The operon has turned itself off.

If both glucose and lactose are available to *E. coli*, glucose, the preferred energy source, is always used up before the bacterium begins to utilize the lactose. This is controlled by the mechanism of *catabolite repression*. Without it, the presence of lactose would cause allolactose to be made, preventing the *lac* repressor protein from binding the operator and allowing the transcription of the *lac* operon. In reality, the availability of glucose represses this process, and when both glucose and lactose are available, RNA polymerase is not able to initiate transcription sufficiently to ensure the levels of production of mRNA, β-galactosidase, and lactose permease reached under the condition of exclusive lactose availability. This failure of transcription initiation is called catabolite repression.

The mechanism of catabolite repression requires a DNA-binding protein, called CAP (for *catabolite activator protein*), which is the product of the *crp* (for cAMP receptor protein) gene. The CAP protein is capable of binding DNA only in the presence of cAMP. This small molecule is a common cellular signal or messenger and is produced by the enzyme adenylate cyclase. When glucose is present, adenylate cyclase does not produce cAMP. When there is no glucose, cAMP is produced. When cAMP binds CAP, the CAP-cAMP complex attaches to the DNA at the *lac* promoter and facilitates the attachment of RNA polymerase at the *lac* promoter. If there is no cAMP present, the CAP protein can't bind to the DNA, the RNA polymerase

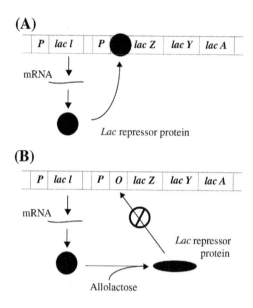

FIGURE 1.2

Panel A. The *lac* repressor protein in action. The *lac* repressor protein binds the *lac* operon at the operator, preventing transcription of the *lac* operon messenger RNA. The operon is Off. **Panel B.** Binding of allolactose to the *lac* repressor causes a conformational change in the repressor, preventing it from binding at the operator. Transcription of the *lac* operon messenger RNA can proceed. The Operon is On. Figure reproduced from *CBE-Life Sciences Education*, Vol. 9, Fall 2010, p. 233.

won't bind at the *lac* promoter, and the *lac* mRNA is not made. No mRNA means no β-galactosidase or lactose permease, and the operon will be off.[1] The key to the entire system is cAMP. If glucose is present, *E. coli* uses the glucose and the catabolism of glucose keeps cAMP levels low. After the glucose is used up, cAMP levels rise. The CAP protein binds cAMP, and the CAP-cAMP complex binds the DNA and facilitates the attachment of RNA polymerase. Figure 1.3 presents a schematic of the whole lac operon regulatory mechanism.

1.3 BOOLEAN NETWORK MODELS OF THE *LAC* OPERON

In this section we design some basic dynamical models of the *lac* operon mechanism, capable of reflecting its biological behavior. At the very minimum, a model should

---

[1] To be more exact, when both lactose and glucose are present, some small amounts of mRNA, β-galactosidase, and lactose permease will always be made (since the repressor protein will be blocked from binding to the operator site). However, those levels will be thousands of times lower than they would be in the absence of glucose.

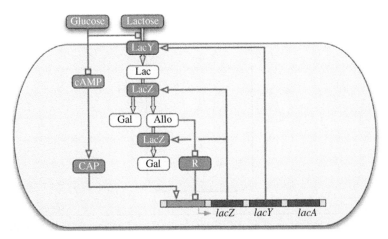

FIGURE 1.3

Schematic of the *lac* operon regulatory mechanism (from Santillan at al. [14]) Protein LacY is a permease that transports external lactose into the cell. Protein LacZ polymerizes into a homotetramer named β-galactosidase. This enzyme transforms internal lactose (Lac) to allolactose (Allo) or to glucose and galactose (Gal). It also converts allolactose to glucose and galactose. Allolactose can bind to the repressor (R) inhibiting it. When not bound by allolactose, R can bind to a specific site upstream of the operon structural genes and thus prevent transcription initiation. External glucose inhibits the production of cAMP that, when bound to the CRP protein (also known as CAP) to form the CAP-cAMP complex, acts as an activator of the *lac* operon. External glucose also inhibits lactose uptake by permease proteins. Reprinted from *Biophysics Journal*, Vol. 92, M. Santillan, M. C. Mackey, and E. S. Zeron, Origin of bistability in the lac operon 3830 - 3842, Copyright (2007), with permission from Elsevier.

imply that when lactose is absent from the medium, the operon is off, and that when lactose is present but glucose is not, the operon is on. We consider Boolean models with various levels of complexity and compare the results. The material can be used as a gentle entry into mathematical modeling, especially for students without Calculus background (see Section 1.5 for more details).

1.3.1   Identifying the Model Variables and Parameters

The modeling process begins with careful examination of the major interactions and components of the system, as depicted in Figure 1.3, and with selecting the model variables and parameters. The model variables are generally chosen to represent the major dynamic elements of the system (quantities that change with time) while the parameters correspond to static descriptors. Different decisions regarding the exclusion or inclusion of any given component or part of the system will lead to different models. The next step is to define a *wiring diagram* (sometimes also called a

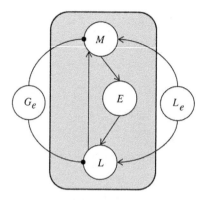

FIGURE 1.4

Wiring diagram for the minimal model. E denotes the *LacZ* polypeptide, M – the mRNA, L- internal lactose. L_e and G_e denote external lactose and glucose, respectively. The nodes in the shaded rectangle represent the model variables while the outside nodes represent the model parameters. Directed links represent influences between the variables: A positive influence is indicated by an arrow; a negative influence is depicted by a circle. Self-regulating links (for nodes that may influence themselves) are not pictured.

dependency graph) for the model that reflects the network topology; that is, it depicts the dependencies between variables and parameters. A wiring diagram is a directed graph in which each node represents a variable or a parameter for the model and the directed links depict influential interactions: if x and y are two nodes of the graph, a directed link from x to y indicates that the quantity x affects the quantity y. Some wiring diagrams contain additional information about the types of interactions between the model components: a positive influence is usually indicated by an arrow, while a negative influence may be depicted by a square (as in Figure 1.3) or by a circle (see Figure 1.4). In many ways, the diagram in Figure 1.3 is similar to a wiring diagram except that it also includes a cartoon of the structural arrangement of the *lac* genes with the promoter and operator regions, and some of the system's elements appear more than once (e.g., the LacZ protein).

It is always best to begin with a simple starter model with a small number of variables and parameters, from which to expand, if necessary. The models of the *lac* operon we will discuss initially follow a "minimal model" approach for choosing variables and parameters after Santillan et al. [14]. The model does not consider the CAP-cAMP positive control mechanism, which is essentially an amplifier for the transcription process. We focus on the following remaining elements (the notations in the parentheses are the mathematical names we will use from now on to denote their concentrations): mRNA (M), β-galactosidase (B), *lac* permease (P), intracellular lactose (L), allolactose (A), external lactose (L_e), and external glucose (G_e). Due to the fact that external conditions for the cell change slowly compared to the lifespan of *E. coli*, the concentrations L_e and G_e remain relatively unchanged with time. We will assume that they are constants and include them in the set of model parameters.

The other quantities ($M, B, P, L,$ and A) will be assumed to vary with time. Some of these variables, however, exhibit related dynamics due to similarities in their underlying biochemical structures and mechanisms of production. Thus, as described next, we can further reduce the number of model variables based on such known dependences.

β-galactosidase is a homo-tetramer made up of four identical LacZ polypeptides. If we denote the LacZ polypeptide by E, the following holds for the concentrations of β-galactosidase and LacZ: $B = E/4$. Further, since the translation rate of the LacY transcript can be assumed to be the same as the rate for the LacZ transcript, the following holds for the concentration of permease: $P = E$. Finally, we can assume that the concentrations of internal lactose (L) and allolactose (A) are proportional, that is $A = pL$, where p indicates the fraction of lactose converted into allolactose and can be determined experimentally. Thus, in our first "minimal" model, we will consider only three variables—M, E, L, and two parameters—L_e and G_e. Knowing the variables M, E, and L would allow us to determine the values of P, B, and A from the equations $B = E/4, P = E$, and $A = pL$. The corresponding dependency graph is depicted in Figure 1.4.

The choice for variables and parameters discussed here is just one possibility among many others. The model by Yildrim and Mackey for instance [15] is based on assumptions leading to a wiring diagram including five nodes corresponding (in our notation) to the variables M, B, P, L, and A, and a node for external lactose as a parameter. In [16], the authors consider a reduction of this five-variable model to a network of three nodes: M, B, and L. These models, together with their Boolean approximations, will be introduced and discussed in Chapter 2 of this volume. Later in this chapter we consider a Boolean model of the *lac* operon with nine variables and two parameters that includes the mechanism of catabolite repression.

Once the model variables have been identified, the decision on the type of mathematical model should be made. As mentioned earlier, various types of mathematical models can be developed (including DE, algebraic, stochastic, and simulation models among others) from the same wiring diagram. In this section, we will focus on developing a Boolean network model.

1.3.2 Boolean Network Models

Boolean variables and Boolean expressions. Boolean models allow only two states, e.g., 0 and 1, indicating the absence or presence of the components represented by the nodes of the wiring diagram. Since in biology trace amounts of various substances may be present at all times, "absence" usually stands for concentrations lower than a certain threshold value separating higher concentrations from the baseline, and "presence" is interpreted as concentrations higher than this threshold. It may appear that because chemical concentrations span a continuous range of values, choosing a single threshold cutoff may not be appropriate—two concentration values may be very close numerically with one of them falling above the threshold and the other one falling below it. Although this may be a legitimate concern in general, it would rarely apply to a model of gene regulation. When the gene is expressed, the concentration

Table 1.1 Tables of values for the basic Boolean operations: Logical AND ($\wedge$), logical OR ($\vee$) and NOT (an overbar identifies the negation of the Boolean variable underneath).

Logical AND: $z = x \wedge y$			Logical OR: $z = x \vee y$			Logical NOT: $z = \bar{y}$	
Input		Output	Input		Output	Input	Output
x	y	z	x	y	z	y	z
0	0	0	0	0	0	0	1
0	1	0	0	1	1	1	0
1	0	0	1	0	1		
1	1	1	1	1	1		

of the protein it makes would be thousands of times higher than the trace amounts that are present when the gene is silent, with a clear distinction between present and absent (1 or 0). Throughout this chapter, we will assume without further mention that a Boolean value of 0 indicates a concentration near the basal level while a Boolean value of 1 signifies a markedly higher concentration.

In Boolean models, the dynamical evolution of the system is described using Boolean functions defined in terms of the model variables and the logical operators AND (denoted by the symbol $\wedge$), OR (denoted by the symbol $\vee$), and NOT (denoted by an overbar on the variable it negates). The tables of values for these Boolean operators are presented in Table 1.1.

In the context of modeling network dynamics, it is often useful to consider the following intuitive interpretation for the operations AND and OR: if the components x and y of the system influence (control) a third component z, then $z = x \wedge y$ means that x and y need to be simultaneously present (have values 1) to affect z; $z = x \vee y$ means that x and y influence z independently and z is affected when x OR y (or both) are present. In the absence of any parentheses, the order of precedence is this: logical NOT has highest precedence, followed by the logical AND, followed by the logical OR.

Sometimes, if we know the value for one of the operands in the AND / OR operations, we may not need to evaluate the other operand to be able to determine the value for the operation. For instance, if $A = 0$, $C = A \wedge B = 0$ regardless of the value of B. In general, if at least one of the operands of the AND operation is zero, the resulting value for the operation is zero no matter what the value of the other operand is. In a similar way, if one of the operands of the OR operation has value 1, the value for the operation is 1 regardless of the value of the other operand. This is known as *short-circuit evaluation*.

Example 1.1.   Assume the Boolean variables A, B, C, and D have values $A = 0$, $B = 1$, $C = 1$, $D = 1$. Determine the value of the Boolean expression $(\bar{D} \vee B) \wedge \bar{A} \wedge C \vee B$.

The expression in the parentheses will be evaluated first and in order to do this, $\bar{D}$ must be computed because NOT has higher precedence than OR. Since $D = 1$, the value of $\bar{D} = 0$. Since $B = 1$, the value of $(\bar{D} \vee B) = 1$. Next, following the rules

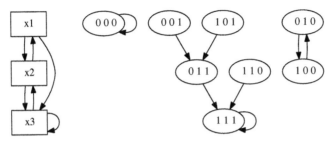

FIGURE 1.5
Wiring diagram and the state space transition diagram for the Boolean dynamical system in Example 1.2 given by Eqs. (1.2). Graphs produced with DVD [12].

of precedence, we compute $\overline{A} = 1$; $(\overline{D} \vee B) \wedge \overline{A} = 1$; $(\overline{D} \vee B) \wedge \overline{A} \wedge C = 1$; and finally $(\overline{D} \vee B) \wedge \overline{A} \wedge C \vee B = 1$.

Exercise 1.1. Consider the Boolean variables w, x, y, z and the Boolean function f of those variables defined by the expression $f(w, x, y, z) = (x \vee y) \wedge (\overline{w \wedge z})$. Determine the values of: (1) $f(1, 1, 0, 1)$; (2) $f(0, 1, 0, 1)$; (3) $f(1, 1, 1, 1)$; (4) $f(0, 0, 0, 0)$. ▽

Modeling the dynamics of a Boolean network: Transition functions. Assume the wiring diagram contains n Boolean model variables (also called *nodes*) denoted by $x_1, x_2, \ldots, x_n$. Each of these n variables can take a value 0 or 1, resulting in a set of n-tuples $V = \{0, 1\}^n = \{(x_1, x_2, \ldots, x_n) | x_i \in \{0, 1\}, i = 1, 2, \ldots, n\}$, containing 2^n elements, representing all possible states for the model variables. Time is discrete and can only take values $t = 0, 1, 2, \ldots$ The values of the nodes $x_1, x_2, \ldots, x_n$ change with time and we write $x_i = x_i(t)$ for the value of the variable x_i at time t. Thus, at each time step t, the system is represented by a binary n-tuple from the set V, where each component stands for the value of the respective Boolean variable at time t. The rules for transitioning between the states at each time step are described by n functions f_{x_i}, $i = 1, 2, \ldots, n$, one function for each model variable. The Boolean expression defining the function f_{x_i}, written in terms of the Boolean operations AND, OR, and NOT, describes in what way the values of the variables $x_1, x_2, \ldots, x_n$ at time t affect the value of the variable x_i at time $t+1$. Thus, for any value of $t = 0, 1, 2, \ldots$, the system "update" for variable x_i from time t to time $t + 1$ is determined by $x_i(t + 1) = f_{x_i}(x_1(t), x_2(t), \ldots x_n(t))$, $i = 1, 2, \ldots, n$. The functions f_{x_i}, $i = 1, 2, \ldots, n$, are the *transition functions* (also called *update rules* or *rules*) for the model. The updates we will be using here are synchronous, meaning that all variables x_i are computed first for time t and then used to evaluate the functions f_{xi}. If we write $x = (x_1, x_2, \ldots, x_n)$ and $f(x) = (f_{x_1}(x), \ldots, f_{x_n}(x))$, the *state space* of the model is defined by the directed graph $\{V, T\}$, where the set $T = \{(x, f(x)) | x \in V\}$ represents the set of edges.

Example 1.2. Assume we have a Boolean network containing three Boolean variables x_1, x_2 and x_3, assume that the wiring diagram for the network is depicted in

Figure 1.5a, and that the transition functions are given by

$$x_1(t+1) = f_{x_1}(x_1(t), x_2(t), x_3(t)) = x_2(t)$$
$$x_2(t+1) = f_{x_2}(x_1(t), x_2(t), x_3(t)) = x_1(t) \lor x_3(t)$$
$$x_3(t+1) = f_{x_3}(x_1(t), x_2(t), x_3(t)) = x_1(t) \land x_2(t) \lor x_3(t). \qquad (1.1)$$

With the understanding that the variables on the right-hand side are always evaluated at time t and that the variables on the left-hand side stand for the values at time $t+1$ we can simplify the notation by removing t and $t+1$ from the equations:

$$x_1 = f_{x_1}(x_1, x_2, x_3) = x_2$$
$$x_2 = f_{x_2}(x_1, x_2, x_3) = x_1 \lor x_3$$
$$x_3 = f_{x_3}(x_1, x_2, x_3) = x_1 \land x_2 \lor x_3. \qquad (1.2)$$

Assume next that at time $t = 0$, the values of the variables x_1, x_2, and x_3 are $x_1 = 0$, $x_2 = 0$, $x_3 = 1$. The values of the model variables at time $t = 0$ are often referred to as *initial conditions*. Using these values to evaluate the transition functions above, we obtain the values of the variables at time $t = 1$:

$$x_1 = f_{x_1}(x_1, x_2, x_3) = f_{x_1}(0, 0, 1) = 0$$
$$x_2 = f_{x_2}(x_1, x_2, x_3) = f_{x_2}(0, 0, 1) = 0 \lor 1 = 1$$
$$x_3 = f_{x_3}(x_1, x_2, x_3) = f_{x_3}(0, 0, 1) = 0 \land 0 \lor 1 = 1.$$

Now take the new values $x_1 = 0$, $x_2 = 1$, $x_3 = 1$. These values are used to evaluate the transition functions f_{x_i} again, producing, for time $t = 2$, the values $x_1 = 1$, $x_2 = 1$, $x_3 = 1$. Plugging these values into the functions again, returns the same values $x_1 = 1$, $x_2 = 1$, $x_3 = 1$ which correspond to the values of the variables at time $t = 3$. Thus, for any future values of t, the values of the model variables will remain $x_1 = 1$, $x_2 = 1$, $x_3 = 1$. We say that we have computed the *trajectory* of the state $(0, 0, 1) : (0, 0, 1) \to (0, 1, 1) \to (1, 1, 1)$. We say that $(1,1,1)$ is a *fixed point* for the Boolean network. Similar considerations show that $(0,0,0)$ is also a fixed point (see Exercise 1.4). Using different starting values for the Boolean variables will lead to different trajectories. For instance, the initial state $(0,1,0)$ generates the following repeating pattern: $(0, 1, 0) \to (1, 0, 0) \to (0, 1, 0) \to (1, 0, 0) \to (0, 1, 0) \to (1, 0, 0)\dots$ (see Exercise 1.5). We say that the states $(0, 1, 0)$ and $(1, 0, 0)$ form a *cycle of length two*. Fixed points can be considered to be cycles of length one.

Since the system is composed of three variables each of which can take values 0 and 1, there are $2^3 = 8$ possible states for the system (see Exercise 1.2). Computing the trajectories initiating from each of these eight initial states and plotting them as in Figure 1.5b visualizes all possible trajectories for the Boolean network, containing all possible transitions between the eight states and, thus, forming the *state space transition diagram* of the Boolean network. Clearly, for a much larger number of variables, computing the trajectories by hand would be impossible and the use of appropriate software is essential. The graphs in Figure 1.5 were created with the freely-available web application *Discrete Visualizer of Dynamics (DVD)* (available at http://dvd.vbi.vt.edu). More details about DVD are provided in Section 1.3.5 below.

Mathematically, the diagram in Figure 1.5b depicts another directed graph. This time, however, the directed graph represents the state-to-state transitions of the system. The edges of the directed graph correspond to transitions between the states as time evolves. Fixed points are recognized as the states that transition to themselves (with edges pointing back to the same state in the diagram). The state space diagram in Figure 1.5b indicates that the three-variable system with wiring diagram given in Figure 1.5a and transition functions described by Eqs. (1.2) has two fixed points: $(0, 0, 0)$ and $(1, 1, 1)$ and a limit cycle of length two. It also shows that the state space graph has three *disconnected components*: The first is $\{(0, 0, 0)\}$ containing only the fixed point $(0, 0, 0)$, the second is $\{(0, 0, 1), (1, 0, 1), (0, 1, 1), (1, 1, 0), (1, 1, 1)\}$, and the third one is $\{(0, 1, 0), (1, 0, 0)\}$. Any trajectories originating from points in the second component will terminate at the fixed point $(1, 1, 1)$. Any trajectory originating from a point in the third component will oscillate, alternating between the two states of the cycle.

Example 1.3. Consider a Boolean network defined by the following set of transition functions describing the interaction dynamics of the three variables x_1, x_2, and x_3:

$$x_1(t + 1) = f_{x_1}(x_1(t), x_2(t), x_3(t)) = x_3(t)$$
$$x_2(t + 1) = f_{x_2}(x_1(t), x_2(t), x_3(t)) = x_1(t)$$
$$x_3(t + 1) = f_{x_3}(x_1(t), x_2(t), x_3(t)) = x_2(t). \qquad (1.3)$$

The wiring diagram for this system and its state space transition diagram are presented in Figure 1.6. The verification is left as an exercise (Exercise 1.7). In addition to the fixed points $(0, 0, 0)$ and $(1, 1, 1)$, the state space diagram contains two cycles of length three: $(0, 0, 1) \rightarrow (1, 0, 0) \rightarrow (0, 1, 0) \rightarrow (0, 0, 1)$ and $(0, 1, 1) \rightarrow (1, 0, 1) \rightarrow (1, 1, 0) \rightarrow (0, 1, 1)$. We say that the state space diagram contains four components, two fixed points, and two cycles of length three.

When the number of variables is small, it is common to use capital letters to denote them instead of using $x_1, x_2, \ldots, x_n$. With such notation, for the example in

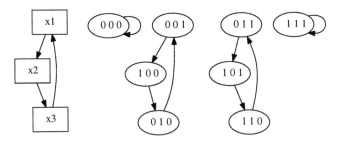

FIGURE 1.6

Wiring diagram and the state space transition diagram for the Boolean dynamical system in Example 1.3 given by Eqs. (1.3). Graphs produced with DVD [12].

Figure 1.4 the model variables are $x_1 = M$, $x_2 = E$, and $x_3 = L$. The functions defining the systems transition will then be denoted as $f_{x_1} = f_M$, $f_{x_2} = f_E$, $f_{x_3} = f_L$.

Deciding on the actual expressions for these Boolean functions is a critical step in the definition of a model. It should take into account all relevant parts of the known biological mechanism of the system being modeled. In the next section we will consider ways of doing this for the *lac* operon.

Exercise 1.2.   How many possible states does a system of n Boolean variables have if:

a. $n = 2$?
b. $n = 4$?
c. $n = 5$?
d. $n = 10$?
e. $n = 100$?   ▽

Exercise 1.3.   For each of the Boolean expressions below, compute the value if $x_1 = 1$, $x_2 = 1$, $x_3 = 0$, $x_4 = 0$, $x_5 = 1$. In the cases where the expressions are not fully parenthesized, be mindful of the operation precedence.

a. $\overline{x_3} \wedge x_5$
b. $(x_3 \vee x_5) \wedge (\overline{x_2} \vee x_1) \vee x_4$
c. $(x_2 \wedge \overline{x_1}) \vee (\overline{x_2} \wedge \overline{x_3}) \wedge x_5$
d. $(x_1 \vee x_2) \wedge \overline{(x_2 \wedge x_3)}$.   ▽

Exercise 1.4.   Show that $(0, 0, 0)$ is a fixed point for the system described by Eqs. (1.1) and (1.2).   ▽

Exercise 1.5.   Show that $(0, 1, 0)$ and $(1, 0, 0)$ form a cycle of length two. That is, show that Eqs. (1.1) and (1.2) lead to the following trajectories that alternate between these two states: $(0, 1, 0) \rightarrow (1, 0, 0) \rightarrow (0, 1, 0) \rightarrow \ldots$ and $(1, 0, 0) \rightarrow (0, 1, 0) \rightarrow (1, 0, 0) \rightarrow \ldots$ ▽

Exercise 1.6.   For the system whose transition functions are given by Eqs. (1.1) and (1.2), compute the trajectory of $(1, 0, 1)$.   ▽

Exercise 1.7.   Verify that the wiring diagram and the state space diagram for the Boolean network defined by Eqs. (1.3) are as depicted in Figure 1.6.   ▽

1.3.3 Creating a Boolean Model of the *Lac* Operon

Once a choice for model variables has been made and the wiring diagram has been constructed, the Boolean transition functions for the model are determined from the wiring diagram and additional available information or assumptions regarding the variable interactions. When multiple quantities impact a third, we need to know if the simultaneous presence of each of these quantities is necessary to exert the effect or if the presence of just one of them would be enough. In the first case, the transition function will use the operator AND, and in the second, it will use the operator OR. We illustrate this by defining a set of transition functions consistent with the wiring diagram in Figure 1.4 for the *lac* operon.

We make the following additional model assumptions. Although some of them may appear too restrictive, keep in mind that they can always be modified or removed after the initial testing and analysis of the model:

- Transcription and translation require one unit of time. This means that when all necessary conditions for the activation of the molecular mechanism are present at time t, the protein production will occur and the product will be available at time $t + 1$.
- Degradation of all mRNA and proteins occurs in one time step. When conditions for making new mRNA or proteins are not met at time t, all previously available amounts of mRNA and proteins have already fallen below the threshold levels and will be absent at time $t + 1$.
- In the presence of β-galactosidase, lactose metabolism occurs in one time step. If lactose and β-galactosidase are available at time t but the conditions for bringing new lactose through the cell membrane from the external medium are not met at time t, by time $t + 1$ all of the lactose is converted to glucose and galactose, and lactose is not available at time $t + 1$.

With this, we define the transition functions (update rules) for the "minimal" *lac* operon model based on the wiring diagram presented in Figure 1.4 as:

$$x_M(t + 1) = f_M(t + 1) = \overline{G_e} \wedge (L(t) \vee L_e(t))$$
$$x_E(t + 1) = f_E(t + 1) = M(t)$$
$$x_L(t + 1) = f_L(t + 1) = \overline{G_e} \wedge ((E(t) \wedge L_e(t)) \vee (L(t) \wedge \overline{E}(t))). \qquad (1.4)$$

We explain these equations next.

Boolean functions for M: The update rule states that for mRNA to be present at time $t + 1$, there should be no external glucose at time t, and either internal or external lactose should be present. When external glucose is present at time t ($G_e(t) = 1$), no mRNA will be available at time $t + 1$ ($M(t + 1) = 0$). Also, when there is no external glucose at time t ($G_e(t) = 0$) and there are high concentrations of lactose inside the cell at time t ($L(t) = 1$) or outside the cell ($L_e(t) = 1$), lactose concentrations in the cell will be sufficiently high (above threshold levels) to cause mRNA production and make mRNA available at time $t + 1$ ($M(t + 1) = 1$).

Boolean function for E: With mRNA available at time t ($M(t) = 1$), the LacZ polypeptide will be produced and available at time $t + 1$ ($E(t) = 1$).

Boolean function for L: When external glucose is available at time t ($G_e(t) = 1$), no lactose will be brought into the cell from the medium and, thus, no internal lactose will be available at time $t + 1$ ($L(t+1) = 0$). When external glucose is absent from the medium at time t ($G_e(t) = 0$), external lactose will be available at time $t + 1$ when at least one of the following conditions is satisfied: (i) External lactose and *lac* permease (as represented by the polypeptide E) are both present at time t ($E(t) \wedge L_e(t) = 1$).

The *lac* permease will then bring the extracellular lactose inside the cell, ensuring the presence of intracellular lactose at time $t + 1$; or (ii) Internal lactose is available at time t but no β-galactosidase (as represented again by the polypeptide E) is available at time t to metabolize it ($L(t) \wedge \overline{E}(t) = 1$). Thus, the internal lactose will still be present at time $t + 1$.

1.3.4 Initial Testing of the Boolean Model of the *Lac* Operon from Eqs. (1.4)

Now that we have defined a model of the *lac* operon, it must be analyzed and validated. Since a model can never be shown to be correct in an absolute sense and is always just an approximation of the actual system, its validation is only appropriate within the context of the questions that the model is developed to help answer. In our case, the simple model we have created should be able to describe the basic qualitative dynamic properties of the *lac* operon. Thus, at a minimum, our model should show that the operon has two steady states, On and Off. When extracellular glucose is available, the operon should be Off. When extracellular glucose is not present and extracellular lactose is, the operon must be On. We next demonstrate that our model satisfies these conditions.

Recall that the operon is On when mRNA is being produced ($M = 1$). When mRNA is present, the production of *lac* permease, and β-galactosidase is also turned on. This corresponds to the fixed-point state $(M, E, L) = (1, 1, 1)$. On the other hand, when mRNA is not made, the operon is Off. This also means no production of *lac* permease, and β-galactosidase. This corresponds to the fixed-point state $(M, E, L) = (0, 0, 0)$.

For the Boolean model of the *lac* operon from Eqs. (1.4), there are four possible combinations for the values L_e and G_e of the model parameters: $L_e = 0, G_e = 0$; $L_e = 0, G_e = 1$; $L_e = 1, G_e = 0$; and $L_e = 1, G_e = 1$. For each one of these pairs of values we can determine the state space transition diagram of the model from the update functions in Eqs. (1.4). The results are shown in Figure 1.7. Notice that according to the model, the operon is On only when external glucose is unavailable and external lactose is present. In all other cases, the operon is Off. These model predictions reflect exactly the expected behavior of the *lac* operon based on the underlying regulatory mechanisms described earlier. This means that our initial model is capable of describing the most fundamental behavior of the *lac* operon system and captures the main qualitative properties of *lac* operon regulation.

Exercise 1.8.   Verify that the space state diagram for the Boolean model described by Eqs. (1.4) is as presented in Figure 1.7b. Notice that for some values of the parameters, the transition functions simplify significantly when we apply short-circuit evaluation for the appropriate Boolean expressions. For instance, when $G_e = 1$, the equations for the transition functions will be: $x_M(t+1) = f_M(t+1) = \overline{1} \wedge (L(t) \vee L_e(t)) = 0$, regardless of the values of L and L_e and $x_L(t + 1) = f_L(t + 1) = \overline{1} \wedge ((E(t) \wedge L_e(t)) \vee (L(t) \wedge \overline{E}(t))) = 0$, regardless of the values of L, E, and L_e. $\triangledown$

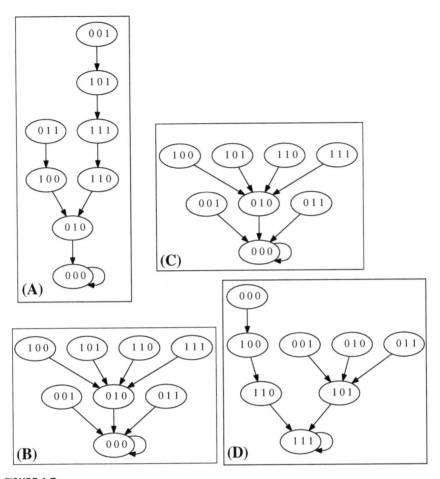

FIGURE 1.7

The state space transition diagram of triples (*M, E, L*) for the Boolean model of the *lac* operon defined in Eqs. (1.4) for the four possible combinations of parameter values. In each of the cases, the operon has a single fixed point and no limit cycles. Panel (A) $L_e = 0$; $G_e = 0$; The operon is Off. Panel (B) $L_e = 0$; $G_e = 1$; The operon is Off. Panel (C) $L_e = 1$; $G_e = 1$; The operon is Off. Panel (D) $L_e = 1$; $G_e = 0$; The operon is On. Graphs obtained using DVD [12].

1.3.5 Using Discrete Visualizer of Dynamics (DVD) to Test a Boolean Model

The *Discrete Visualizer of Dynamics* (DVD) is a suite of freely available web applications (http://dvd.vbi.vt.edu) that takes the transition functions of the Boolean model

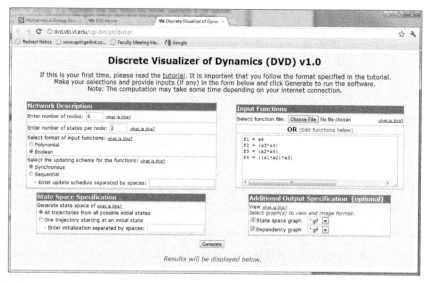

FIGURE 1.8

DVD v1.0 main page screenshot.

as input and returns the wiring diagram and information about the state space of the network including fixed points, number of components, cycles, and cycle lengths. For systems with a relatively small number of variables, DVD can also compute and display the entire space transition diagram. For larger networks, given any initial state, DVD can compute and display the trajectory of this initial state. DVD is fairly intuitive, well documented, and easy to use. In this chapter we will use DVD v1.0 available under the link with the same name from DVDs main page (Figure 1.8). A link to an online tutorial is available on the same page (see [12] for more details). Our next example illustrates the use of DVD.

Example 1.4.   Assume we want to analyze the behavior of a Boolean network with four nodes with the following set of transition functions:

$$x_1(t+1) = f_{x1}(x_1(t), x_2(t), x_3(t), x_4(t)) = x_2(t)$$
$$x_2(t+1) = f_{x2}(x_1(t), x_2(t), x_3(t), x_4(t)) = x_3(t) \wedge x_4(t)$$
$$x_3(t+1) = f_{x3}(x_1(t), x_2(t), x_3(t), x_4(t)) = x_2(t) \wedge x_4(t)$$
$$x_4(t+1) = f_{x4}(x_1(t), x_2(t), x_3(t), x_4(t)) = x_1(t) \wedge x_2(t) \wedge x_3(t).$$

In DVD, the number of nodes under "Network Description" should be set to four; the number of states for each node should be two (since only the Boolean values 0 and 1 are allowed); the format for the input functions should be set to Boolean, and the updating schedule should be set to synchronous (see Figure 1.8).

To analyze the network, we enter the transition functions in the Input Functions text area as follows:

$$f1 = x4$$
$$f2 = (x3 * x4)$$
$$f3 = (x2 * x4)$$
$$f4 = ((x1 * x2) * x3). \tag{1.5}$$

DVD uses the following symbols for the basic logical operations: $*$ for AND, $+$ for OR, and $\sim$ for NOT. Notice the multiple sets of parentheses. DVD requires that all Boolean expressions are *fully parenthesized*, meaning that every single operation should be enclosed in parentheses. Multiple sets of parentheses for the same operations *should not be used*, as this may lead to incorrect results. In DVD the nodes are always labeled x1, x2, and so on; their respective update rules are denoted f1, f2, and so on. If your model uses different variable names, they need to be changed before using DVD to conform with this requirement.

Exercise 1.9. Open DVD v1.0 and enter and run the example from Eqs. (1.5). Check the appropriate boxes in the bottom right panel "Additional Output Specification" to generate the state space transition graph and the dependency graph. Clicking on the "Generate" button displays the characteristics of the Boolean network. Follow the links to the space graph and dependency graph to examine them. Answer the following questions: (1) How many fixed points does the model have? (2) How many components does the state space graph have? (3) Are there any limit cycles?   ▽

Exercise 1.10. For the example from Eqs. (1.5), use DVD to find the trajectory of the state (0, 1, 1, 1). To do so, select the radio button "One Trajectory…" in the "State Space Specification" panel and enter the components of the space separated by spaces: 0 1 1 1. Clicking on the "Generate" button will display the path.   ▽

DVD does not provide a specific option for designating selected nodes as parameters. Thus, parameter values should either be entered explicitly as 0s or 1s in the model equations or they should be treated as "variables" that retain their constant values for all time steps. To do the latter, for each parameter P, we add an update rule of the form $f_P = P$ (usually at the end of the model, after the update rules for all of the variables). This ensures that P retains its initial value for all time steps t. Of course, with this approach, the DVD output should be interpreted carefully and with the understanding that different sets of initial values for the parameters should be considered as separate outputs from the model.

Example 1.5. We will show how the *lac* operon model from Eqs. (1.4) can be analyzed using DVD. First, the variable names will need to change to names accepted by DVD. We will use x1 for M, x2 for E, and x3 for L. Two different approaches are possible for the model parameters:

1. Enter and run the model four different times for each of the four combinations of the parameters L_e and G_e (see Exercise 1.11). As an example, for $L_e = 1$ and

Table 1.2 DVD output for the update functions in Eqs. (1.6) See the text for details.

ANALYSIS OF THE STATE SPACE [m = 2, n = 5]

There are 4 components and 4 fixed point(s)

Components	Size	Cycle Length
1	8	1
2	8	1
3	8	1
4	8	1

TOTAL: $32 = 2^5$ nodes
Printing fixed point(s).
[0 0 0 0 0] lies in a component of size 8.
[0 0 0 0 1] lies in a component of size 8.
[0 0 0 1 1] lies in a component of size 8.
[1 1 1 1 0] lies in a component of size 8.

$G_e = 0$, we should enter the model into DVD as

$$f1 = ((\sim 0) * (x3 + 1))$$
$$f2 = x1$$
$$f3 = ((\sim 0) * ((x2 * 1) + (x3 * (\sim x2)))).$$

2. Introduce two new variables, $x4$ for L_e and $x5$ for G_e. The update rules for these variables are $f4 = x4$ and $f5 = x5$ and the entire model is

$$f1 = ((\sim x5) * (x3 + x4))$$
$$f2 = x1$$
$$f3 = ((\sim x5) * ((x2 * x4) + (x3 * (\sim x2))))$$
$$f4 = x4$$
$$f5 = x5. \tag{1.6}$$

Taking the second approach and running the model in DVD produces the output in Table 1.2 and the state space diagram in Figure 1.9. There are four fixed points, each one of which corresponds to a different combination of the parameter values for L_e and G_e. Each such combination corresponds to a component in the state space transition graph in Figure 1.9. The last two values of each state encode the values of the parameters L_e and G_e, respectively. Viewed this way, Figure 1.9 is identical to the analysis of the *lac* operon model presented in Figure 1.7.

Exercise 1.11. Use DVD to analyze the model of the *lac* operon from Eqs. (1.4), running it four times for the four different values of the parameters. Compare the results with those in Table 1.2, Figure 1.7 and Figure 1.9. ▽

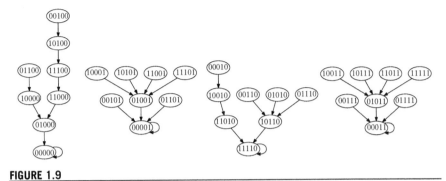

FIGURE 1.9

State space diagram for the model defined by the update functions in Eqs. (1.6). The order of components for each state is (M, E, L, L_e, G_e). See the text for further details.

1.3.6 How to Recognize a Deficient Model

In the minimal model introduced and discussed above, the testing and analysis showed that the model can adequately represent the behavior of the operon to be On or Off. One should remember, however, that this model is just one of many possible models that can be created and that every model relies on a set of assumptions made during the modeling process. For the minimal Boolean model above, we made several assumptions both during the process of selecting the model variables and at the stage of writing down the transition functions that determine the dynamical behavior of the system. Different choices and assumptions would lead to different models, some of which may work and some of which may not. In this section we present a "model" (and the quotation marks here are meant to indicate that we will ultimately show that it is not a good model) that may look legitimate at first. The initial testing, however, will show that the model does not adequately capture the regulatory behavior of the *lac* operon.

The model is based on the following five variables: mRNA (M), β-galactosidase (B), lac permease (P), intracellular lactose (L), and allolactose (A). This certainly appears to be a reasonable choice, as it includes the primary components of the *lac* operon regulatory mechanism. As in the previous model, and for the same reasons as before, the CAP-cAMP positive control mechanism is excluded from the modeling effort. The following transition functions are proposed for describing the underlying biology:

$$f_M = A$$
$$f_B = M$$
$$f_A = A \vee (L \wedge B)$$
$$f_L = P \vee (L \wedge \overline{B})$$
$$f_P = M. \tag{1.7}$$

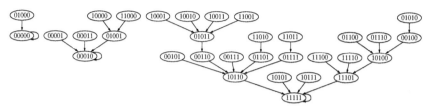

FIGURE 1.10

The state space transition diagram of the points (*M, B, A, L,P*) for the Boolean model defined by Eqs. (1.7).

Here, as for the minimal model from Eqs. (1.4), it is assumed that translation and transcription require one unit of time, protein and mRNA degradation require one unit of time, and lactose metabolism requires one unit of time. This model does not involve any parameters: instead, it assumes that extracellular lactose is always available, and extracellular glucose is always unavailable. The rationale and interpretation for the functions proposed by Eqs. (1.12) comes from the biological interactions between the main regulatory elements. For instance, the transition function f_L indicates that at time $t + 1$ lactose (*L*) will be available if permease (*P*) is available at time t to bring the extracellular lactose into the cell or, in case lactose (*L*) is already available at time t, there is no β-galactosidase at time t ($\overline{B}$) to metabolize the lactose into glucose and galactose.

Exercise 1.12.   Sketch the wiring diagram (the dependency graph) for the model described by Eqs. (1.7).   ▽

As for our earlier model, one way to initially test the model from Eqs. (1.7) is to examine its fixed points and determine if they correspond to states that are biologically feasible. The state space graph for the model defined by Eqs. (1.7) is presented in Figure 1.10. It has three fixed points (*M, B, A, L, P*): (0, 0, 0, 0, 0), (1, 1, 1, 1, 1), and (0, 0, 0, 1, 0). The first two correspond to the operon being Off and On, respectively. The third one, however, corresponds to a biological scenario under which the bacterium does not metabolize the intracellular lactose, which is unrealistic (recall that the model assumed that extracellular lactose is always available and that there is no extracellular glucose).

The fact that the state (*M, B, A, L, P*) = (0, 0, 0, 1, 0) is a fixed point for the model indicates that the model does not represent accurately the most important qualitative behavior of the *lac* operon regulation. Thus, the model fails the initial testing and is in need of modification.

Exercise 1.13.   Consider the transition functions in Eqs. (1.7) and criticize the model. Can you find reasons to question the definitions of the transition functions? The way to approach this is by examining each function and asking if it accurately reflects the underlying biology and/or the model assumptions.   ▽

Exercise 1.14.   Consider possible modifications of the transition functions in Eqs. (1.7), aimed at eliminating the biologically infeasible fixed point. Give the rationale for your modification and specify the biological mechanism or model assumptions that justify the change. Use DVD to analyze the modified model. For each of your modifications, use the number of fixed points to decide if they correspond to biologically realistic situations. Note that there should be no limit cycles.   ▽

1.3.7 **A More Refined Boolean Model of the *Lac* Operon**

The models we have considered so far attempted to reproduce features of the *lac* operon mechanisms by including a relatively small number of variables. The catabolite repression mechanism (the CAP-cAMP positive control loop) was excluded, as was the explicit modeling of the presence of low levels of lactose and allolactose. For proteins and enzymes, since basal concentrations are always nonzero, we used the Boolean value 0 to refer to basal levels and 1 for concentrations that are much higher. This does not apply to lactose and allolactose, the concentrations of which may be truly zero. Thus, in the case of lactose and allolactose, it is justified to look for ways to model low but nonzero concentrations separately.

There are several possible options to do so. One of them is to consider discrete models in which the variables take values from a set S of three or more values, corresponding to ranges in the respective concentrations. If $S = \{0, 1, 2\}$, the value 0 may be used to represent a concentration near zero, 1 to represent a low concentration that is not near zero, and 2 to represent a high concentration. This approach can no longer be implemented using Boolean networks and leads to discrete models with multiple states. When the set S has a prime number of elements, it can be shown that the transition functions of the model have a representation as polynomials of the model variables. The theory of polynomial dynamical systems is then applicable to the analysis of such networks and there are many interesting mathematical questions arising in this context (see, e.g., [17–21]).

Another possible approach, which we will examine here, is to stay within the framework of Boolean networks and introduce additional Boolean variables to allow for a separation into three concentration ranges instead of two. The model that follows comes from Stigler and Veliz-Cuba [22] and utilizes this approach. It also includes the CAP-cAMP positive control mechanism of the *lac* operon that was not considered in the models discussed so far.

We begin by identifying the model variables and parameters. As in the minimal model, L_e and G_e are model parameters, denoting the extracellular lactose and the extracellular glucose. The variables L_l and A_l (the index stands for low concentration) are introduced to facilitate the ability of the model to distinguish between "no lactose" and "some lactose" and similarly for allolactose. This is an improvement over our initial model because, unlike for proteins and enzymes, it would not be warranted to assume that baseline levels of lactose or allolactose are always present. When L_l and A_l have value 1, this means that at least low concentrations of lactose and allolactose, respectively, are available in the cell. As before, $L = 1$ and $A = 1$ stand for high levels of lactose and allolactose. High levels of lactose or allolactose at any given time t imply at least low levels for the next time step $t + 1$. The complete list of model variables is:

$M = lac$ mRNA	$L =$ high concentration of intracellular lactose
$P = lac$ permease	$A =$ high concentration of allolacose (inducer)
$B = \beta$-galactosidase	$L_l =$ (at least) low concentration of intracellular
$C =$ catabolite activator protein CAP	lactose
$R =$ repressor protein lacI	$A_l =$ (at least) low concentration of allolactose

The model assumptions are:

- Transcription and translation require one unit of time. This means that if all necessary conditions for the activation of the molecular mechanism are present at time t, the protein production will be happening in time $t + 1$.
- Degradation of all mRNA and proteins occurs in one time step.
- High levels of lactose or allolactose at any given time t imply at least low levels for the next time step $t + 1$.

The Boolean transition functions for the model reflect the dependencies between variables according to the regulatory mechanisms of the lac operon from Section 2. We provide justification for two of the functions; the rest are left as exercises. The corresponding wiring diagram is depicted in Figure 1.11.

$$
\begin{aligned}
f_M &= \overline{R} \wedge C \\
f_P &= M & f_B &= M \\
f_C &= \overline{G_e} & f_R &= \overline{A} \wedge \overline{A_l} & \quad (1.8) \\
f_A &= L \wedge B & f_{A_l} &= A \vee L \vee L_l \\
f_L &= \overline{G_e} \wedge P \wedge L_e & f_{L_l} &= \overline{G_e} \wedge (L \vee L_e).
\end{aligned}
$$

Boolean function for R : For the concentration of the repressor protein to be high ($R = 1$), there should be no allolactose present; that is $A = A_l = 0$.

Boolean function for M: In order for production of mRNA to be high, there should be no repressor protein ($R = 0$) and the concentration of CAP should be high ($C = 1$).

Exercise 1.15.   Consider the rest of the transition functions from Eqs. (1.8). Give justification for the definition of each function and provide any additional assumptions that the definition may imply.   ▽

Exercise 1.16.   There is no variable to represent cAMP in the Boolean model with the wiring diagram depicted in Figure 1.11 and defined by Eqs. (1.8). Could you justify this decision? How would the model change if a cAMP variable is introduced? Do you think this change will impact the qualitative behavior of the model?   ▽

As with the earlier Boolean models we have examined, the next logical step in the process is to find the fixed points of the model and determine if it accurately reflects the ability of the lac operon to be On or Off. The model now has nine variables,

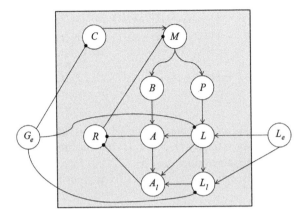

FIGURE 1.11
The wiring diagram for the Boolean model defined by Eqs. (1.8). The nodes inside the square region represent model variables, while the outside nodes correspond to model parameters. Arrows indicate positive interactions, circles indicate negative interactions. (Figure adapted from Stigler and Veliz-Cuba [22]).

leading to a state space for the Boolean network model of size $2^9 = 512$ for each of the four combinations of the parameter values. It would be nearly impossible to compute the individual trajectories of each of these points by hand and determine the fixed points this way but we may think that a computer would be able to easily do that. However, for networks with large numbers of nodes, such a "brute force" approach based on considering all possible trajectories becomes impossible even for the fastest computers. As an example, a Boolean model of T cell receptor signaling developed in [23] contains 94 nodes. This means that the state space contains 2^{94} states (approximately equal to $2 * 10^{28}$, a number that is about 3,000,000 times larger than the estimated number of stars in the observable universe [24]). This means that years of computing time would be needed even on the fastest computers. Clearly, this is not practical, implying that DVD and other similar software applications must use a different, computationally efficient, way to determine the fixed points of a Boolean network. This justifies the following question: Given a Boolean network model, how can we determine its fixed points without having to compute the entire state space transition diagram?

1.4 DETERMINING THE FIXED POINTS OF BOOLEAN NETWORKS

The dynamical behavior of a Boolean network model with nodes $x_1, x_2, \ldots, x_n$ is described by the transition functions $f_{x_1}, f_{x_2}, \ldots, f_{x_n}$, namely

$$x_1(t+1) = f_{x_1}(x_1(t), x_2(t), \ldots, x_n(t))$$
$$x_2(t+1) = f_{x_2}(x_1(t), x_2(t), \ldots, x_n(t))$$
$$\cdots$$
$$x_n(t+1) = f_{x_n}(x_1(t), x_2(t), \ldots, x_n(t)). \tag{1.9}$$

A point $(p_1, p_2, \ldots, p_n)$ is *a fixed point* for the set of Eqs. (1.9), when plugging the values $(p_1, p_2, \ldots, p_n)$ for $x_1(t), x_2(t), \ldots, x_n(t)$ into the functions $f_{x_1}, f_{x_2}, \ldots f_{x_n}$, from Eqs. (1.9) returns the same values $(p_1, p_2, \ldots, p_n)$ for $x_1(t+1), x_2(t+1), \ldots, x_n(t+1)$. In other words, $(p_1, p_2, \ldots, p_n)$ is a fixed point for the set of Eqs. (1.9) when

$$p_1 = f_{x_1}(p_1, p_2, \ldots, p_n)$$
$$p_2 = f_{x_2}(p_1, p_2, \ldots, p_n)$$
$$\cdots$$
$$p_n = f_{x_n}(p_1, p_2, \ldots, p_n). \tag{1.10}$$

We seek a method for determining all points $(p_1, p_2, \ldots, p_n)$ that satisfies the condition in Eqs. (1.10).

The brute force approach in this context would be to test all possible states for the fixed-point property defined by Eqs. (1.10). However, since the size of the state space grows exponentially with the number of nodes in the network, we already decided that going over the entire state space to test all of its elements is not practical (see Exercise 1.2). The approach we describe next makes it possible to rephrase the problem of finding all fixed points $(p_1, p_2, \ldots, p_n)$ of a Boolean network as a problem of solving systems of polynomial equations and determining a computationally efficient way of finding the solutions. This method uses a computational algebra approach and is based on some fundamental results in abstract algebra and algebraic geometry and on the use of Groebner bases for solving systems of polynomial equations [25].

Equations (1.10) state that the fixed points of the model are the solutions $(x_1, x_2, \ldots, x_n) = (p_1, p_2, \ldots, p_n)$ of the system of equations

$$x_1 = f_{x_1}(x_1, x_2, \ldots, x_n) \qquad\qquad f_{x_1}(x_1, x_2, \ldots, x_n) - x_1 = 0$$
$$x_2 = f_{x_2}(x_1, x_2, \ldots, x_n) \quad \text{or, equivalently,} \quad f_{x_2}(x_1, x_2, \ldots, x_n) - x_2 = 0$$
$$\cdots \qquad\qquad\qquad\qquad\qquad\qquad \cdots$$
$$x_n = f_{x_n}(x_1, x_2, \ldots, x_n) \qquad\qquad f_{x_n}(x_1, x_2, \ldots, x_n) - x_n = 0, \tag{1.11}$$

where, the functions $f_{x_1}, f_{x_2}, \ldots, f_{x_n}$, are Boolean functions.

Table 1.3 The table of values for Boolean functions $x_1 \wedge x_2$; $x_1 \vee x_2$, and $\overline{x_1}$ and their equivalents in polynomial form.

x_1	x_2	$x_1 \wedge x_2$	$x_1 x_2$	$x_1 \vee x_2$	$x_1 + x_2 + x_1 x_2$	$\overline{x_1}$	$x_1 + 1$
0	0	0	0				
0	1	0	0				
1	0	0	0				
1	1	1	1				

The general theory we will be applying here describes a method for solving such systems of equations in the special case where the functions $f_{x_1}, f_{x_2}, \ldots, f_{x_n}$, are polynomials of the variables $x_1, x_2, \ldots, x_n$. This may appear to be a serious restriction since in the case of Boolean networks the transition functions do not seem to satisfy this condition. However, as we will see next, each Boolean function can be rewritten as a *Boolean polynomial*, where a Boolean polynomial of the (Boolean) variables $x_1, x_2, \ldots, x_n$ is a sum of terms of the form $x_1^{b_1} x_2^{b_2} \ldots x_n^{b_n}$ with $b_i \in \{1, 0\}$, $i = 1, 2, \ldots, n$.

Example 1.6. $f(x_1, x_2, x_3, x_4) = x_1 x_3 x_4 + x_2 x_3 + x_1 x_3$ is a Boolean polynomial of the variables x_1, x_2, x_3, x_4. For the first term, we have $b_1 = 1, b_2 = 0, b_3 = 1$, and $b_4 = 1$. For the second and third terms, $b_1 = 0, b_2 = 1, b_3 = 1$, and $b_4 = 0$, and $b_1 = 1, b_2 = 0, b_3 = 1$, and $b_4 = 0$, respectively.

To convert any Boolean function into a Boolean polynomial, the Boolean operations AND, OR, and NOT need to be converted to polynomials as follows:

1. $x_1 \wedge x_2 = x_1 x_2$
2. $x_1 \vee x_2 = x_1 + x_2 + x_1 x_2$
3. $\overline{x_1} = x_1 + 1$.

To verify that the Boolean and polynomial functions (1)–(3) are identical, we need to compare their values and verify that the two functions return the same values for the same inputs. Table 1.3 contains the proof that $x_1 \wedge x_2 = x_1 x_2$.

Since addition and multiplication are performed over the field $\{0,1\}$, the following addition and multiplication rules apply for any value $a \in \{0, 1\} : a + a = 0$ and $a * a = a$.

Exercise 1.17.   Fill in the remaining columns of Table 1.3 to show that the equalities (2) and (3) above hold.   ▽

Exercise 1.18.   Show that for any $a, b \in \{0, 1\}, a - b = a + b$. As a special case, over the field $\{0, 1\}, -1 = 1$.   ▽

Example 1.7.   We next translate the Boolean model of the *lac* operon defined by Eqs. (1.8) into polynomial form. We will give the translation for the function $f_{A_l} = A \vee L \vee L_l$ and leave the rest of the functions as an exercise:

$$f_{A_l} = A \vee L \vee L_l = ((A \vee L) \vee L_l) = ((A + L + AL) \vee L_l)$$
$$= (A + L + AL) + L_l + (A + L + AL)L_l$$
$$= A + L + AL + L_l + AL_l + LL_l + ALL_l.$$

Thus, in polynomial form, $f_{A_l} = ALL_l + AL_l + LL_l + AL + A + L + L_l$.

Exercise 1.19.   Translate the remaining eight functions in Eqs. (1.8) into polynomials and simplify as much as possible to obtain the polynomial form of the Boolean model defined by the set of Eqs. (1.8). Write down the whole model (it has nine variables and two parameters!) and save it. You will need this polynomial form of the model again in Example 1.10 and in Exercise 1.20 below.   ▽

Once the system of Boolean Eqs. (1.10) is translated into a system of polynomial equations, finding the fixed points becomes a problem of solving a polynomial system of equations. This can be done by employing a technique from computational algebra involving Groebner bases. With this technique, it is generally possible to rewrite a system of polynomial equations in a much simpler form while preserving the set of solutions. Under some additional conditions, the reduced form of the system of equations is guaranteed to have a form that makes finding the solutions possible by back-substitution. Readers familiar with the method of Gaussian elimination for systems of linear equations will notice that finding a Groebner basis that diagonalizes the system of equations can be viewed as a generalization of the Gaussian elimination method for polynomial functions.

For readers with appropriate mathematical background in abstract algebra, including the basic theory of rings, fields, ideals, and varieties, an outline of the theory of Groebner bases, as it relates to solving systems of polynomial equations, is presented in the online Appendix 1. For what follows, however, we do not assume familiarity with this theory. Instead, we illustrate how knowing the diagonalized forms of the polynomial systems (obtained by determining the Groebner basis with the use of computational software) can help to solve the system of polynomial equations and determine the fixed points.

Various computer systems including Macaulay 2, MAGMA, CoCoA, SINGULAR, and others, have the capability of computing Groebner bases. For our needs, using the web-based *SAGE* interface for Macaulay 2 for determining the Groebner basis for a diagonalized reduction is easy and convenient (http://www.sagemath.org/). We suggest that you use the online option first. Select "Try *SAGE* Online" and register for a *SAGE* notebook account. We consider a few examples below. The first two illustrate how the methods can be used to solve systems of polynomial equations over the real numbers, a problem that the reader is likely to be more familiar with. We then use the method to determine the fixed points of the Boolean model of the *lac* operon defined by Eqs. (1.8).

Example 1.8.   Consider the system of polynomial equations below where x, y, and z are real numbers. We want to find the solutions of the system of equations.

$$x^2 + y^2 + z^2 = 1$$
$$x^2 + z^2 = y$$
$$x - z = 0.$$

To use Groebner bases to diagonalize the system, we first need to rewrite the equations to ensure that the right-hand side is zero:

$$x^2 + y^2 + z^2 - 1 = 0$$

$$x^2 + z^2 - y = 0 \qquad (1.12)$$

$$x - z = 0.$$

Next, we need to find the Groebner basis for the functions that form the left-hand sides of the equations: $h_1 = x^2 + y^2 + z^2 - 1$; $h_2 = x^2 + z^2 - y$; $h_3 = x - z$. To compute the Groebner basis for these functions in *SAGE*, we use the following commands (click on "evaluate" after entering each line)[2]:

```
P.<x,y,z> = PolynomialRing(RR, 3, order='lex')
I = ideal(x^2+y^2+z^2-1, x^2+z^2-y, x-z)
B = I.groebner_basis()
```

In the first command, the variables in the system are listed on the left. On the right, RR denotes the real numbers (meaning we want to work over the reals) and three is the number of variables. For our purposes, we will always need to use `order='lex'`.

After evaluating the last command, *SAGE* returns the Groebner basis:

```
[x-z, y - 2*z^2, z^4 + 1/2*z^2 - 1/4].
```

It follows from the general theory outlined in the online Appendix 1 that the systems of Eqs. (1.12) has a solution set equivalent to the solution set of the system

$$z^4 + \frac{1}{2}z^2 - \frac{1}{4} = 0$$

$$y - 2z^2 = 0$$

$$x - z = 0.$$

Notice that this system is "diagonal" in the sense that if we solve the first equation for z, we can use the values in the second equation to solve for y and then in the third equation for x. After doing this, we obtain the solutions $z = \pm\sqrt{\frac{-1+\sqrt{5}}{4}}$; $y = 2z^2$; $x = z$ for the system, which leads to the solutions $z = \sqrt{\frac{-1+\sqrt{5}}{4}}$; $y = \frac{-1+\sqrt{5}}{2}$; $x = \sqrt{\frac{-1+\sqrt{5}}{4}}$ and $z = -\sqrt{\frac{-1+\sqrt{5}}{4}}$; $y = \frac{-1+\sqrt{5}}{2}$; $x = -\sqrt{\frac{-1+\sqrt{5}}{4}}$.

---

[2]Note that exponentiation is indicated with ^ and multiplication must be indicated with *.

Example 1.9.   Repeat Example 1.8 with the following system of polynomial equations.

$$x^2y - z^3 = 0$$
$$2xy - 4z - 1 = 0$$
$$z - y^2 = 0$$
$$x^3 - 4zy = 0.$$

After entering the appropriate commands as in Example 1.8 for this system of equations, *SAGE* returns the Groebner basis [1] for the functions $f_1 = x^2y - z^3$; $f_2 = 2xy - 4z - 1$; $f_3 = z - y^2$; $f_4 = x^3 - 4zy$.

This means that the solution set of the system of equations above is equivalent to the solution set of the equation $1 = 0$, indicating that the solution set is empty. The system does not have a real-valued solution.

Example 1.10.   Consider again Exercise 1.19 where you wrote the polynomial form of the Boolean model of the *lac* operon given by Eqs. (1.7). If in those equations we rename the variables M, P, B, C, R, A, A_l, L, L_l to $x1$, $x2$, $x3$, $x4$, $x5$, $x6$, $x7$, $x8$, $x9$ (in this order) and the parameters L_e and G_e to a and g, respectively, and rewrite the equations in a form where the right-hand side is zero, we will obtain[3]

$$x1 + x4 * x5 + x4 = 0$$
$$x1 + x2 = 0,$$
$$x1 + x3 = 0$$
$$x4 + (g + 1) = 0$$
$$x5 + x6 * x7 + x6 + x7 + 1 = 0$$
$$x6 + x3 * x8 = 0 \qquad\qquad (1.13)$$
$$x6 + x7 + x8 + x9 + x8 * x9 + x6 * x8 + x6 * x9 + x6 * x8 * x9 = 0$$
$$x8 + (g + 1) * a * x2 = 0$$
$$x9 + (g + 1) * (x8 + a * x8 + a) = 0.$$

We now need to solve this system four times, using all four combinations of $a = 0,1$ and $g = 0,1$ for the parameter values. Using *SAGE* we need to "evaluate" the following command lines:

```
P.<x1,x2,x3,x4,x5,x6,x7,x8,x9>=PolynomialRing(GF(2),
9, order = 'lex').
```

We use GF(2) here since we want to find the solutions of the system of equations from Eqs. (1.13) over the field $\{0, 1\}$, often referred to as the Galois field of two elements (hence the notation). We also use nine to indicate the number of variables. After setting $a = g = 0$ in the polynomials from Eqs. (1.13), we enter the set of functions:

---

[3]To get the system in this form we used the facts that, over the field $\{0,1\}$, $1 + 1 = 0$, $-1 = 1$, and $a - b = a + b$.

```
I=ideal(x1+x4*x5+x4,x1+x2,x1+x3,x4+1,x5+x6*x7+x6+x7
+1,x6+x3*x8,
x6+x7+x8+x9+x8*x9+x6*x8+x6*x9+x6*x8*x9, x8, x9+x8)
I.groebner_basis().
```

Evaluating, *SAGE* returns the following Groebner basis for the set of functions in the left-hand sides of Eqs. (1.13):

```
[x1, x2, x3, x4 + 1, x5 + 1, x6, x7, x8, x9].
```

Thus, the system of equations from Eqs.(1.13) has the same solution set as the system of equations

$$x1 = 0 \quad x4 + 1 = 0 \quad x7 = 0$$
$$x2 = 0 \quad x5 + 1 = 0 \quad x8 = 0$$
$$x3 = 0 \quad \quad x6 = 0 \quad x9 = 0.$$

This gives the steady state $(M, P, B, C, R, A, A_l, L, L_l) = (x1, x2, x3, x4, x5, x6, x7, x8, x9) = (0, 0, 0, 1, 1, 0, 0, 0, 0)$ for $L_e = a = 0, G_e = g = 0$.

Exercise 1.20.   Use *SAGE* to continue the computations from Example 1.10 and show that the Boolean model of the *lac* operon from Eqs. (1.8) has the following fixed points for the remaining combinations of parameter values: (1) For : $a = 0, g = 1$: $(M, P, B, C, R, A, A_l, L, L_l) = (0, 0, 0, 0, 1, 0, 0, 0, 0)$; (2) For : $a = 1, g = 1$: $(M, P, B, C, R, A, A_l, L, L_l) = (0, 0, 0, 0, 1, 0, 0, 0, 0)$; (3) For : $(a = 1, g = 0$: $(M, P, B, C, R, A, A_l, L, L_l) = (1, 1, 1, 1, 0, 1, 1, 1, 1)$.   ▽

The results from Example 1.10 and Exercise 1.20 show that the Boolean model of the *lac* operon from Eqs. (1.8) has the right qualitative behavior, predicting that the operon is On only when external lactose is available and external glucose is not. In this case all variables of the model, except for the repressor protein, are present. When glucose is available, the operon is Off. All fixed points are biologically feasible.

Now that we understand how fixed points may be found as solutions of systems of polynomial equations, we can use DVD again to analyze the model and compare the results to those from Example 1.10 and from Exercise 1.20. In DVD you can either enter and run the model four times for the four different combinations of the parameter values or use the approach we took in Example 1.5 (2) and include an additional "variable" for each of the parameters with an update rule that keeps that variable constant for all time steps. The analysis in DVD confirms the fixed points from Example 1.10 and from Exercise 1.20 and affirms that the state space does not contain limit cycles.

1.5 CONCLUSIONS AND DISCUSSION

Utilization of lactose by the bacterium *E.coli* requires the production of two proteins, the transporter lactose permease and the enzyme β-galactosidase. Efficient use of cellular energy and materials requires that the bacterium only make these two proteins when needed. The regulatory mechanism of the *lac* operon ensures the repression of these two proteins when glucose, the preferred energy source, is present and the coordinated production of these two proteins when lactose alone is present. In this chapter we show how Boolean networks can be used to model the control mechanisms of the *lac* operon system. The chapter can be used as an introduction to mathematical modeling without calculus. A brief primer of Boolean algebra is included, so virtually no mathematical prerequisites are required. We introduce specialized web-based software for testing and analyzing the models, which is a convenient way to separate theory from applications. For readers interested in the theoretical underpinnings, the chapter provides an online appendix (Appendix 1) that outlines the use of Groebner bases as it pertains to solving systems of polynomial equations.

The models considered in this chapter are relatively simple and capture only the most significant qualitative behavior of the *lac* operon – its ability to turn on its lactose utilization mechanism when glucose is absent from the external medium and lactose is present. It is clear even from these simple models, that threshold values separating concentrations with Boolean value 0 from those with Boolean value 1 need to be selected carefully based on the biology. We generally assume that a value of 0 and 1 indicate, respectively, "low" and "high" concentrations, where "low" does not always mean a concentration of zero. For concentrations of mRNA, *lac* permease, and β-galactosidase, as well as for other proteins and enzymes involved in the *lac* operon control, which increase by a factor of thousands when the operon is turned on in comparison with their baseline concentrations, such threshold values can easily be selected. The concentrations of lactose, allolactose, and glucose, however, depend on the external environment and change more gradually. For those, "low" could truly mean "absent" and medium-level concentrations are biologically completely feasible. In the model from Eqs. (1.8), the authors take care to introduce designated Boolean variables for low lactose and allolactose to ensure that "low" means "some but not zero" [22]. Other models (see, e.g., [26,27] and Chapter 2 of this volume) consider Boolean frameworks within which it is possible to distinguish between low, medium, and high lactose concentrations.

Focusing on the medium range of lactose concentration is of particular interest since the lactose operon has been shown to exhibit *bistability* for medium levels of lactose. Bistability is the ability of a system to settle in one of two different fixed points under the same set of external conditions. Which of these fixed points the system will reach depends on its history. It has been known since the 1950s [28] that for medium lactose concentrations, both induced and uninduced cells can be observed in a population of *E.coli*. Cells grown in an environment poor in extracellular lactose will likely remain uninduced for medium lactose levels while cells grown in a lactose-rich environment will likely retain their induced state. Chapter 2 of this volume examines Boolean network models of the *lac* operon that can capture the bistability property of the system.

Boolean models are important from an educational perspective since they require only a minimal mathematics background. At the introductory level, the construction of a simple model may essentially amount to translating the system's interactions represented by a biology "cartoon" into a directed graph (wiring diagram), followed by a subsequent translation into logical expressions (the update rules). This makes

Boolean models ideal for an early (below-calculus level) introduction to mathematical models, removing the need for calculus or other mathematical prerequisites. For mathematics students, such models can be introduced in low-level finite mathematics or discrete mathematics courses and used to provide an early demonstration of the important link between mathematics and biology.

At the more advanced mathematics level, Boolean models can be generalized to finite dynamical systems (FDS) and used to provide an introduction to some serious theoretical mathematical questions or as a path to questions appropriate for student research projects. In this chapter we examined the questions of determining the fixed points of FDS. This leads to a question of solving systems of polynomial equations over a finite field, for which the theory of Groebner bases provides a practical solution. The algorithm is essentially a generalization of the well-known process of Gaussian elimination for solving systems of linear equations.

The actual implementation of the method requires the use of specialized software (as even the verification that a given set of polynomials is a Groebner basis for an ideal is labor intensive and virtually impossible to do by hand). Although there are several open-source computational algebra systems that compute Groebner bases (e.g., Macaulay 2, MAGMA, CoCoA, SINGULAR, and others), most such systems require download and installation. For the purposes of this chapter the web-based SAGE interface to Macaulay 2 is appropriate, as it requires only a few straightforward commands. The students can then focus on the output and its interpretation with regard to the question of solving polynomial systems of equations.

For mathematics students, we see the use of the chapter material to be threefold. On one hand, it introduces them to a new modeling approach that is currently not taught in any of the mainstream undergraduate mathematics courses. On the other hand, FDS models provide links to important mathematical theory and results in abstract algebra and algebraic geometry that can be further pursued in advanced-level courses or as independent student research projects. Finally, the topic provides evidence for the important connections between modern biology and modern mathematics, and can be used to highlight mathematical and systems biology as career paths for mathematics majors.

For biology students, the chapter can be used as an introduction to mathematical modeling without calculus prerequisites. Our experience indicates that the "just-in-time" approach for developing the necessary mathematical concepts as a way to formalize specific aspects of the biology works well for Boolean models. It allows students to focus on the logical links that determine the variable interactions instead of on the detailed kinetics needed for calculus-based models. Concurrent or subsequent introduction to such models in calculus or differential equations courses will allow students to reinforce the conceptual framework, further improve their mathematical sophistication, and solidify the retention of basic ideas.

Acknowledgments

The authors gratefully acknowledge the support of the National Science Foundation under the Division of Undergraduate Education award 0737467.

1.6 SUPPLEMENTARY MATERIALS

The online appendix, additional files, and computer code associated with this article can be found, in the online version, at http://dx.doi.org/10.1016/B978-0-12-415780-4.00021-1 and from the volume's website http://booksite.elsevier.com/9780124157804

References

[1] Kauffman S. Metabolic stability and epigenetics in randomly constructed gene nets. J Theor Biol 1969;22:437–467.

[2] Jacob F, Perrin D, Sanchez C, Monod J. L'Operon: groupe de gène à expression par un operatour. C.R. Seances Acad Sci 1960;250:1727–1729.

[3] Jacob F, Monod J. Genetic regulatory mechanisms in the synthesis of proteins. J Mol Biol 1961;3:318–356.

[4] Cheng B, Fournier RL, Relue PA. The inhibition of Escherichia coli lac operon gene expression by antigene oligonucleotidesmathematical modeling. Biotechnol Bioeng 2000;70:467–472.

[5] Doi A, Fujita S, Matsuno H, Nagasaki M, Miyano S. Constructing biological pathway models with hybrid functional Petri nets. In Silico Biology 2004;4:271–91.

[6] Farina M, Prandini M. A mathematical model for genetic regulation of the lactose operon. In: Bemorad, A. et al., editors. Hybrid systems: computation and control, Lecture Notes in Computer Science, Vol. 4416; 2007 p. 693–7.

[7] Romero-Campero FJ, Pérez-Jiménez MJ. Modelling gene expression control using P systems: the lac operon, a case study. Biosystems 2008;91:438–457.

[8] Setty Y, Mayo AE, Surette MG, Alon U. Detailed map of a cis-regulatory input function. Proc Natl Acad Sci USA 2003;100:7702–07.

[9] Tian, T., Burrage, K. A mathematical model for genetic regulation of the lactose operon. In: Gervasi, O. et al. editors. Computational science and its applications—ICCSA 2005, Lecture Notes in Computer Science, vol. 3481; 2005, p. 1245–53.

[10] van Hoek M, Hogeweg P. The effect of stochasticity on the lac operon: an evolutionary perspective. PLoS Comput Biol 2007;3:e111.

[11] Muller-Hill B. The lac operon: a short history of a genetic paradigm. Berlin: Walter De Gruyter, Inc.; 1996.

[12] Vastani H, Jarrah AS, Laubenbacher R, Visualization of dynamics for biological networks. http://dvd.vbi.vt.edu/dvd.pdf.

[13] Russell PJ. iGenetics. Benjamin Cummings, San Francisco: A Molecular Approach; 2006.

[14] Santillán M, Mackey MC, Zeron E. Origin of bistability in the lac operon. Biophys J 2007;92:3830–3842.

[15] Yildrim N, Mackey MC. Feedback regulation in the lactose operon: a mathematical modeling study and Comparison with Experimental Data. Biophys J 2003;84:2841–51.

[16] Yildirim N, Santillán M, Horike D, Mackey MC. Dynamics and bistability in a reduced model of the lac operon. Chaos 2004;14:279–292.

[17] Laubenbacher, R. Algebraic models in systems biology. In: Anai H, Horimoto K, editors. Algebraic Biology, Universal Academy Press: Tokyo; 2005, p. 33–40.

[18] Allen E, Fetrow J, Daniel L, Thomas S, John D. Algebraic dependency models of protein signal transduction networks from time-series data. J Theor Biol 2006;238:317–330.

[19] Delgado-Eckert E. An algebraic and graph theoretic framework to study monomial dynamical systems over a finite field. Complex Systems 2009;19:307–328.

[20] Jarrah AS, Laubenbacher R, Veliz-Cuba A. The dynamics of conjunctive and disjunctive Boolean network models. Bull Math Biol 2010;72:1425–47.

[21] Veliz-Cuba A, Jarrah AS, Laubenbacher R. Polynomial algebra of discrete models in systems biology. Bioinformatics 2010;26:1637–1643.

[22] Stigler B, Veliz-Cuba A, Network topology as a driver of bistability in the lac operon. 2008, arXiv:0807.3995. http://arxiv.org/abs/0807.3995

[23] Saez-Rodriguez J, Simeoni L, Lindquist JA, Hemenway R, Bommhardt U, Arndt B, Haus U, Weismantel R, Gilles ED, Klamt S, Schraven B. A logical model provides insights into T cell receptor signaling. PLOS Comp Biol 2007;3:e163.

[24] Astronomers count the stars. BBC News, July 22, 2003. http://news.bbc.co.uk/2/hi/science/nature/3085885.stm. Retrieved June 25, 2012.

[25] Laubenbacher R, Sturmfels B. Computer algebra in systems biology. Am Math Monthly 2009;116:882–891.

[26] Hinkelmann F, Laubenbacher R. Boolean models of bistable biological systems. Discrete and continuous dynamical systems 2011;4:1442–56.

[27] Veliz-Cuba A, Stigler B. Boolean models can explain bistability in the lac operon. J Comput Biol 2011;18:783–94.

[28] Novick A, Weiner M. Enzyme induction as an all-or-none phenomenon. Proc Natl Acad Sci U S A 1957;43:553–566.

Bistability in the Lactose Operon of *Escherichia coli*: A Comparison of Differential Equation and Boolean Network Models

2

Raina Robeva* and Necmettin Yildirim[†]

*Department of Mathematical Sciences, Sweet Briar College, Sweet Briar, VA 24595, USA
[†]Division of Natural Sciences, New College of Florida, FL 34243, USA

2.1 INTRODUCTION

Cellular systems are complex and involve many interacting components, which are ultimately responsible for cellular functions. Such complexity is capable of producing interesting dynamic behaviors. Among such behaviors, bistability is extremely important and is a recurring pattern in biological systems. A system is called *bistable* if it is capable of resting in two stable steady states separated by an unstable steady state. A bistable system can reside in one of the two states for a long time unless a strong external perturbation kicks in. Bistability also provides a perfect discontinuous switching among the stable steady states and can convert graded inputs into switch-like outputs.

Bistability can be either reversible or irreversible. A system is *irreversibly bistable* if it cannot go back to its earlier state once it settles down to a state after a perturbation. However, a reversible bistable system can go back and forth between the stable states as the system parameters vary. Figure 2.1a shows a reversible bistable switch and Figure 2.1b is for an irreversible bistable switch. As seen in Figure 2.1a, the value of the parameter p for the system to switch from a lower steady state to a higher steady state is different from the value of the parameter p for which a switch from a higher steady state to a lower steady state occurs. This is called hysteresis. Bistability comes with hysteresis, which can be defined as dependence of the current state of a system not only on its current state but also on its past state. To predict future state of a bistable system, both its current state and its history must be known.

Mathematical Concepts and Methods in Modern Biology. http://dx.doi.org/10.1016/B978-0-12-415780-4.00002-8

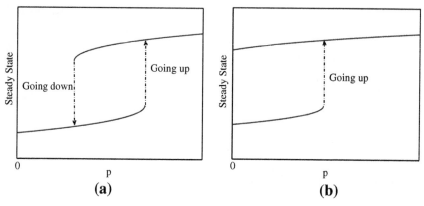

FIGURE 2.1

Bistability and hysteresis emerging in cellular reaction networks. Bistability can be reversible (a) or irreversible (b). In reversible bistable systems, the steady states can change from low to high or high to low state as the parameter p is tuned. However, the steady state changes from a low state to a high state at a higher value of p than that for which the change from a high steady state to a low steady state occurs. In an irreversible bistable system, once a system shifts from a low state to a high state, it stays there even when the value of the parameter p is set back to a value much lower than that where an earlier shift took place.

In the case of the lactose (*lac*) operon in *Escherichia coli* (*E. coli*), it has been known since the 1950s [1] that it exhibits a hysteresis effect: In the absence of glucose, the operon is uninduced for low concentrations below a threshold L_1 of extracellular lactose and fully induced at high concentrations above a threshold of L_2. In the interval (L_1, L_2), both induced and uninduced cells can be observed and their status depends on the cell history (the system response is hysteresis). The interval (L_1, L_2) is defined as a region of bistability and is referred to as *maintenance concentration*. Cells grown in an environment poor in extracellular lactose will have low levels of internal lactose and allolactose and may remain uninduced for lactose levels in the interval (L_1, L_2). In contrast, cells grown in a lactose-rich environment remain induced for concentrations in the interval (L_1, L_2). A recent discussion of the bistability feature of the *lac* operon together with green-fluorescence and inverted phase-contrast images of a cell population showing a bimodal distribution of *lac* expression levels can be found in [2]. A survey of the quantitative approaches to the study of bistability in the *lac* operon is given in [3].

In this chapter we examine two types of mathematical models of the *lac* operon: differential equation models and Boolean network models. In both cases the focus will be on bistability and on the capability of the models to capture the bistable behavior of the system. The chapter is organized as follows: Section 2.2 presents a short description of the regulatory mechanism of the *lac* operon. In Section 2.3 we give a step-by-step tutorial on how to model biochemical reactions using differential equations. These approaches are then used in Section 2.4 to justify several widely used differential equation models of the *lac* operon. Section 2.5 extends the material from Chapter 1 on Boolean networks, describing Boolean models with varying elimination times and Boolean models involving delays. In Section 2.6 we build Boolean model approximations of the differential equation models from Section 2.4 and demonstrate that those models capture the bistability of the system. Section 2.7 contains some closing comments and conclusions.

2.2 THE LACTOSE OPERON OF *ESCHERICHIA COLI*

The *lac* operon is a well-known example of an inducible genetic circuit. It has been serving as a model system for understanding many aspects of gene regulations since its discovery in the late 1950s. The *lac* operon encodes the genes for internalization of extracellular lactose and then its conversion to glucose. A cartoon that depicts the major regulatory components of this system is shown in Figure 2.2. The *lac* operon consists of a promoter/operator region (*P* and *O*) and three structural genes *lacZ*, *lacY*, and *lacA*. A regulatory gene *lacI* (*I*) preceding the *lac* operon is responsible for producing a repressor (*R*) protein. The molecular mechanism of this operon is as follows: In the presence of allolactose, a complex between allolactose and the

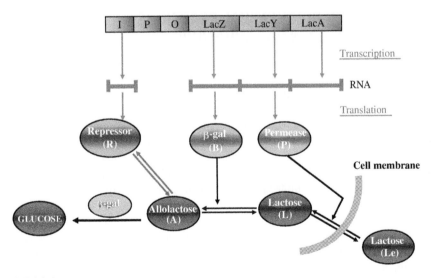

FIGURE 2.2

Schematic representation of the lactose operon regulatory system.
See the text for details. Reprinted from *Methods in Enzymology*, v. 484, Yildirim, N. and Kazanci, C., Deterministic and stochastic simulation and analysis of biochemical reaction networks the lactose operon example, p. 371–395, (2011), with permission from Elsevier.

repressor is formed. This complex is unable to bind to the operator region, which allows the RNA polymerase to bind to the promoter site, starting initiation of transcription of the structural genes to produce $mRNA(M)$. However, in the absence of allolactose (A) the repressor protein R binds to the operator region (O) and blocks the RNA polymerase from transcribing the structural genes. The translation process takes place once the mRNA has been produced. The *lacZ* gene encodes the portion of the mRNA for the production of the enzyme β-galactosidase (B) and translation of the *lacY* gene produces the mRNA that is needed for the production of the enzyme permease (P). The final portion of mRNA produced by the *lacA* gene is responsible for the production of the enzyme thiogalactoside transacetylase, which does not play a role in the regulation of the *lac* operon [4].

A bistable system often involves a positive feedback loop or a double negative feedback. Such motifs are very common in control mechanisms in gene networks. The lactose operon and the arabinose operon of *Escherichia coli* both have positive feedback mechanisms [5,6]. In the lactose operon, once external lactose is converted into allolactose, the allolactose feeds back and forms a complex with the lactose repressor, which allows the transcription process to start. This eventually leads to synthesis of more enzymes to transport more external lactose into the cell and then convert it into allolactose. More information about this mechanism and about the *lac* operon in general can be found in Chapter 1.

2.3 MODELING BIOCHEMICAL REACTIONS WITH DIFFERENTIAL EQUATIONS

Differential equations are widely used to study reaction kinetics. In this section we outline the basics of reaction modeling with differential equations. This approach will be applied in Section 2.4 to justify two well-known differential equation models of the *lac* operon.

Two or more species can react if they come together and collide. Collision alone is not enough for a reaction to occur since the molecules have to collide the right way and with enough energy. Even if the molecules are orientated properly when they collide, a reaction may not happen if the molecules do not have enough energy. On the other hand, if the molecules collide with enough energy but their orientations are not appropriate, then the reaction still may not occur. Reactions with a single species, such as degradation reactions, do not require orientation of collisions. Only if they have enough energy will those reactions take place. Most of the reactions involve one or two species. Reactions with three or more species are extremely uncommon.

The rate of a reaction reflects how fast or slow it takes place. There are different approaches and methodologies to studying reaction rates. Mass-action kinetics is a microscopic approach to reaction kinetics and it is one of the most common methods. Modeling based on mass-action kinetics results in a system of differential equations as a mathematical description of the reaction rates. This approach is fully deterministic and it is only appropriate under certain conditions, such as constant temperature, well-mixed solution, large number of molecules, etc. In this section, we briefly describe how to write differential equation models from a reaction mechanism.

Reactions can be classified based on their reactants. The reaction $A \rightarrow P$ is called uni-molecular, since one reactant A becomes a product P. The reaction $A + B \rightarrow P$ is called bi-molecular, since there are two reactants A and B becoming a product P. A tri-molecular reaction looks like $A + B + C \rightarrow P$, and has three reactants A, B, and C.

According to the mass-action kinetics, a reaction rate is proportional to the probability of collision of the reactants involved. At higher concentration, the collisions occur more often. This probability can be taken to be proportional to each reactant concentration.

Now consider a uni-molecular reaction in which A becomes P,

$$A \xrightarrow{k} P. \tag{2.1}$$

Here k is called a kinetic rate constant describing how likely it is for this reaction to occur and produce the product P. According to mass action kinetics, the rate of this reaction can be written as

$$v = \frac{d[P]}{dt} = k[A].$$

Here $[A]$ and $[P]$ represent concentrations of A and P, respectively; k is a first order rate constant. Units for $[A]$ and $[P]$ are concentrations and unit of $\frac{d[P]}{dt}$ is concentration per time. Therefore, unit of the rate constant k has to be time^{-1}.

A bi-molecular reaction looks like,

$$A + B \xrightarrow{k} P. \tag{2.2}$$

In this reaction the reactants A and B react and become a product P with a rate constant k. The rate constant quantifies how likely it is that the collision of A and B ends up with the product P. The rate of this reaction is

$$v = \frac{d[P]}{dt} = k[A][B]. \tag{2.3}$$

Here k is a second order rate constant and its unit becomes $(\text{concentration} \times \text{time})^{-1}$.

The reactions given in both Eq. (2.1) and Eq. (2.2) are uni-directional reactions. In theory, all chemical reactions are reversible. Now let's consider a two-directional reaction like

$$A + B \underset{k_2}{\overset{k_1}{\rightleftharpoons}} P. \tag{2.4}$$

In this reaction, A and B react and become P with an associated rate constant k_1. On the other hand, P can break down and produce A and B. The rate constant for the backward reaction is k_2. Here P is produced at a rate $v_f = k_1[A][B]$ and consumed at a rate $v_b = k_2[P]$. Hence, how the concentration of P changes over time is given by the difference of these two rates:

$$v = \frac{d[P]}{dt} = v_f - v_b \tag{2.5}$$

$$= k_1[A][B] - k_2[P]. \tag{2.6}$$

In this equation k_1 is a second order rate constant and k_2 is a first order rate constant.

2.3.1 Enzymatic Reactions and the Michaelis-Menten Equation

Enzymes are specific proteins that catalyze reactions. An enzyme can increase the rate of a reaction up to 10^{12}-fold [7], compared to the spontaneous reaction without the enzyme. The enzyme first binds to its substrate (reactant), forms a complex (an enzyme-substrate complex), and performs a chemical operation on it. Then it releases from the complex, resulting in conversion of the substrate into a product. Enzymes stay unchanged after the reaction. Some enzymes bind to a single substrate while others can bind to multiple substrates and combine them to produce a final product.

Consider now the enzyme catalyzed reaction in Eq. (2.7), which involves three individual reactions.

$$E + S \underset{k_2}{\overset{k_1}{\rightleftharpoons}} ES \overset{k_3}{\longrightarrow} P + E. \tag{2.7}$$

The first reaction is $E + S \overset{k_1}{\longrightarrow} ES$, where an enzyme E binds to a substrate S and forms an enzyme-substrate complex ES with an associated rate constant k_1. Since this is a reversible reaction, ES can break down into E and S ($ES \overset{k_2}{\longrightarrow} E + S$). The associated rate constant for this backward reaction is k_2. The third reaction is $ES \overset{k_3}{\longrightarrow} P + E$ in which the enzyme E releases from ES, producing a product P with a rate constant k_3.

The differential equation describing the dynamics of the concentration of the enzyme-substrate complex ES is the difference between the gain and loss terms. ES is produced with the first reaction and consumed with the second and third reactions. Hence, we have

$$\frac{d[ES]}{dt} = k_1[E][S] - (k_2 + k_3)[ES]. \tag{2.8}$$

The dynamics of the product P are modeled by Eq. (2.9). This equation has a single term, since P is produced by the third reaction and is not consumed by any of the reactions.

$$\frac{d[P]}{dt} = k_3[ES]. \tag{2.9}$$

If the total concentration of the enzyme stays constant over the duration of this reaction, we can write,

$$E_0 = [E] + [ES], \tag{2.10}$$

where E_0 represents the initial enzyme concentration.

Equations (2.8)–(2.9) now describe the dynamics of the single enzyme single substrate reaction in Eq. (2.7). However, with an additional assumption, these two equations can further be simplified. In an enzymatic reaction, the enzyme-substrate complex reaches a steady state much earlier than the product does. This allows us to take $\frac{d[ES]}{dt} \approx 0$. Therefore,

$$\frac{d[ES]}{dt} = k_1[E][S] - (k_2 + k_3)[ES] = 0. \tag{2.11}$$

Solving Eq. (2.11) for $[E]$ yields

$$[E] = \frac{(k_2 + k_3)[ES]}{k_1[S]}. \tag{2.12}$$

After plugging $[E]$ from Eq. (2.12) into the enzyme conversation equation given by Eq. (2.10) and solving the resultant equation for $[ES]$, we obtain $[ES]$ in terms of only $[S]$:

$$[ES] = \frac{E_0[S]}{\frac{k_2+k_3}{k_1} + [S]}. \tag{2.13}$$

Substituting $[ES]$ given by Eq. (2.13) into Eq. (2.9) yields

$$\frac{d[P]}{dt} = \frac{V_{\max}[S]}{K_m + [S]}, \tag{2.14}$$

where

$$V_{\max} = k_3 E_0, \qquad K_m = \frac{k_2 + k_3}{k_1} \tag{2.15}$$

are two positive parameters. This equation is called the *Michaelis-Menten* equation. The right-hand side of this equation $f([S]) = \frac{V_{\max}[S]}{K_m+[S]}$ is an increasing function of the substrate concentration with two properties: (i) $\lim_{[S] \to \infty} f([S]) = V_{\max}$, and (ii) $f([S] = K_m) = \frac{V_{\max}}{2}$. From a biological point of view, this reaction never occurs at a rate greater than $V_{\max}$. K_m is the substrate concentration at which the reaction rate is equal to half of its maximum value $V_{\max}$. Furthermore, since $V_{\max} = k_3 E_0 > 0$, the reaction rate is a linear increasing function of the initial enzyme concentration E_0, which means that higher initial enzyme concentration makes the reaction go faster and this relationship between the rate and E_0 is linear.

If the total substrate concentration is conserved throughout the course of the reaction, we can write

$$S_0 = [S] + [ES] + [P]. \tag{2.16}$$

Since the initial concentration of the enzyme is usually much smaller than the initial substrate concentration, the concentration of *ES* is negligibly small compared to either the substrate or the product concentration. Hence, (2.16) simplifies to

$$S_0 = [S] + [P]. \tag{2.17}$$

Differentiating both sides of this equation with respect to t yields $\frac{-d[S]}{dt} = \frac{d[P]}{dt}$. Hence, Eq. (2.14) becomes

$$\frac{d[S]}{dt} = -\frac{V_{\max}[S]}{K_m + [S]}. \tag{2.18}$$

The evolution of the substrate concentration can be simulated by solving this differential equation after fixing $V_{\max}$ and K_m and assigning a value to the initial concentration of the substrate. Then, from Eq. (2.17), one can compute the progress curve for the concentration of the product.

Exercise 2.1.   Consider the reactions where two substrates S_1 and S_2 compete for binding to an enzyme E to produce two different products P_1 and P_2. Under the assumption that each reaction follows the Michaelis-Menten kinetics, derive rate equations for P_1 and P_2 in this system and explain the effects of the competition occurring.

$$E + S_1 \underset{k_2}{\overset{k_1}{\rightleftharpoons}} ES_1 \overset{k_3}{\longrightarrow} P_1 + E.$$

$$E + S_2 \underset{p_2}{\overset{p_1}{\rightleftharpoons}} ES_2 \overset{p_3}{\longrightarrow} P_2 + E.$$

$\triangledown$

2.3.2 Multi-Molecule Binding and Hill Equations

Some enzymes react with more than one substrate molecule, e.g., the reaction in Eq. (2.19), in which n-molecules of a substrate react with an enzyme E. For simplicity, let's assume that the binding sides on the enzyme are all identical, the bindings take place simultaneously and the bindings of n-molecules are independent. Otherwise, the influence of each binding step to other binding steps has to be considered.

$$E + nS \underset{k_2}{\overset{k_1}{\rightleftharpoons}} ES_n \overset{k_3}{\rightarrow} P + E. \tag{2.19}$$

This reaction consists of three individual reactions. In the first reaction, $E + nS \overset{k_1}{\longrightarrow} ES_n$, n-molecules of a substrate S interact with one molecule of an enzyme E to form an enzyme-substrate complex ES_n with a rate constant k_1. $ES_n \overset{k_2}{\longrightarrow} E + nS$ is the second reaction, in which the enzyme-substrate complex ES_n can break down into the enzyme E and n-substrate S molecules with a rate constant k_2. In the third reaction, $ES_n \overset{k_3}{\longrightarrow} E + P$, the enzyme E releases unchanged from the complex and a product P is produced with a rate constant k_3. The differential equation modeling dynamics of the concentration of the enzyme-substrate complex ES_n is the difference between the production and consumption rates of this complex. ES_n is produced with the first reaction and it is used up with both the second and the third reactions. Hence,

$$\frac{d[ES_n]}{dt} = k_1[E][S]^n - (k_2 + k_3)[ES_n]. \tag{2.20}$$

Since the product P is produced by the third reaction and it is not consumed, the dynamics of the product P concentration can be modeled as in Eq. (2.21).

$$\frac{d[P]}{dt} = k_3[ES_n].$$ (2.21)

Suppose that the complex ES_n reaches a steady state faster than the product P. Then, Eq. (2.20) becomes

$$\frac{d[ES_n]}{dt} = k_1[E][S]^n - (k_2 + k_3)[ES_n] = 0.$$ (2.22)

Solving Eq. (2.22) for $[E]$ gives

$$[E] = \frac{(k_2 + k_3)[ES_n]}{k_1[S]^n}.$$ (2.23)

Assuming that the total amount of enzyme is conserved throughout the reaction, we have

$$E_0 = [E] + [ES_n],$$ (2.24)

where E_0 stands for the total enzyme concentration. Substituting Eq. (2.23) into Eq. (2.24) and then solving it for $[ES_n]$, we obtain

$$[ES_n] = \frac{E_0[S]^n}{\frac{k_2+k_3}{k_1} + [S]^n}.$$

Then the dynamics of the concentration of P given by Eq. (2.21) becomes

$$\frac{d[P]}{dt} = \frac{V_{\max}[S]^n}{K_m + [S]^n},$$ (2.25)

where n, $V_{\max}$ and K_m are positive constants. The parameters $V_{\max}$ and K_m are as in Eq. (2.15), and n is a *Hill coefficient*. This equation is known as the *Hill* equation. Unlike the Michaelis-Menten equation, the Hill equation has three adjustable parameters. The right-hand side of the equation $f([S]) = \frac{V_{\max}[S]^n}{K_m[S]^n}$ is a strictly increasing function of $[S]$ and satisfies the following two conditions: (i) $\lim_{[S] \to \infty} f([S]) = V_{\max}$, and (ii) $f([S] = K_m^{1/n}) = \frac{V_{\max}}{2}$. Biologically, this reaction can never take place at a rate larger than $V_{\max}$. The value $K_m^{1/n}$ is the substrate concentration necessary for this reaction to occur at a rate equal to half of the maximum rate, $V_{\max}$. Since $V_{\max} = k_3 E_0$, increasing the initial enzyme concentration linearly increases the reaction rate. When $n \gg 1$, the Hill function behaves like a step function. For example, for large n values, $f([S]) \approx 0$ when $[S]$ is small. On the other hand, $f([S]) = V_{\max}$ when $[S]$ is big (see Figure 2.3).

When the initial substrate concentration is conserved throughout the course of the reaction and the initial enzyme concentration is much smaller than that of the substrate

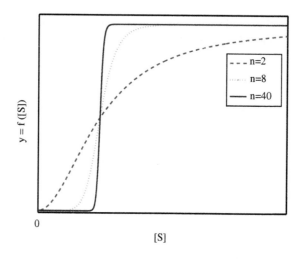

FIGURE 2.3

Representative curves from the Hill equation given by Eq. (2.25) for various values of the Hill coefficient $n \geqslant 2$ when both K_m and V_{max} are constant. The curves are all sigmoidal and they behave like a step function when n is big (solid curve).

concentration, Eq. (2.25) can be written in terms of the substrate consumption rate. That is, when

$$S_0 = [S] + [P],$$
(2.26)

where S_0 is the initial substrate concentration, Eq. (2.25) becomes

$$\frac{d[S]}{dt} = -\frac{V_{max}[S]^n}{K_m + [S]^n}.$$

Notice that the Hill function reduces down to a Michaelis-Menten function when $n = 1$. Both the Hill function and the Michaelis-Menten function are commonly used in modeling of biochemical reaction networks.

Exercise 2.2. The Hill equation is an approximation for multi-molecule binding and it assumes simultaneous binding of n-molecules of a substrate S to the enzyme E above. Now suppose that two molecules of the substrate S are undergoing a reaction with an enzyme in an ordered manner as follows:

$$E + S \underset{k_2}{\overset{k_1}{\rightleftharpoons}} ES + S \underset{k_4}{\overset{k_3}{\rightleftharpoons}} ES_2 \overset{k_5}{\rightarrow} P + E.$$
(2.27)

Derive a rate equation under the steady state assumption and compare it with the Hill equation given in Eq. (2.25). When do these two equations become roughly the same?

$\triangledown$

2.4 THE YILDIRIM-MACKEY DIFFERENTIAL EQUATION MODELS FOR THE LACTOSE OPERON

In this section, we talk about two delay differential equation models [8,9] recently developed to study the *lac* operon. Both models employ experimentally estimated parameters and are compared against independently collected, published experimental data from this genetic circuit. The first model is a 3 variable model and only takes into account mRNA (*M*), β-galactosidase (*B*), and allolactose (*A*) dynamics. Internal lactose (*L*) is assumed to be readily available and is included in the model as a parameter [8]. In the second model, besides these three dynamic variables, the internal lactose (*L*) and permease (*P*) are also explicitly modeled [9].

2.4.1 Model Justification

We need a few more preliminaries before presenting the differential equation models.

Modeling dilution in protein concentration due to bacterial growth: When developing a mathematical model for bacterial systems like the *lac* operon, it is important to know how fast the bacteria actually grow. Depending on the environmental conditions, an *E. coli* population can double in size as fast as every 20 min [10]. This increase in the cellular volume results in a dilution in the protein concentrations, which can be critical in developing reliable mathematical models. Dilution in protein concentrations due to bacterial growth can be modeled as follows: Suppose that *V* is the average volume of a bacterial cell and *x* represents the number of molecules of a protein *X* in that cell. Assume that the cell volume increases exponentially in time. Then we have,

$$\frac{dV}{dt} = \mu V \tag{2.28}$$

where $\mu > 0$ denotes the growth rate. Also, assume that the degradation of X is exponential. Then we can write,

$$\frac{dx}{dt} = -\beta x, \tag{2.29}$$

where $\beta > 0$ represents this decay rate. Concentration of the protein *X* is equal to the number of molecules *x* of *X* divided by volume of the cell, namely

$$[x] = \frac{x}{V}, \tag{2.30}$$

where [*x*] stands for concentration of the protein *X*. Differentiating both sides of Eq. (2.30) with respect to time *t* results in

$$\frac{d[x]}{dt} = \left(\frac{dx}{dt} V - \frac{dV}{dt} x \right) \frac{1}{V^2}. \tag{2.31}$$

Substituting Eq. (2.28) and Eq. (2.29) into Eq. (2.31) and making necessary simplifications gives us

$$\frac{d[x]}{dt} = -\left(\mu + \beta\right)[x].\qquad(2.32)$$

As seen in this equation, an increase in the volume over time can successfully be modeled by adding the growth rate to the protein degradation rate.

Modeling of the lactose repressor dynamics: The repressor protein plays a major role in the regulation of the *lac* operon. In both absence and presence of the external lactose, the repressor protein is produced at a constant rate. A binary complex between allolactose and the repressor protein is formed. This complex cannot bind to the operator site (O). Yagil and Yagil [11] have modeled the dynamics of the repressor protein (R) as follows:

$$R + nA \overset{K_1}{\rightleftharpoons} RA_n.\qquad(2.33)$$

Here n represents the effective number of allolactose molecules required to form this complex. If we assume this reaction is at equilibrium, we have

$$K_1 = \frac{[RA_n]}{[R][A]^n},\qquad(2.34)$$

where K_1 is the association constant for this reaction. The repressor molecules can also bind to the operator region (O) in the absence of allolactose and block the transcription process. We assume this interaction is of the following form:

$$O + R \overset{K_2}{\rightleftharpoons} OR.\qquad(2.35)$$

At equilibrium, we have

$$K_2 = \frac{[OR]}{[R][O]}.\qquad(2.36)$$

Here K_2 is also an association constant of this reaction. Let O_{tot} be the total operator concentration. It is plausible to take the total concentration of the operator as constant. Therefore,

$$O_{tot} = [O] + [OR].\qquad(2.37)$$

From Eq. (2.36) and (2.37), we can write

$$\frac{[O]}{O_{tot}} = \frac{1}{1 + K_2[R]}.\qquad(2.38)$$

Now, let R_{tot} be the total concentration of the repressor protein. Since the repressor protein is not regulated by the extracellular lactose, we can assume that the total concentration of this protein also stays unchanged. Hence,

$$R_{tot} = [R] + [OR] + [RA_n].\qquad(2.39)$$

Since there are at most a few molecules of the operator site per cell, it is reasonable to assume $[OR] \ll \max\{[R], [RA_n]\}$. Then Eq. (2.39) simplifies to

$$R_{\text{tot}} = [R] + [RA_n]. \tag{2.40}$$

Putting Eq. (2.34) into Eq. (2.40) and solving the resultant equation for $[R]$, we obtain

$$[R] = \frac{R_{\text{tot}}}{1 + K_1[A]^n}. \tag{2.41}$$

By plugging the repressor protein concentration from Eq. (2.41) into Eq. (2.38), we obtain the portion of free operator site concentration in terms of the allolactose concentration as

$$\frac{[O]}{O_{\text{tot}}} = \frac{1 + K_1[A]^n}{K + K_1[A]^n}, \tag{2.42}$$

where $K = 1 + K_2 R_{\text{tot}}$. This equation describes the available operator concentration for RNA polymerase binding and for initiating the transcription of the structural genes. The right-hand side of the function in Eq. (2.42) is a sum of two functions: (i) a decreasing Hill function, $\frac{1}{K+K_1[A]^n}$, and (ii) an increasing Hill function $\frac{K_1[A]^n}{K+K_1[A]^n}$ of the form given in Eq. (2.25). Denote $f([A]) = \frac{1+K_1[A]^n}{K+K_1[A]^n}$ and notice that f takes a positive value $1/K$ when $[A] = 0$. Since f is an increasing function of $[A]$ and the concentration of A is always non-negative, this is the lowest value f can take. Biologically this can be interpreted as the process of ongoing transcription of the structural genes occurring at a constant rate in the absence of allolactose, and providing basal concentrations of the proteins encoded by the structural genes. On the other hand, when allolactose is abundant, $\lim_{[A] \to \infty} f([A]) = 1$. This can be interpreted as having all of the operator sites available to mRNA polymerase for transcription of the structural genes, with transcription occurring at the maximal possible rate. Yagil and Yagil [11] have experimentally shown that Eq. (2.42) can accurately fit their observations and estimated the parameters K, K_1, and n from the data.

The Yildirim-Mackey models of the lac operon: The dynamics of mRNA is modeled with the following equation

$$\frac{dM}{dt} = \alpha_M \frac{1 + K_1 \left(e^{-\mu \tau_M} A_{\tau_M} \right)^n}{K + K_1 \left(e^{-\mu \tau_M} A_{\tau_M} \right)^n} - \widetilde{\gamma_M} M, \tag{2.43}$$

where $A_{\tau_M} = A(t - \tau_M)$ and $\widetilde{\gamma_M} = \gamma_M + \mu$. The rate of change in mRNA concentration is the difference between its gain and loss terms. We assume that the production rate of mRNA is proportional to the fraction of free operators as described in Eq. (2.42). Here α_M represents such a proportionality constant. A_{τ_M} represents the allolactose concentration at time $(t - \tau_M)$. The production of mRNA from DNA via transcription is not an instantaneous process. In fact, it requires a period of time τ_M for RNA polymerase to transcribe the first ribosomes binding site. Hence, the production rate of mRNA is not a function of the available allolactose concentration at time t, but a function of the available concentration of allolactose at time $t - \tau_M$. The exponential prefactor $e^{-\mu \tau_M}$ accounts for the dilution of the allolactose through bacterial growth during the transcriptional period, which can be derived as follows:

Suppose that a protein P decays exponentially. If p represents the concentration of this protein, then we have

$$\frac{dp}{dt} = -\mu p, \tag{2.44}$$

where $\mu > 0$ quantifies how fast this degradation happens. When this equation is integrated from $(t - \tau_M)$ to t, we obtain

$$p(t) = e^{-\mu \tau_M} p_{\tau_M}, \tag{2.45}$$

where $p_{\tau_M} = p(t - \tau_M)$. This derivation explains the prefactor $e^{-\mu \tau_M}$ in Eq. (2.43).

Exercise 2.3.   Derive the function $p(t)$ in Eq. (2.45) from Eq. (2.44).   ▽

In the 5-variable model, the mRNA equation has an extra parameter Γ_0 that accounts for the basal transcription rate in the absence of the extracellular lactose. Therefore, we use Eq. (2.46) for the dynamics of mRNA concentration in the 5 variable model.

$$\frac{dM}{dt} = \alpha_M \frac{1 + K_1 \left(e^{-\mu \tau_M} A_{\tau_M}\right)^n}{K + K_1 \left(e^{-\mu \tau_M} A_{\tau_M}\right)^n} + \Gamma_0 - \widetilde{\gamma_M} M. \tag{2.46}$$

The basal transcription rate for mRNA is not explicitly modeled in Eq. (2.43) but is included in the parameters α_M and K, instead. The second term in Eq. (2.43) models the loss in mRNA concentration. It is a sum of two terms and given by $\widetilde{\gamma_M} M = \gamma_M M + \mu M$. The term $\gamma_M M$ accounts for the loss due to mRNA degradation and the term μM is the effective loss due to the bacterial growth, as explained in Eq. (2.32).

Equation (2.47) models the dynamics of the β-galactosidase concentration. It can be assumed that the rate of production of β-galactosidase is directly proportional to the mRNA concentration at time $(t - \tau_B)$ with a proportionality constant α_B.

$$\frac{dB}{dt} = \alpha_B e^{-\mu \tau_B} M_{\tau_B} - \widetilde{\gamma_B} B. \tag{2.47}$$

Here τ_B is the time required for the translation of mRNA, $\widetilde{\gamma_B} = \gamma_B + \mu$ and $M_{\tau_B} = M(t - \tau_B)$.

Equation (2.48) governs the allolactose dynamics. The first term on the right-hand side of this equation gives the gain in allolactose due to the conversion of lactose mediated by β-galactosidase. We assume that this conversion follows the Michaelis-Menten equation as derived in Eq. (2.14).

$$\frac{dA}{dt} = \alpha_A B \frac{L}{K_L + L} - \beta_A B \frac{A}{K_A + A} - \widetilde{\gamma_A} A. \tag{2.48}$$

The second term denotes the loss of allolactose due to its conversion to glucose and galactose by β-galactosidase, which is also assumed to follow a Michaelis-Menten kinetics. The last term accounts for the loss in the allolactose concentration due to its degradation and dilution. The equations for the 3-variable model are presented in Table 2.1.

Table 2.1 The 3 variable Yildirim-Mackey Model (from [8]).

$$\frac{dM}{dt} = \alpha_M \frac{1 + K_1 \left(e^{-\mu \tau_M} A_{\tau_M}\right)^n}{K + K_1 \left(e^{-\mu \tau_M} A_{\tau_M}\right)^n} - \widetilde{\gamma}_M M$$

$$\frac{dB}{dt} = \alpha_B e^{-\mu \tau_B} M_{\tau_B} - \widetilde{\gamma}_B B \tag{2.49}$$

$$\frac{dA}{dt} = \alpha_A B \frac{L}{K_L + L} - \beta_A B \frac{A}{K_A + A} - \widetilde{\gamma}_A A$$

In addition to the equations in Table 2.1, the 5-variable model includes specific equations for the dynamics of permease and internal lactose. The equation for the dynamics of the permease is similar to the β-galactosidase equation and it is given in Eq. (2.50).

$$\frac{dP}{dt} = \alpha_P e^{-\mu(\tau_B + \tau_P)} M_{\tau_B + \tau_P} - \widetilde{\gamma}_P P. \tag{2.50}$$

Here $M_{\tau_B + \tau_P} = M(t - (\tau_B + \tau_P))$ and $\widetilde{\gamma}_P = \gamma_P + \mu$. The first term models the gain in the permease concentration due to the transcriptional induction of *lac*Y gene by the external lactose, and it is assumed to be proportional to the mRNA concentration at a time $\tau_B + \tau_P$ ago. Here α_P is the proportionality constant for this process, and τ_B and τ_P are the times required for the translation of functional β-galactosidase and permease, respectively. The term $e^{-\mu(\tau_B + \tau_P)}$ models the dilution in mRNA concentration due to cellular growth during the translation of such enzymes. The last term is for the loss in the permease concentration due to cellular growth and degradation of this enzyme.

The equation governing the dynamics of internal lactose is given in Eq. (2.51). The first three terms in this equation are in the Michaelis-Menten form. The first term models the gain in the concentration of lactose due to the transport of external lactose by the permease.

$$\frac{dL}{dt} = \alpha_L P \frac{L_e}{K_{L_e} + L_e} - \beta_{L_1} P \frac{L}{K_{L_1} + L} - \alpha_A B \frac{L}{K_L + L} - \widetilde{\gamma}_L L. \tag{2.51}$$

Here L_e denotes the concentration of external lactose, which is a parameter for the 5-variable model. The permease-mediated transport of lactose through the cellular membrane is reversible [12]. The second term in Eq. (2.51) accounts for the loss in the internal lactose concentration due to the reversible nature of this transport process. The third term describes the loss in the lactose level due to its conversion to allolactose, which is mediated by β-galactosidase. The last term models the loss in the lactose concentration due to the cellular growth and its degradation. The equations for the 5-variable model are presented in Table 2.2.

Table 2.2 The 5 variable Yildirim-Mackey Model (from [9]).

$$\frac{dM}{dt} = \alpha_M \frac{1 + K_1 \left(e^{-\mu \tau_M} A_{\tau_M}\right)^n}{K + K_1 \left(e^{-\mu \tau_M} A_{\tau_M}\right)^n} + \Gamma_0 - \widetilde{\gamma_M} M$$

$$\frac{dB}{dt} = \alpha_B e^{-\mu \tau_B} M_{\tau_B} - \widetilde{\gamma_B} B$$

$$\frac{dA}{dt} = \alpha_A B \frac{L}{K_L + L} - \beta_A B \frac{A}{K_A + A} - \widetilde{\gamma_A} A \tag{2.52}$$

$$\frac{dP}{dt} = \alpha_P e^{-\mu(\tau_B + \tau_P)} M_{\tau_B + \tau_P} - \widetilde{\gamma_P} P$$

$$\frac{dL}{dt} = \alpha_L P \frac{L_e}{K_{L_e} + L_e} - \beta_{L_1} P \frac{L}{K_{L_1} + L} - \alpha_A B \frac{L}{K_L + L} - \widetilde{\gamma_L} L$$

2.4.2 Numerical Simulation of the Yildirim-Mackey Models and Bistability

In this section, we will perform steady state analysis and numerical simulations for both models. The models consist of delay differential equations with discrete time delays due to the transcription and translation processes. Besides delay parameters, they also involve a number of unknown parameters that need to be fixed in order to perform steady state analysis and numerical simulations. Yildirim et al. [9,8] did an extensive literature search to estimate these parameters from published articles. The values and details on estimation of the parameters can be found in [9,8,13]. In our simulations in this chapter, we have used two different μ values, as was done in the original papers. For simulations with the 3 variable model we took $\mu = 3.03 \times 10^{-2}$ min^{-1}. For the simulations with the 5 variable model, a smaller μ value was used, $\mu = 2.26 \times 10^{-2}$ min^{-1}. These values were estimated by fitting the differential equation models to experimental data.

Experimental results showing that the *lac* operon in *Escherichia coli* is capable of showing bistable behavior for a range of extracellular lactose concentrations have been available since the late 1950s [1,14] and more recently in [2]. To see if the 3 variable model can show bistability, steady state analysis is performed, which can be studied by setting the left-hand side of each equation in the 3 variable model given in Table 2.1 to zero and solving it for a range of L concentrations after keeping all the other parameters fixed at their estimated values. The result is shown in (Figure 2.4). The 3 variable model predicts that there is a range for the internal lactose concentration, which corresponds to the S-shaped curve in the figure. When the lactose concentration is in this range, the *lac* operon can have three coexisting steady states.

Figure 2.5 shows how the bistable behavior arises in the time series simulation of the 3 variable model. For this simulation, all the parameters are kept constant at their estimated values when $L = 50 \times 10^{-3}$ mM. As seen from the steady state curve in (Figure 2.4), there are three distinct steady states for this particular concentration of L.

Table 2.3 The steady state values calculated by solving the nonlinear system of equations by setting time derivatives to zero in the 3 variable model when $L = 50 \times 10^{-3}$ mM for which there exist three steady states.

Steady States	A* (mM)	M* (mM)	B*(mM)
I	4.27×10^{-3}	4.57×10^{-7}	2.29×10^{-7}
II	1.16×10^{-2}	1.38×10^{-6}	6.94×10^{-7}
III	6.47×10^{-2}	3.28×10^{-5}	1.65×10^{-5}

We calculated these steady state values numerically as in Table 2.3. The steady state values were calculated by solving the nonlinear system of equations obtained by setting time derivatives to zero in the model equations in Table 2.1 and solving it for the concentrations. Then six distinct initial values were chosen for the protein concentration around the unstable steady state concentration (steady state II in Table 2.3). Three of them were below the unstable steady state and the other three were above it. As expected, three initials converged to the lower steady state (steady state I in Table 2.3), the other three ended up settling at the higher stable steady state (steady state III in Table 2.3).

Exercise 2.4.   The Lac repressor protein is a tetramer of identical subunits [11]. It has been shown experimentally that two allolactose molecules on average bind to this protein and effectively block it so that transcription of new proteins can take place

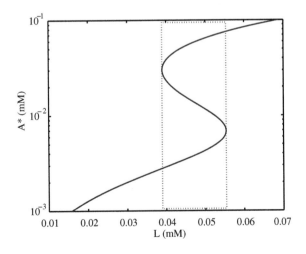

FIGURE 2.4

Bistability arises in $(L; A^*)$ space in the 3 variable *lac* operon model (y-axis is in logarithmic scale). For a range of L concentrations there are three coexisting steady states for the allolactose concentration. We estimated this range to be $(0.039, 0.055)$ mM of L concentration.

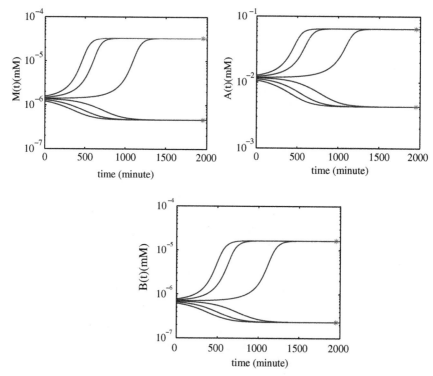

FIGURE 2.5

Time series simulation of the mRNA, β-galactosidase and allolactose concentrations. These plots were produced by numerically solving the 3 variable model when $L = 50 \times 10^{-3}$ mM. For this value of the internal lactose concentration, there exist three coexisting steady states (see Figure 2.4). The (∗)'s in these plots represent the location of the low and high stable steady states. See the text for selection of the initials.

in the presence of external lactose. Now suppose that three molecules of allolactose are needed for effective blockage of the repressor protein. Numerically study how this will affect the bistability range in this system. Use the 3 variable model and the parameter values given in Table I in the paper [8]. You should take the Hill coefficient as $n = 3$.

MATLAB starter code is provided for this exercise. Open the file *Code_for_Ex_2_4_Starter.m* and add the appropriate lines of code. Note that this exercise requires the use of the MATLAB's Global Optimization Toolbox. ▽

The 5 variable model was analyzed in a similar way. Figure 2.6 shows a plot of the allolactose steady state values as responses to the external lactose concentration in the 5 variable model. To produce this plot, the system of nonlinear equations in Table 2.2 was solved for a range of L_e values after setting each of the time derivatives

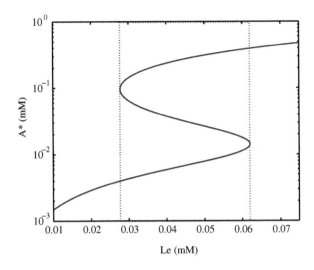

FIGURE 2.6

Bistability arises in $(L_e; A^*)$ space in the 5 variable *lac* operon model. Notice that the y-axis is in the logarithmic scale and there exists a range of L_e concentrations for which there are three coexisting steady states for the allolactose concentration. We estimated this range to be (0.027, 0.062) mM of $[L_e]$.

Table 2.4 The steady state values calculated from the 5 variable model when $L_e = 50 \times 10^{-3}$ mM for which there exist three steady states (see Figure 2.4).					
Steady States	**A^*(mM)**	**M^*(mM)**	**B^*(mM)**	**L^*(mM)**	**P^*(mM)**
I	7.85×10^{-3}	2.48×10^{-6}	1.68×10^{-6}	1.69×10^{-1}	3.46×10^{-5}
II	2.64×10^{-2}	7.58×10^{-6}	5.13×10^{-6}	2.06×10^{-1}	1.05×10^{-4}
III	3.10×10^{-1}	5.80×10^{-4}	3.92×10^{-4}	2.30×10^{-1}	8.09×10^{-3}

to zero. As seen in (Figure 2.6), there is a range of L_e concentrations for which there are three coexisting steady states for the allolactose concentration.

Table 2.4 shows the three steady states calculated from the 5 variable model when the extracellular lactose concentration is $L_e = 50 \times 10^{-3}$. To compute those steady states, the time derivatives in the model equations given in Table 2.2 are set to zero and the system of nonlinear equations is solved numerically while all the parameters are kept constant at their estimated values.

Figure 2.7 shows time-series simulations for the time evolution of mRNA, β-galactosidase, allolactose, lactose, and permease concentrations in the 5 variable model when the extracellular lactose concentration is $L_e = 50 \times 10^{-3}$ mM. There are three coexisting steady states for this value of L_e (see Figure 2.6 and Table 2.4), two

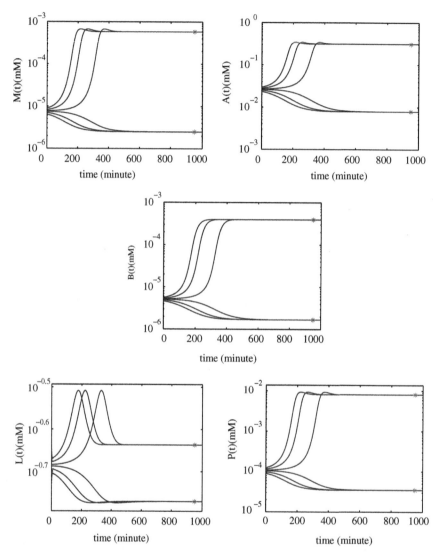

FIGURE 2.7

Time series simulation of the mRNA, β-galactosidase, allolactose, lactose and permease concentrations produced from the 5 variable model when the extracellular lactose concentration is $L_e = 50 \times 10^{-3}$ mM. For this value of the external lactose concentration, there are three coexisting steady states (see Figure 2.6). Two of the three states are locally stable and the third one is unstable [15, 9]. In the plots, ($*$)'s represent the location of the low and high stable steady states. See the text for selection of the initials and details for this simulation.

of which are locally stable and the third one is unstable [9, 15]. For this simulation, we have chosen six different initial values for the protein concentrations in the neighborhood of the unstable steady state (steady state II in Table 2.4) when all the parameters were held at their estimated values in Table 2.4. Since we started our simulations from a point close to the unstable steady state (but not exactly from it), three of the initials converged to the higher steady state (steady state III in Table 2.4) and and the other three initials converged to the lower steady state (steady state I in Table 2.4).

Exercise 2.5.   Consider the 5 variable model and the parameter values given in Table I in the paper [9] except for the bacterial growth rate μ. Bacterial growth rate can change depending on the environmental condition. In our analysis, we estimated this parameter to be about 30 min. Assume that you have a bacteria population that can double in size every 100 min then compute numerically the range for the extracellular lactose concentration and produce the bistability plot in (L_e, A^*) space. Furthermore, compute and estimate the value for the bacterial growth rate from this model for which the lactose operon is no longer capable of showing bistable behavior.

MATLAB starter code is provided for this exercise. Open the file *Code_for_Ex_2_5_Starter.m* and add the appropriate lines of code. Note that this exercise requires the use of the MATLAB's Global Optimization Toolbox. $\triangledown$

2.5 BOOLEAN MODELING OF BIOCHEMICAL INTERACTIONS

Boolean models were introduced and discussed in detail in Chapter 1 of this volume and we refer the reader to Chapter 1 for a primer on Boolean networks. We repeat some of the basics here for easy reference. In a Boolean model, all model variables are discretized to take values 1 or 0 (we also say that a Boolean variable can be On or Off, respectively). A certain threshold of discretization is chosen for each variable and a value of 0 typically represents the case when only "trace" (baseline) values of a substance are available. A value of 1 refers to concentrations larger than the threshold level.

As in the differential equation models, each variable in a Boolean model represents the concentration of a molecular species (e.g., enzyme, substrate, protein, or mRNA). A Boolean model of n-variables $x_1, x_2, \ldots, x_n$ consists of n transition Boolean functions (also called update functions) $f_{x_1}, f_{x_2}, \ldots, f_{x_n}$ describing the dynamical evolution of the model variables, where $f_{x_j}(x_1, x_2, \ldots, x_n) : \{0, 1\}^n \rightarrow \{0, 1\}$, $j = 1, 2, \ldots, n$. Equations (2.53) present a simple example.

$$
\begin{aligned}
f_{x_1} &= f_1(x_1, x_2, x_3, x_4) = x_3, \\
f_{x_2} &= f_1(x_1, x_2, x_3, x_4) = x_3 \wedge x_4, \\
f_{x_3} &= f_1(x_1, x_2, x_3, x_4) = x_2 \wedge x_3, \\
f_{x_4} &= f_1(x_1, x_2, x_3, x_4) = x_1 \wedge x_2 \wedge x_3.
\end{aligned}
\tag{2.53}
$$

Boolean models are time-discrete and the dynamical evolution of the model is determined by iterating the transitions defined by the update functions. Starting from

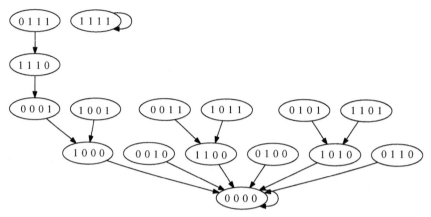

FIGURE 2.8

The state space diagram for the Boolean model from Eqs. (2.53).

an initial condition $(x_1^0, x_2^0, \ldots, x_n^0)$ at time $t = 0$, the state of the system at time $t = 1$ will be $(x_1^1, x_2^1, \ldots, x_n^1)$ where $x_j^1 = f_{x_j}(x_1^0, x_2^0, \ldots, x_n^0)$ for $j = 1, 2, \ldots, n$. The transition between time t and $t + 1$ is given by $x_j^{t+1} = f_{x_j}(x_1^t, x_2^t, \ldots, x_n^t)$, $j = 1, 2, \ldots, n$. As an example, consider the set of transition functions defined in Eqs. (2.53), with the initial condition $(x_1^0, x_2^0, x_3^0, x_4^0) = (0, 0, 1, 1)$. Substituting these values into Eqs. (2.53) yields

$$
\begin{aligned}
x_1^1 &= f_{x_1}(x_1^0, x_2^0, x_3^0, x_4^0) = f_1(0, 0, 1, 1) = 1, \\
x_2^1 &= f_{x_2}(x_1^0, x_2^0, x_3^0, x_4^0) = f_2(0, 0, 1, 1) = 1 \wedge 1 = 1, \\
x_3^1 &= f_{x_3}(x_1^0, x_2^0, x_3^0, x_4^0) = f_3(0, 0, 1, 1) = 0 \wedge 1 = 0, \qquad (2.54)\\
x_4^1 &= f_{x_4}(x_1^0, x_2^0, x_3^0, x_4^0) = f_4(0, 0, 1, 1) = 0 \wedge 0 \wedge 1 = 0.
\end{aligned}
$$

Next, the values $(x_1^1, x_2^1, x_3^1, x_4^1) = (1, 1, 0, 0)$ are used to evaluate the transition functions again, producing $(x_1^2, x_2^2, x_3^2, x_4^2) = (0, 0, 0, 0)$. A subsequent iteration of the functions f_{x_j} returns the same values $(x_1^3, x_2^3, x_3^3, x_4^3) = (0, 0, 0, 0)$. We say that we have computed the trajectory $(0, 0, 1, 1) \to (1, 1, 0, 0) \to (0, 0, 0, 0) \to (0, 0, 0, 0)$, and that $(0,0,0,0)$ is a *fixed point* for the Boolean network. A visual representation of this process for all 16 different sequences $(x_1^0, x_2^0, x_3^0, x_4^0)$ composed of 0s and 1s as initial states is given by the directed graph in Figure 2.8. This graph depicts the *transition diagram* of the Boolean model. Loops of length larger than one on the transition diagram correspond to limit cycles. Figure 2.8 shows that the Boolean model from Eqs. (2.53) has two fixed points (0,0,0,0) and (1,1,1,1) and no limit cycles. *The fixed points in a Boolean model are equivalent to the steady states of a differential equation model.*

In Chapter 1 of this volume we considered several models of the *lac* operon, specifically examining their transition functions, state-space diagrams and fixed points. A common feature of those models was the assumption that all major processes in the system's regulation (e.g., dilution and degradation, translation and transcription, and the effects of interaction between the system's components) take place in a single time step. Although models constructed under such assumptions can play an important role in the qualitative analysis of a system, such simple models lack the capability of describing delayed interactions and systems for which not all degradation rates are equal. Further, since those models assume only high and low levels of internal or external lactose, they effectively bypass the bistability phenomenon occurring at intermediate lactose concentrations.

We next consider Boolean models that can incorporate time delays, account for different degradation rates, and can exhibit bistability. The approach we outline below follows in most parts the general methodology proposed by Hinkelmann et al. in [16]. We begin with some background considerations.

1. Assume that a variable Y regulates the production of X.[1] As in Chapter 1, we use $X(t)$ and $Y(t)$ to denote the values of the Boolean variables X and Y at time t, where t is discrete, $t = 0, 1, 2, \ldots$ Assume that $Y(t) = 1$ implies $X(t+1) = 1$; that is, the presence of Y at time t implies that X will be present at time $t+1$. Assume that the loss of X due to dilution and degradation occurs over the course of several time steps. To account for this process, n additional Boolean variables $X_{\text{old}(1)}, X_{\text{old}(2)}, \ldots, X_{\text{old}(n)}$, are introduced (where n is a positive integer) with the property that:

    i. If $Y(t) = 0$ and $X(t) = 1$, then $X_{\text{old}(1)}(t+1) = 1$. A value of 1 for $X_{\text{old}(1)}(t+1)$ indicates that the amount of X present at time $t+1$ is already reduced once by dilution and degradation.

    ii. If $Y(t) = 0$ and $X_{\text{old}(i-1)}(t) = 1$, for some $i = 2, 3, \ldots, n$, then $X_{\text{old}(i)}$ $(t+1) = 1$. A value of 1 for $X_{\text{old}(i)}(t+1)$ indicates that the amount of X present at time $t+1$ is already reduced i times by dilution and degradation.

    iii. The number of required "old" variables is determined by the number of time steps needed to reduce the concentration of X below the discretization threshold when there is no new production of X. For instance, assume that in the absence of new production, the concentration of X needs to be diluted n times before falling below the discretization threshold. Thus, $X(t+1) = 1$ when either $Y(t) = 1$ (that is, a new amount of X will be produced by time $t+1$) or when $(X(t) \wedge \overline{X_{\text{old}(n)}}(t)) = 1$ (that is, previously available amounts of X are still available at time t and have not yet been reduced n-fold). In other words, $X(t+1) = Y(t) \vee (X(t) \wedge \overline{X_{\text{old}(n)}}(t))$. We should note that our approach here differs somewhat from that in [16]. We highlight the differences in Section 2.7.

2. To be able to model bistability we need to be able to distinguish between low, medium, and high concentrations of lactose L. This can be achieved by introducing an additional variable L_{high} such that $L_{\text{high}} = 1$ implies $L = 1$. In this

---

[1] In the models we consider later, Y will usually be a compound Boolean expression describing multiple regulating factors for X.

framework, $L = 1$ means that the concentration of L is at least medium, and when $L = 1$ and $L_{high} = 0$, the concentration of L is only medium. $L = 0$ and $L_{high} = 0$ stands for low concentration of L; $L = 1$ and $L_{high} = 1$ represents high concentration of L.

3. In a Boolean model, delays can be incorporated by introducing additional variables. The number of additional variables depends on the magnitudes of the delays and on the choice of the time step for the Boolean model. Assume a variable R regulates the production of X but the effect that R exerts on X is delayed by time τ.[2] If the delay τ is commensurable with n time steps in the model, n additional Boolean variables $R_i, i = 1, 2, \ldots, n$, can be introduced to represent the delayed regulation. The following motif will then be present in the set of transition functions: $R_1(t + 1) = R(t)$, $R_2(t + 1) = R_1(t)$, ..., $R_n(t + 1) = R_{n-1}(t)$, $X(t + 1) = R_n(t)$. Expressed in terms of the transition functions, the same can be stated as $f_X = R_n, f_{R_1} = R_2, \ldots, f_{R_n} = R_{n-1}$.

We now use these techniques to create Boolean models that approximate the Yildirim-Mackey differential equation models from Section 2.4.

2.6 BOOLEAN APPROXIMATIONS OF THE YILDIRIM-MACKEY MODELS

2.6.1 Boolean Variants of the 3-Variable Model

We want to build a Boolean model based on the assumptions used in the 3-variable differential equation model from Table 2.1. Recall that $A_{\tau_M} = A(t - \tau_M)$, $M_{\tau_B} = M(t - \tau_B)$. The delays are estimated in [8] to be $\tau_M = 0.10$ min, $\tau_B = 2.00$ min. Various estimates are available from the literature for the loss rates $\widetilde{\gamma_M}$, $\widetilde{\gamma_B}$, and $\widetilde{\gamma_A}$ and in some cases the range for the estimates may be rather wide. For instance in Yildirim and Mackey [9], the degradation rate of A is estimated at $\gamma_A = 0.52$ min^{-1}, in Yildirim et al. [8] the estimate is $\gamma_A = 1.35 \times 10^{-2}$ min^{-1}, and in Wong et al. [17] an estimate as low as $\gamma_A = 1.8 \times 10^{-4}$ min^{-1} is considered. The degradation rates γ_M and γ_B are estimated in both [8,9] to be $\gamma_M = 0.411$ min^{-1}, $\gamma_B = 8.3 \times 10^{-4}$ min^{-1}. The effective loss in the concentrations due to dilution is proportional to the growth rate μ and is estimated to be between $\mu_{min} = 4.5 \times 10^{-3}$ min^{-1} and $\mu_{max} = 3.47 \times 10^{-2}$ min^{-1}.

As Boolean models are qualitative in nature, the actual values of these constants are not essential and the only consideration of importance is the comparative order of their magnitudes. For our first Boolean model below, we have selected the following values by considering middle-of-the-range estimates: $\mu = 3 \times 10^{-2}$ min^{-1} and $\gamma_A = 1.35 \times 10^{-2}$ min^{-1}. Thus the loss terms in the 3-variable model in Table 2.1 are estimated to be $\widetilde{\gamma_M} = \gamma_M + \mu = 0.441$, $\widetilde{\gamma_B} = \gamma_B + \mu = 0.031$, and $\widetilde{\gamma_A} = \gamma_A + \mu = 0.044$. From here the times needed to reduce the concentrations of

---

[2] In the models we consider below, R will usually be just one of the regulating factors for X but the idea remains the same.

M, B, and A by half are calculated to be $\widetilde{h}_M = 1.572$ min, $\widetilde{h}_B = 22.360$ min, and $\widetilde{h}_A = 15.753$ min.

Combining this information with the methodology outlined in Section 2.5 leads to the following choice of variables for the Boolean model:

1. We will model the dynamics of the Boolean variables M, B, and A, denoting mRNA, β-galactosidase, and allolactose.

2. We assume that glucose is absent and intracellular lactose is present at all times. The intracellular lactose concentration L is a parameter for the model. From Section 2.4, we know that the model in Table 2.1 has multiple steady states when the internal lactose concentration is in a certain intermediate (maintenance) range, estimated numerically to be between $(0.039, 0.055)$ mM (see Figure 2.4). To distinguish between low, medium, and high lactose concentrations in the Boolean case, we introduce an additional Boolean parameter L_{high}. The value $L = 1$ implies intracellular lactose concentrations are within the maintenance range or higher while $L = 0$ implies intracellular lactose concentrations are below the maintenance range. When $L_{\text{high}} = 1$, internal lactose concentration is higher than the maintenance range.

3. We select a discrete time step of about 10 min for the Boolean model. The delays τ_M, τ_B can then be ignored since they are much smaller than the time step. Similarly, since $\widetilde{h}_M \ll 10$ min, in the absence of new mRNA production, any available amounts of mRNA would be reduced below the discretization threshold in one time step. Thus there is no need for introducing M_{old}.

4. We define additional variables A_{old}, $B_{\text{old}(1)}$, and $B_{\text{old}(2)}$ to model the different degradation rates of allolactose and β-galactosidase.

We emphasize that the choices regarding the size of the discrete time step and the exact number of "old" variables chosen for this model represent just one of many possibilities. Alternative models are considered later in the chapter and in the exercises.

Combining these assumptions with the assumption that translation and transcription happen in one time step, we obtain the following Boolean model:

$$f_M = A \qquad\qquad f_{B_{\text{old}(2)}} = \overline{M} \wedge B_{\text{old}(1)}, \qquad (2.55)$$

$$f_B = M \vee (B \wedge \overline{B_{\text{old}(2)}}) \quad f_A = (B \wedge L) \vee L_{\text{high}} \vee ((A \wedge \overline{A_{\text{old}}}) \wedge \overline{B}),$$
$$f_{B_{\text{old}(1)}} = \overline{M} \wedge B \qquad f_{A_{\text{old}}} = ((\overline{B} \vee \overline{L}) \wedge \overline{L_{\text{high}}}) \wedge A.$$

A justification for the transition functions in Eqs. (2.55) follows:

Transition Equation for M: In the presence of allolactose at time t, mRNA will be produced and present at the next time step $t + 1$. Availability of M at time t does not affect its availability at time $t + 1$ since M is assumed to degrade completely within one time step.

Transition Equation for B: When mRNA is available at time t, translation and transcription of β-galactosidase will take place and $B = 1$ at step $t + 1$. If amounts of B produced earlier are still available in high enough concentrations to not fall below the discretization threshold by time $t + 1$ (that is, $B \wedge \overline{B_{old(2)}} = 1$), B will still be available at time $t + 1$.

Transition Equation for $B_{old(1)}$: When no mRNA is available at time t, no newly produced β-galactosidase will be available at time $t + 1$. By time $t + 1$, the amounts of B produced at time t will be reduced once due to dilution and degradation and $B_{old(1)}(t + 1) = 1$.

Transition Equation for $B_{old(2)}$: When no mRNA is available at time t, no newly produced β-galactosidase will be available at time $t + 1$. By time $t + 1$, the amounts of B, already reduced once at time t, will be further reduced due to dilution and degradation and $B_{old(2)}(t + 1) = 1$.

Transition Equation for A: There are three possible ways for A to be available at time $t + 1$: (i) At time t, β-galactosidase above the discretization threshold is available together with at least medium concentration of lactose. Under those conditions, β-galactosidase will convert lactose into allolactose by time $t + 1$. (ii) At time t, the high concentration of intracellular lactose ensures that available trace amounts of β-galactosidase will convert enough lactose molecules into allolactose to bring the concentration of allolactose at time $t + 1$ above the discretization threshold. (iii) At time t, the concentration of A is high enough not to be reduced below the threshold level at time $t + 1$ due to dilution and it will not be lost via conversion into glucose and galactose (mediated by β-galactosidase).

Transition Equation for A_{old}: At time t, the conditions for producing A by time $t + 1$ are not met. Amounts of A available at time t will be reduced once by time $t + 1$ due to dilution and degradation and $A_{old}(t + 1) = 1$.

We analyze the Boolean model defined by Eqs. (2.55) using the web-based DVD software [18], considering all four possible combinations for the parameter values. When $M = 1$, $A = 1$, and $B = 1$, the cell is producing all necessary proteins for the metabolism of lactose. In this case we say that the operon is On. We say that the *lac* operon is Off when those proteins are not being produced ($M = 0$, $A = 0$, and $B = 0$). The results are presented in Table 2.5. Regardless of the parameter values, the system has only fixed points and no limit cycles. As expected, low concentration of lactose ($L = 0$, $L_{high} = 0$) drives the system to a single steady state in which the operon is Off, while for high concentrations of lactose ($L = 1$, $L_{high} = 1$) the system settles in a fixed point at which the operon is On. For intermediate lactose concentrations ($L = 1$, $L_{high} = 0$) the model approximates the bistable nature of the operon, qualitatively replicating the bistability results from Section 2.4. If we start at the fixed point 1 in Table 2.5 (corresponding to low inducer concentration and an Off state for the operon) and increase the lactose concentration to medium, the operon remains Off (fixed point 3). If we start at fixed point 1 and increase the lactose concentration to high, the operon turns On (fixed point 4). In a similar way, if we start at fixed point 2 (corresponding to high inducer concentration and an On state for the operon) and reduce the lactose concentration to medium, the operon remains On (fixed point 4). Thus, in intermediate inducer concentrations the long-term behavior of the system depends on its history (hysteresis).

By considering the middle-of-the-range value for γ_A, we essentially made the assumption that allolactose degrades much slower than mRNA and that the delays τ_M and τ_B are negligible in comparison with the time for degradation and dilution of allolactose. However, considering estimates for γ_A among the largest reported in the literature (e.g., $\gamma_A = 0.52$ min^{-1}), a different Boolean model would be more appropriate since in that case the half-life for A, estimated at $\tilde{h}_A = 1.260$ min, is similar to the half-life of mRNA. This situation calls for assumptions different from those used to build the model given by Eqs. (2.55): (i) If a larger time step is considered (e.g., $t = 15$ min), we can eliminate the variables A_{old} and $B_{old(2)}$. (ii) If we consider a much smaller time step (e.g., $t = 1$ min), more additional variables will be needed in order to account for the system delays and for the fact that multiple time steps will be needed for dilution and degradation to bring M and A below the discretization threshold levels.

The Boolean model defined by Eqs. (2.56) would be appropriate under assumption (i).

$$f_M = A \qquad\qquad f_{B_{old}} = \overline{M} \wedge B,$$
$$f_B = M \vee (B \wedge \overline{B_{old}}) \quad f_A = (B \wedge L) \vee L_{high}. \qquad (2.56)$$

Exercise 2.6. (a) Justify the model presented by Eqs. (2.56) by explaining the logical expression defining each of the transition functions as was done above for the Boolean model in Eqs. (2.55); (b) Explain why the model would be consistent with choosing a time step $t = 15$ min.; (c) Use DVD to analyze the model and obtain the state space diagram presented in Figure 2.9; (d) Does the result from (c) imply bistability for the system? ▽

The Boolean model defined by Eqs. (2.57) on the other hand would be appropriate under the assumption (ii) above ($t = 1$ min). New variables M_1 and M_2 are introduced to model the delayed effect (by τ_B) of mRNA on the production of β-galactosidase and A_1 is needed to represent the delayed action of A on the production of mRNA by $\tau_M = 0.1$ min. As before, M_{old} and A_{old} track the loss of M and A due to dilution and degradation. Since the loss of B is much slower, several "old" variables are needed. We have used two such variables here but one would think that a much larger number of such variables will be needed to represent the time scales accurately. In the discussion

Table 2.5 Fixed points for the Boolean model from Eqs. (2.55). The system always settles in a fixed point. There are no limit cycles. Two steady states are possible for medium lactose concentrations (the system is bistable).

	Inducer Level	L	L_{high}	M	B	$B_{old(1)}$	$B_{old(2)}$	A	A_{old}	Operon is
1	Low lactose	0	0	0	0	0	0	0	0	Off
2	High lactose	1	1	1	1	0	0	1	1	On
3	Medium lactose	1	0	0	0	0	0	0	0	Off
4	Medium lactose	1	0	1	1	0	0	1	0	On

we argue that this is unnecessary.

$$
\begin{aligned}
f_M &= A_1 \vee (M \wedge \overline{M_{\text{old}}}) & f_{B_{\text{old}(1)}} &= \overline{M_2} \wedge B \\
f_{M_1} &= M & f_{B_{\text{old}(2)}} &= \overline{M_2} \wedge B_{\text{old}(1)} \\
f_{M_2} &= M_1 & f_A &= (B \wedge L) \vee L_{\text{high}} \vee ((A \wedge \overline{A_{\text{old}}}) \wedge \overline{B}) \quad (2.57) \\
f_{M_{\text{old}}} &= \overline{A_1} \wedge M & f_{A_1} &= A \\
f_B &= M_2 \vee (B \wedge \overline{B_{\text{old}(2)}}) & f_{A_{\text{old}}} &= ((\overline{B} \vee \overline{L}) \wedge \overline{L_{\text{high}}}) \wedge A.
\end{aligned}
$$

Analysis of the long-term behavior of the models from Eqs. (2.56) and (2.57) leads to results similar to those for the Boolean model from Eqs. (2.55). Since the state space for the model defined by Eqs. (2.56) is relatively small, we have presented it in Figure 2.9 (see Exercise 2.6). Regardless of the initial conditions, for low lactose ($L = 0, L_{\text{high}} = 0$) the system settles in a state where $M = 0$, $A = 0$, and $B = 0$ and the operon is Off. For high lactose ($L = 1, L_{\text{high}} = 1$) the system settles in a state where $M = 1$, $A = 1$, and $B = 1$ and the operon is On. For intermediate levels

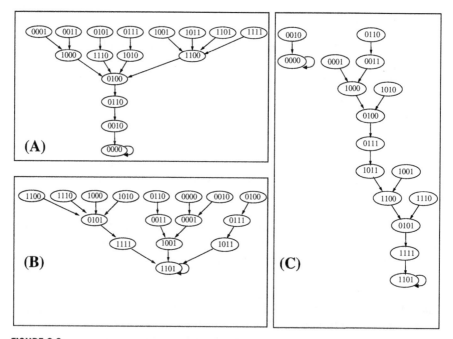

FIGURE 2.9

The state space diagram displaying the points (M,B,B$_{\text{old}}$, A) for the Boolean model from Eqs. (2.56). Panel A: Low level of internal lactose, corresponding to $L = 0$; $L_{\text{high}} = 0$. Single fixed point, the operon is Off; Panel B: High level of lactose, corresponding to $L = 1$; $L_{\text{high}} = 1$. Single fixed point, the operon is On; Panel C: Medium level of lactose, corresponding to $L = 1$; $L_{\text{high}} = 0$. Two possible fixed points exist, the operon can be On or Off.

of lactose ($L = 1$, $L_{\text{high}} = 0$), there are two fixed points and the state of the operon depends on the initial conditions (its history): If the lactose concentration is raised to medium for cells grown in low lactose concentrations (those cells have thus settled in the Off state), the operon remains Off. If the concentration of lactose is decreased to medium for cells grown in lactose-rich environment (those cells have thus settled in the On state) the operon remains On.

For the model defined by Eqs. (2.57) the entire state space is too large to display but we present its fixed points in Table 2.6.

Exercise 2.7.   Verify that the fixed points for the Boolean model from Eqs. (2.57) are as presented in Table 2.6.   ▽

So far, we have used designated "old" variables to separate the time scales of the dilution and degradation processes from those of synthesis. As the next model shows, this could also be done implicitly through the logical assumptions built into the transition functions. If we choose to work with the estimate for $\gamma_A = 1.8 \times 10^{-4} \text{min}^{-1}$ from Wong et al. [17], the degradation time for A will be the slowest among the three variables M, B, and A. The transition function for A in the next model accounts for this by allowing for $A = 1$ either because of new production or because previously produced amounts of A are still present and have not been lost due to conversion to glucose and galactose.

$$f_M = A, \tag{2.58}$$
$$f_B = M,$$
$$f_A = (B \wedge L) \vee L_{\text{high}} \vee (A \wedge \overline{B}).$$

Exercise 2.8.   Justify and analyze the model from Eqs. (2.58). Show that its long-term behavior captures the bistability of the *lac* operon in medium lactose concentrations.   ▽

2.6.2 Boolean Variants of the 5-Variable Model

We next consider Boolean variants of the model from Table 2.2 that uses five dynamic variables, M, B, A, L, and P, representing mRNA, β-galactosidase, allolactose, intracellular lactose, and *lac* permease. Recall that extracellular glucose is assumed to be absent and extracellular lactose (L_e) present at all times. L_e is a parameter for the model.

The degradation constants for L and P are estimated to be $\gamma_L = 0.0 \text{ min}^{-1}$ (meaning the dilution loss of L is fully due to the growth rate) and $\gamma_P = 0.65 \text{ min}^{-1}$; $\tau_P = 0.83$ min is an estimate for the delay τ_P in the equations for P [9].

As before, we need to be able to distinguish between high, medium, and low concentrations of external lactose, where medium concentration would roughly correspond to the maintenance range estimated to be $(0.027, 0.062)$ mM (see Figure 2.6). We introduce an additional parameter $L_{e_{\text{high}}}$. Assuming that $L_e = 1$ stands for at least medium external lactose, the combination $L_e = 1$; $L_{e_{\text{high}}} = 0$ indicates medium

Table 2.6 Fixed points for the Boolean model from Eqs. (2.57). The system always settles in a fixed point. There are no limit cycles. Two steady states are possible for medium lactose concentrations (the system is bistable).

	Inducer Level	L	L_{high}	M	M_1	M_2	M_{old}	B	$B_{old(1)}$	$B_{old(2)}$	A	A_1	A_{old}	State
1	Low lactose	0	0	0	0	0	0	0	0	0	0	0	0	Off
2	High lactose	1	1	1	1	1	0	1	0	0	1	1	0	On
3	Medium lactose	1	0	0	0	0	0	0	0	0	0	0	0	Off
4	Medium lactose	1	0	1	1	1	0	1	0	0	1	1	0	On

external lactose, while $L_e = 0$; $L_{e_{high}} = 0$ and $L_e = 1$; $L_{e_{high}} = 1$ represent low and high external lactose, respectively.

As in the case of the 3 variable differential equation model from Table 2.1, it was shown in [9] that the delays in the equations do not affect the bistable behavior of the system. With this motivation, our first Boolean model for the 5-variable system ignores the delays. We also choose to introduce a single "old" variable for each of the model variables (see Section 2.7 for more on this). The transition functions for this Boolean model are:

$$f_M = A \vee (M \wedge \overline{M_{old}})$$
$$f_{M_{old}} = \overline{A} \wedge M$$
$$f_A = (B \wedge L) \vee (L \wedge L_{e_{high}}) \vee ((A \wedge \overline{A_{old}}) \wedge \overline{B})$$
$$f_{A_{old}} = ((\overline{B} \vee \overline{L}) \wedge (\overline{L} \vee L_{e_{high}}) \wedge A$$
$$f_L = ((P \wedge L_e) \vee L_{e_{high}}) \vee ((L \wedge \overline{L_{old}}) \wedge (\overline{B} \wedge \overline{P}))$$
$$f_{L_{old}} = ((\overline{P} \vee \overline{L_e}) \wedge \overline{L_{e_{high}}}) \wedge L$$

$$f_B = M \vee (B \wedge \overline{B_{old}})$$
$$f_{B_{old}} = \overline{M} \wedge B$$
$$f_P = M \vee (P \wedge \overline{P_{old}})$$
$$f_{P_{old}} = \overline{M} \wedge P$$

$$(2.59)$$

The justification for the transition functions of the "old" variables is the same as before: no new quantities are made and previously available amounts have not already been reduced once by dilution and degradation. Also note that the justifications for the transition functions for M and B are as those for the Boolean models from the previous section.

Transition Equation for A: There are three possible ways for A to be available at time $t + 1$: (i) At time t, β-galactosidase is available together with at least medium concentration of lactose. Under those conditions β-galactosidase will convert lactose into allolactose by time $t + 1$. (ii) At time t, internal lactose is already present and the concentration of extracellular lactose is high; this will ensure that by time $t + 1$, the intracellular lactose concentration is high enough to find available trace amounts of β-galactosidase and initiate enough conversions into allolactose to bring the concentration of allolactose above the discretization threshold. (iii) At time t the concentration of A is high enough not to be reduced below the threshold level at time $t + 1$ due to dilution or degradation and that it will not be lost to conversion into glucose and galactose (mediated by β-galactosidase) by time $t + 1$.

Transition Equation for L: There are three possible ways for L to be available at time $t + 1$: (i) At time t, permease above the discretization threshold is available together with at least medium levels of external lactose. Under those conditions, the permease will facilitate the transport of external lactose into the cell. (ii) At time t, the concentration of P may be below the discretization threshold but the high concentration of external lactose at time t will guarantee that the external lactose enters the cell and brings, by time $t + 1$, internal lactose above the threshold. (iii) At time t there is still high enough concentration of L that will not be lost due to the reversible nature of permease-facilitated lactose transport and also not lost to conversion into allolactose mediated by B.

Transition Equation for P: When mRNA is available at time t, translation and transcription of permease will take place to make P available at time $t + 1$. If high enough amounts of P produced earlier are still available by the time t (that is, by time t they have not already been reduced enough to be below the threshold at time $t + 1$) P will still be available at $t + 1$.

When the variables M, A, L, B, and P all have values 1, the operon is On. When they all have values 0, the operon is Off. Analyzing this model with DVD shows no limit cycles, a single fixed point for the cases of low and high concentrations of external lactose and two fixed points for medium concentrations. Regardless of the initial conditions, the operon is On for high external lactose and Off for low external lactose. When the external lactose concentration is medium, the operon can be On or Off based on the initial conditions (Table 2.7).

Exercise 2.9. Assuming a time step $t = 1$ min, expand the Boolean model from Eqs. (2.59) to include additional variables M_1, M_2, M_3, and A_1 accounting, as in the model from Eqs. (2.57), for the delays in the 5-variable model from Table 2.2. Use DVD to show that introducing the delays does not affect bistability. ▽

Exercise 2.10. Reduce the Boolean model from Eqs. (2.59) by removing all "old" variables except for L_{old}. Explain why this would be justified. Use DVD to find the fixed points of the system and show that the qualitative bistable behavior of this model remains the same as for the Boolean model from Eqs. (2.59). ▽

Exercise 2.11. In Santillán et al., 2007 [3], the authors consider a minimal ODE model of the *lac* operon involving three variables: the intracellular concentration of mRNA (M), *lacZ* polypeptide (E), and intracellular lactose (L). Since β-galactosidase is a homo-tetramer made up of four identical *lacZ* polypeptides and the translation rate of the *lacY* transcription is assumed to be the same as the rate for the *lacZ* transcript, the following holds for the intracellular concentrations of β-galactocidase (B) and permease (Q): $Q = E$ and $B = E/4$. The model also assumes that the concentrations of internal lactose (L) and allolactose (A) are the same. The three ODEs for the model are given by Eqs. (2.60). External lactose (L_e) and external glucose (G_e) are parameters for the model. The full model is presented and justified in [3].

$$\frac{dM}{dt} = Dk_M \mathcal{P}_D(G_e)\mathcal{P}_R(A) - \gamma_M M,$$

$$\frac{dE}{dt} = k_E M - \gamma_E E,$$

$$\frac{dL}{dt} = k_L \beta_L(L_e)\beta_G(G_e)Q - 2\phi_M \mathcal{M}(L)B - \gamma_L L. \qquad (2.60)$$

In Chapter 1 of this volume we examined and analyzed a Boolean model built under the same assumptions and defined by the set of transition functions in Eqs. (2.61). Since this Boolean model does not have the ability to distinguish between

Table 2.7 Fixed points for the Boolean model from Eqs. (2.59). The system always settles in a fixed point. There are no limit cycles. Two steady states are possible for medium lactose concentrations (the system is bistable).

	Inducer Level	L_e	$L_{e_{high}}$	M	M_{old}	B	B_{old}	A	A_{old}	L	L_{old}	P	P_{old}	State
1	Low ext. lactose	0	0	0	0	0	0	0	0	0	0	0	0	Off
2	High ext. lactose	1	1	1	0	1	0	1	0	1	0	1	0	On
3	Medium ext. lactose	1	0	0	0	0	0	0	0	0	0	0	0	Off
4	Medium ext. lactose	1	0	1	0	1	0	1	0	1	0	1	0	On

low, medium, and high levels of lactose, it does not exhibit bistability.

$$f_M = \overline{G_e} \wedge (L \vee L_e),$$
$$f_E = M,$$
$$f_L = \overline{G_e} \wedge ((E \wedge L_e) \vee (L \wedge \overline{E})). \qquad (2.61)$$

1. Give justification for the transition functions presented by Eqs. (2.61). Use DVD to calculate the system's fixed points and present a table with the fixed points for all possible combinations of the parameter values.
2. Now assume that in the model from Eqs. (2.61), L_e stands for external concentration of lactose that is at least medium. Then introduce a new parameter $L_{e_{high}}$ to denote high levels of external lactose. The combination of parameter values $L_e = 1$ and $L_{e_{high}} = 0$ now stands for medium external lactose. Modify the transition functions from Eqs. (2.61) to make the modified model exhibit bistable behavior for medium lactose concentrations.   ▽

2.7 CONCLUSIONS AND DISCUSSION

In this chapter we developed and compared a number of mathematical models of the lactose regulatory mechanism of E. coli. The focus was on bistability. The continuous models were previously published in [8,9], while the Boolean models we have considered are new. Since the discovery of the *lac* operon by Jacob and Monod in 1960 ([19,20]), both differential equation models and Boolean models have been introduced to describe operon system dynamics, beginning with the work of Goodwin [21] and Kauffman [22], respectively. Bistability for the *lac* operon system has been experimentally observed and simulated by a number of continuous models, including those considered in this chapter. However, establishing that Boolean models are capable of capturing the system's bistability has only been done recently [16,23]. A necessary condition for bistability when modeling with Boolean networks is to make possible distinguishing between at least three levels of inducer concentrations: low, medium, and high. Both deterministic models (e.g., [16]) and models involving stochasticity (e.g., [23]) have been proposed.

In this chapter we confirmed that Boolean models can approximate delayed differential equation models, preserving critical qualitative features such as bistability. It was noted in [8] that the model in [9] does not consider inducer exclusion or catabolite repression (both are external-glucose-dependent mechanisms), indicating that bistability is independent from the presence of glucose in the extracellular medium. The 3-variable model from Table 2.1 in [8] ignores the lactose permease in the operon regulation and considers only the role of β-galactosidase. This shows that of all feedback loops in the system, the β-galactosidase regulation is the one responsible for the bistable behavior of the operon. The fact that the Boolean models that approximate this continuous model preserve bistability further underscores this result. In [8,9] the authors also show that bistability of the system is preserved when the delays in the regulatory equations from Tables 2.1 and 2.2 are ignored. The Boolean models replicate this finding, indicating once again that bistability is a robust phenomenon that arises from the network structure.

The Santillán model from Exercise 2.11 [3], considers external lactose as a parameter of the model and the function $\mathcal{P}_R(L, G_e) = \mathcal{P}_D(G_e)\mathcal{P}_R(A)$ (an increasing function of L and a decreasing function of G_e) in the equation for M in Eqs. (2.60) is a possible way to model the dependence of the system on these parameters. In [24] the authors provide a detailed comparison of a number of ODE models of the *lac* operon that differ in the numbers of variables and type (stochastic or deterministic), including the three ODE models considered in this chapter. Many of those models differ in the ways the dependence of mRNA on external glucose and internal lactose is modeled. It is noted in [24] that in some cases heuristic reasoning is used to propose Hill-type equations for the function $\mathcal{P}_R(L, G_e)$ with a different level of detail. Establishing that Boolean approximations of the minimal model from Eqs. (2.60) can describe and explain bistability indicates that such differences are nonessential with regard to bistability and do not impact this feature of the system.

It has been shown that bistability requires a direct positive feedback loop or an indirect positive feedback loop, such as a double negative feedback loop [25]. However, a feedback loop alone is not enough for a system to exhibit bistable behavior. It must also possess some type of nonlinearity within the feedback circuit. That is, some of the proteins in the feedback circuit must respond to their upstream regulators in an ultrasensitive manner [26–28]. The Hill coefficient is used to quantify the steepness of this response. A Hill coefficient larger than one is considered to be the ultrasensitive response [29]. In Boolean models this condition is automatically satisfied since the On-Off switches in such models can be viewed as responses with very large Hill coefficients.

The analysis of the Boolean models was done using the web-based software DVD [18]. When the number of variables is small, DVD can be used to produce the entire state space for the model. When the number of variables grows, obtaining the entire state space is not feasible, but DVD computes and returns the model's fixed points together with the number of connected components in state space. DVD's successor ADAM [30], which handles a broader class of discrete models, including logical models and Petri nets in addition to Boolean and polynomial models, can also be used.

We found that introducing multiple "old" variables into the Boolean models does not appear to change the long-term behavior of the system and to impact bistability. This is consistent with findings reported in [31] where the author presents a method for reducing Boolean networks and their wiring diagrams while preserving the set of fixed points. The reduction is done by deleting vertices with no self-loops (that is, vertices whose transition functions do not include them as inputs). If a vertex is to be removed from the network, its transition function is substituted for the variable representing this vertex in all of the other transition functions in the model. The idea is to reduce the size of the network by providing a way to delete a vertex and "pass on" its functionality to other variables in the network. A reduced model with the same fixed points as the original model would be indicative of characteristics that emerge not as a result of some specific system interactions but, instead, from the core network topology. For instance, in [23] Veliz-Cuba and Stigler present examples of Boolean models of the *lac* operon incorporating inducer inclusion and catabolite repression. The networks for those models are then reduced by this method to smaller models that no longer include those regulatory mechanisms but preserve the bistable behavior of the system.

In the Boolean models proposed in this chapter we use "old" variables to track the dilution and degradation of concentrations that require multiple steps of time for reduction below the discretization threshold. In the context of the notation used in Section 2.5, if a variable Y regulates the production of X, which (if no new amounts are produced) would degrade below threshold levels in n steps, the Boolean model will include the following motif:

$$f_X = Y \lor (X \land \overline{X_{\text{old}(n)}}) \qquad \cdots \qquad (2.62)$$
$$f_{X_{\text{old}(1)}} = \overline{Y} \land X \qquad\qquad f_{X_{\text{old}(n-1)}} = \overline{Y} \land X_{\text{old}(n-2)}$$
$$f_{X_{\text{old}(2)}} = \overline{Y} \land X_{\text{old}(1)} \qquad\qquad f_{X_{\text{old}(n)}} = \overline{Y} \land X_{\text{old}(n-1)}$$
$$\cdots \qquad\qquad\qquad f_Y = \ldots,$$

We can reduce the network by eliminating $X_{\text{old}(n)}$. To do this, substitute its transition function in place of $X_{\text{old}(n)}$ in the transition function of X: $f_X = Y \lor (X \land \overline{\overline{Y} \land X_{\text{old}(n-1)}})$, which simplifies to $f_X = Y \lor (X \land \overline{X_{\text{old}(n-1)}})$, resulting in exactly the same structural motif with one less "old" variable. Variables $X_{\text{old}(n-1)}, \ldots, X_{\text{old}(2)}$ can be eliminated similarly, leading to the reduction $f_X = Y \lor (X \land \overline{X_{\text{old}(1)}})$, $f_{X_{\text{old}(1)}} = \overline{Y} \land X$, $f_Y = \ldots$. When applied to the Boolean models in this chapter, this reduction process indicates that the bistability property of the system does not depend on the number of "old" variables used in the model. We finally note that if we choose to also eliminate $X_{\text{old}(1)}$ from the model, this leads to the transition equations $f_X = Y \lor X$, $f_Y = \ldots$, reflecting a situation in which X is considered completely stable (a situation in which the combined degradation and dilution rate for X is infinitely small and thus, the growth rate is negligible), which is biologically unrealistic. Thus, at least one "old" variable is needed.

Exercise 2.12.   Confirm that $Y \lor (X \land \overline{\overline{Y} \land X_{\text{old}(n-1)}}) = Y \lor (X \land \overline{X_{\text{old}(n-1)}})$. ∇

The use and treatment of "old" variables here differ from the approach introduced in [16]. We stipulate that an "old" variable has value 1 at time $t + 1$ only when conditions for new production are not met at time t <u>and</u> when previously produced amounts available at time t have not already been reduced by a certain factor due to dilution and degradation. In [16], an "old" variable has value 1 at time $t + 1$ when conditions for new production are not met at time t. Our approach provides a mechanism to track the level of reduction of X: since only one $X_{\text{old}(k)}$ could be equal to 1 at each time step, $X_{\text{old}(k)}(t) = 1$ means that at time t the concentration of X has already been reduced exactly k times ($k = 1, 2, \ldots, n$).

Thus, when delay variables are added to our Boolean model from Eqs. (2.59) (see Exercise 2.9) the resulting model provides a Boolean approximation of the 5-variable differential equation model from Table 2.57 that differs from the Boolean approximation of the same model presented in [16]. Both models capture the bistable behavior of the lactose operon but the model in [16] generates several fixed points that are not biologically meaningful. Another difference in comparison with the Boolean model in [16] is that the Boolean models in this chapter factor in the bacterial growth rate μ.

The Boolean modeling carried out here is strictly deterministic in nature and uses synchronous updates for all variables. It would be interesting to examine what impact stochasticity and asynchronous update schedules may have on the model outcomes.

Acknowledgment

Raina Robeva gratefully acknowledges the support of NSF under the DUE award 0737467.

2.8 SUPPLEMENTARY MATERIALS

All supplementary files and/or computer code associated with this article can be found from the volume's website http://booksite.elsevier.com/9780124157804

References

[1] Novick A, Wiener M. Enzyme induction as an all-or-none phenomenon. Proc Natl Acad Sci USA 1957;43:553–566.

[2] Ozbudak EM, Thattai M, Lim HN, Shraiman BI, Oudenaarden AV. Multistability in the lactose utilization network of *Escherichia coli*. Nature 2004;427:737–740.

[3] Santillán M, Mackey MC, Zeron E. Origin of bistability in the lac operon. Biophys J 2007;92:3830–3842.

[4] Beckwith J. The lactose operon in *Escherichia coli* and *Salmonella*: Cellular and molecular biology. In: Neidhardt FC, Ingraham JL, Low KB, Magasanik B, Umbarger HE, editors. American Society for Microbiology vol. 2. Washington, DC; 1987. p. 1444–52.

[5] Schleif R. Regulation of the L-arabinose operon of *Escherichia coli*. Trends in Genetics 2000;16(12):559–565.

[6] Lewin Benjamin. Genes Jones and Bartlett Publishers 9th ed. 2008.

[7] Klipp E., Herwig R., Kowald A., Wierling C., Lehrach H. Systems Biology in Practice. Wiley-VCH first ed. 2005.

[8] Yildirim N, Santillán M, Horike D, Mackey MC. Dynamics and bistability in a reduced model of the lac operon. Chaos 2004;14(2):279–292.

[9] Yildirim N, Mackey MC. Feedback Regulation in the Lactose Operon: A Mathematical Modeling Study and Comparison with Experimental Data. Biophysical J 2003;84:2841–2851.

[10] Watson JD. Molecular Biology of the Gene. New York: W.A. Benjamin Inc. third ed; 1977.

[11] Yagil G, Yagil E. On the relation between effector concentration and the rate of induced enzyme synthesis. Biophys J 1971;11:11–27.

[12] Saier MH, Ramseier TM, Reizer J, Regulation of carbon utilization in *Escherichia coli* and *Salmonella*: Cellular and molecular biology. Neidhardt FC, Curtiss R, Ingraham JL, et al., editors. American Society for Microbiology. vol. 1. p. 1325–1343Washington, D.C.: 1996.

[13] Mackey MC, Santillán M, Yildirim N. Modeling operon dynamics: the trypto-phan and lactose operons as paradigms. C R Biol 2004;327:211–224.

[14] Cohn M, Horibata K. Inhibition by glucose of the induced synthesis of the β-galactosidase-enzyme system of *Escherichia coli*: Analysis of maintenance. J Bacteriol 1959;78:613–623.

[15] Yildirim N, Kazanci C. Deterministic and stochastic simulation and analysis of biochemical reaction networks the lactose operon example. Methods Enzymol 2011;487:371–395.

[16] Hinkelmann F, Laubenbacher R. Boolean Models of Bistable Biological Systems. Discrete and Continuous Dynamical Systems 2011;4:1443–1456.

[17] Wong P, Gladney S, Keasling JD. Mathematical model of the lac operon: Inducer exclusion, catabolite repression, and diauxic growth on glucose and lactose. Biotechnol Prog 1997;13:132–143.

[18] Vastani H, Jarrah A, Laubenbacher. Visualization of Dynamics for Biological Networks. http://dvd.vbi.vt.edu/dvd.pdf.

[19] Jacob F, Perrin D, Sanchez C, Monod J. L'Operon: groupe de gène à expression par un operatour. C. R. Seances Acad Sci 1960;250:1727–1729.

[20] Jacob F, Monod J. Genetic regulatory mechanisms in the synthesis of proteins. J Mol Biol 1961;3:318–356.

[21] Goodwin B. Oscillatory behaviour in enzymatic control process. Adv Enz Regul 1965;3:425–438.

[22] Kauffman S. Metabolic stability and epigenetics in randomly constructed gene nets. J Theor Biol 1969;22:437–467.

[23] Veliz-Cuba A, Stigler B. Boolean models can explain bistability in the lac operon. J Comput Biol 2011;18:783–794.

[24] Santillán M, Mackey M. Quantitative approaches to the study of bistability in the lac operon of *Escherichia coli*. J R Soc Interface 2008;5:S29–S39.

[25] Laurent M, Kellershohn N. Multistability: a major means of differentiation and evolution in biological systems. Trends Biochem Sci 1999;24:418–422.

[26] Koshland D. E., Goldbeter A., Stock JB. Amplification and adaptation in regulatory and sensory systems. Science (New York, N.Y.) 1982;217:220–225.

[27] Ferrell JE. Self-perpetuating states in signal transduction: positive feedback, double-negative feedback and bistability. Curr Opin Cell Biol 2002;14:140–148.

[28] Ferrell JE. Tripping the switch fantastic: how a protein kinase cascade can convert graded inputs into switch-like outputs. Trends Biochem Sci 1996;21:460–466.

[29] Ferrell JE. Building a cellular switch: more lessons from a good egg. Bioessays 1999;21: 866–870.

[30] Hinkelmann F, Brandon M, Guang B, et al. ADAM: Analysis of discrete models of biological systems using computer algebra. BMC Bioinformatics 2011;12:437–467.

[31] Veliz-Cuba A. Reduction of Boolean network models. J Theor Biol 2011;289: 167–172.

Inferring the Topology of Gene Regulatory Networks: An Algebraic Approach to Reverse Engineering

3

Brandilyn Stigler and Elena Dimitrova

Southern Methodist University, Dallas, TX 75275-0156, USA
Clemson University, Clemson, SC 29634-0975, USA

3.1 INTRODUCTION

3.1.1 Gene Regulatory Networks in Molecular Biology

In everyday language the word "genes" has become synonymous with heredity. Nowadays children learn early that who we, and the living world around us, are is encoded in each and every cell of the organism. This, of course, is generally true.[1] Indeed genes hold the information to build and maintain an organism's cells and pass genetic traits to offspring. However, thinking of a genome simply as a book full of facts about an organism paints a rather incomplete static picture. It hides the fact that the "book" also contains instructions for the mechanisms through which genetic information is extracted and plays a role in cellular processes, such as controlling the response of a cell to environmental signals and replication of the DNA preceding the cell division. This process of genetic information extracting and utilizing is part of what is know as *gene regulation*. Gene regulation is essential for prokaryotic and eukaryotic cells, as well as for viruses, as it increases the cell's flexibility in responding to the environment, allowing it to synthesize gene products (which are most often proteins) when needed.

Gene regulation is an intricate process whose complexity makes it an extreme challenge for mathematical modeling. For example, proteins synthesized from genes may control the flow of genetic information from DNA to mRNA, function as enzymes catalyzing metabolic reactions, or may be components of signal transduction pathways. Gene regulation also drives the processes of cellular differentiation and morphogenesis, leading to the creation of different cell types in multicellular organisms where the different types of cells may possess different gene expression profiles though they all possess the same genome sequence. The degradation of proteins

---

[1] If we overlook the distinction between genes and their alleles and the fact that some organelles are self-replicating and are not coded for by the organism's DNA.

Mathematical Concepts and Methods in Modern Biology. http://dx.doi.org/10.1016/B978-0-12-415780-4.00003-X

and the immediate DNA products can also be regulated in the cell. The proteins involved in the regulatory functions are produced by other genes. This gives rise to a *gene regulatory network* (GRN) consisting of regulatory interactions between DNA, RNA, proteins, and small molecules. A simple GRN consists of one or more input genes, metabolic and signaling pathways, regulatory proteins that integrate the input signals, several target genes, and the RNA and proteins produced from target genes.

The first discovery of a GRN is widely considered to be the identification in 1961 of the *lac* operon, discovered by Jacques Monod [1], in which proteins involved in *lac* metabolism are expressed by *Escherichia coli* and some other enteric bacteria only in the presence of lactose and absence of glucose. Since its discovery, the *lac* operon has often been used as a model system of gene regulation. Most mathematical models of the *lac* operon have been given as systems of differential equations (see Chapter 2 for examples) but discrete modeling frameworks are increasingly receiving attention for their use in offering global insights [2–4]. For examples of discrete mathematical models of a GRN, see the Boolean network models of the *lac* operon in *E. coli* in Chapter 1.

Boolean networks are particularly useful in the case where one is interested in qualitative behavior. However, allowing each variable to assume only 0 and 1 as its values (often interpreted as 'absent' and 'present', or 'off' and 'on') does not always allow for sufficient flexibility in the model construction and can cause the loss of important information. Furthermore, few GRNs are as well understood as the *lac* operon, which prohibits model construction that is purely based on knowledge of the GRN. Technological advances in the life sciences, however, have triggered an enormous accumulation of experimental data representing the activities of the living cell. As a result, the type of mathematical modeling which starts from experimental data and builds a model with little additional information has become of interest. Such modeling methods are sometimes referred to as *reverse engineering* and are the topic of the following section.

3.1.2 Reverse Engineering of Gene Regulatory Networks

Reverse engineering is an approach to mathematical modeling in which a model is constructed from observations of a system in response to stimulus. The process of reverse engineering is similar to playing the parlor game Twenty Questions, in which one tries to discover a secret word by asking yes/no questions. While one can ask all possible feature-type questions (is it alive, is it a physical object), one can typically get away with far fewer questions. In fact, one wins if fewer than 20 questions are asked. In molecular biology, a biological network, such as our immune system's response to infection, plays the role of the secret word and laboratory experiments take the place of the questions. The goal then is to discover the network through the experiments with the hope of gaining deeper insight into a particular phenomenon, such as seasonal effects on immunity. For reasons of cost, it is advantageous to be able to discover the network with as few experiments as possible.

Reverse engineering is a critical step in the systems biology paradigm that has pervaded the biological sciences in recent history. The modeling process starts with a biological system under study. Given a question or hypothesis about the system, the researcher designs experiments to probe the system in hopes of addressing the question or hypothesis. A model is then built from the data resulting from the experiments. Predictions or simulations are extracted from the model and then compared against the original system. The step "from data to a model" is one that uses reverse engineering. This paradigm differs from the previous "reductionist" paradigm that dominated twentieth century biology: the parts of a system were of importance for study in reductionism, whereas the system itself is of interest in systems biology.

Generally speaking, reverse engineering in systems biology aims to recover the network topology and dynamics of a network from observations. A model is built to fit the given observations and the topology and/or the dynamics of the network can be inferred from the model. *Network topology* refers to the physical structure of the network, that is, how the components in the network are connected. It is often encoded as a directed graph, or a wiring diagram, where vertices represent the components of the network (genes, proteins, signaling molecules, etc.) and a directed edge is drawn between two vertices if there is an interaction between the associated components (regulation, synthesis, activation, etc.). *Dynamics* refer to the behavior of the network over time, that is the time evolution of the network processes. The dynamics of a network is also depicted as a graph; in the case of finite dynamical systems, which are considered here, the graph is finite and directed. The so-called state space graph is comprised of vertices representing network states (n-tuples) and a directed edge is drawn between two vertices A and B if the network advances from state A to state B. See Section 3.2 for definitions.

Gene regulatory networks are often modeled using collections of mathematical functions in which each molecular component is assigned a function that formalizes the dynamics of the component [5,6]. From these functions the wiring diagram and state space can be constructed and the structure and behavior of the GRN can be analyzed. A challenge for molecular geneticists is to identify causal links in the network. Identification and control of these links are important first steps in repairing defects in regulation. While there is a growing amount of data being collected from such networks, control of GRNs requires knowledge of the topology and dynamics of the GRN.

Mathematical methods to reverse engineer GRNs are diverse and draw from statistics, graph theory, network theory, computational algebra, and dynamical systems [7,8]. The performance of these methods intrinsically depends on the amount and quality of data provided [9]. In practice, there is insufficient data to uniquely infer a model for a GRN and the number of models that fit the data may be considerably large. An area of continual growth is the development of methods to select biologically feasible or likely models from a pool of candidate models. There are numerous strategies for model selection. For example, some methods restrict the space of likely wiring diagrams to those that have few inputs per vertex or whose in-degree distribution follows a power law, features which are consistent with what is believed about

GRNs [10,11]. As it has also been observed that GRNs tend to be ordered systems with a limited number of regimes, there are methods which filter their results to only those state spaces with few steady states and short limit cycles (see Section 3.2 for definitions). Other aspects of GRNs, such as the negative or positive feedback loops and oscillatory behavior, are also considered in model selection.

3.2 POLYNOMIAL DYNAMICAL SYSTEMS (PDSs)

Different mathematical frameworks have been proposed for the modeling of GRNs, with continuous models using differential equations being the most common approach. While the behavior of a biological system may be seen as continuous in that change of concentration of biochemicals can be modeled with continuous functions, the technology to record observations of the system is not continuous and the available experimental data consist of collections of discrete instances of continuous processes. Furthermore, discrete models where a variable (e.g., a gene) could be in one of a finite number of states are more intuitive, phenomenological descriptions of GRNs and, at the same time, require less data to build. The framework of finite dynamical systems (FDS) provides the opportunity to model a GRN as a collection of variables that transition discretely from one state to the next.

Definition 3.1.   A *finite dimensional system* of dimension n is a function $F = (f_1, \ldots, f_n) : S^n \rightarrow S^n$, where each $f_i : S^n \rightarrow S$ is called a *local* (or *transition*) function, and S is a finite set.

For each variable x_i, its local function f_i determines the state of x_i in the next iteration based on the current state of all variables, including possibly x_i itself. If we further require that the state set S be a finite field, then a result in [12] guarantees that the local functions of an FDS can be expressed as polynomial functions. Working exclusively with polynomials facilitates the modeling process significantly, as we shall see in later sections. Fortunately, from basic abstract algebra we know that a restriction on the size of S can turn it into a *finite field* by requiring that the cardinality of S be a power of a prime integer. An example of a finite field is the set of integers modulo p, denoted $\mathbb{Z}_p$, where p is prime. The elements of $\mathbb{Z}_p$ are the equivalence classes of remainders upon division by p, namely $[0], [1], \ldots, [p-1]$, and addition and multiplication of these classes corresponds to addition and multiplication of representatives from the classes modulo p.

Example 3.1.   The finite field $\mathbb{Z}_5$ has elements $[0], [1], [2], [3]$, and $[4]$. We see that $[2] + [4] = [1]$ since $2 + 4 = 6 \equiv 1 \bmod 5$. Similarly, $[2] \cdot [4] = [3]$ since $2 \cdot 4 = 8 \equiv 3 \bmod 5$.

For simplicity, we will write m instead of $[m]$ to represent the elements of a finite field.

If the state set for an FDS F is a finite field $\mathbb{F}$, we call F a *polynomial dynamical system* (PDS).

Example 3.2. Let $\mathbb{F} = \mathbb{Z}_3$ and $F : \mathbb{F}^2 \rightarrow \mathbb{F}^2$ be a PDS with transition functions

$$f_1 = f_1(x_1, x_2) = 2x_2,$$
$$f_2 = f_2(x_1, x_2) = x_1 + x_1^2.$$

The dimension of F is 2, where each of the two variables x_1 and x_2 can be in one of three states, i.e., 0, 1, or 2.

A key issue in the study of GRNs is understanding the relationships among genes, proteins, metabolites, etc. in the network. Thus an important characteristic of a PDS is the way its variables are "connected." To visualize the network topology, a graph called a wiring diagram is used, where the vertices are labeled by the PDS variables and the directed edges signify the direction of interaction between two variables.

Definition 3.2. Let F be an n-dimensional PDS on variables $x_1, \ldots, x_n$. The *wiring diagram* (also known as a *dependency graph*) of F is a directed graph with vertex set $V = \{x_1, \ldots, x_n\}$. A directed edge is drawn from x_i to x_j if, and only if, x_i is present in at least one polynomial term (with a nonzero coefficient) of the local function f_j.

Figure 3.1(a) shows the wiring diagram of the PDS from Example 3.2.

While the wiring diagram of a PDS provides only a *static* snapshot of the network topology with no reference to strength and timing of the interactions, such a graph carries very useful information about the biological system and several methods have been developed for its inference. Sections 3.5.1 and 3.5.2 present two such methods.

Just as a phase portrait is used to plot the simultaneous change of two or three variables in a system of differential equations [13], the *dynamic* properties of a PDS are captured in a graph called a *state space graph*. In fact, the dynamics of a PDS $F : \mathbb{F}^n \rightarrow \mathbb{F}^n$ is uniquely represented by its state space graph, which has $|\mathbb{F}|^n$ vertices.

Definition 3.3. The *state space graph* of a PDS $F : \mathbb{F}^n \rightarrow \mathbb{F}^n$ on variables $x_1, \ldots, x_n$ is a directed graph whose vertex set is $\mathbb{F}^n$ and contains an edge directed from $\mathbf{u}$ to $\mathbf{v}$ if, and only if, $F(\mathbf{u}) = \mathbf{v}$, where $\mathbf{u}, \mathbf{v} \in X^n$.

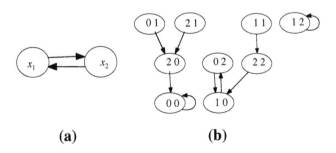

(a)   (b)

FIGURE 3.1

Wiring diagram (a) and state space graph (b) of the PDS from Example 3.2.

The state space graph of the PDS from Example 3.2 is shown in Figure 3.1(b). For example, there is an arrow from state $(2, 1)$ to state $(2, 0)$ because $F((2, 1)) = (f_1(2), f_2(1)) = (2, 0)$ using modulo 3 arithmetic (since the field is $\mathbb{Z}_3$). The graph consists of $3^2 = 9$ vertices arranged into three connected components. Notice that $F((0, 0)) = (0, 0)$ and $F((1, 2)) = (1, 2)$. Such points are called *fixed points* and these two are the only fixed points for this PDS. Also notice that $F \circ F((0, 2)) = F(F((0, 2))) = F((1, 0)) = (0, 2)$, or graphically, $(0, 2) \leftrightarrows (1, 0)$. This is called a *limit cycle* and its length is two. Analogously, for any positive integer m, a *limit cycle of length m* arises whenever $F^m(\mathbf{v}) = \mathbf{v}$ (and m is the smallest positive integer with this property) for some $\mathbf{v} \in \mathbb{F}^n$. Fixed points can be considered as limit cycles of length one. For this PDS, $(0, 2) \leftrightarrows (1, 0)$ is the only limit cycle of length larger than one.

Exercise 3.1. Construct by hand the state space graphs and wiring diagrams for the following PDSs. Can a vertex in the state space graph have an out-degree greater than one, that is, can the vertex have more than one directed edge coming out of it? How about a vertex in the wiring diagram?

1. $F = (f_1, f_2) : \mathbb{Z}_2^2 \rightarrow \mathbb{Z}_2^2$, where

$$f_1 = x_1 + x_2,$$
$$f_2 = x_1 x_2.$$

2. $F = (f_1, f_2, f_3) : \mathbb{Z}_3^3 \rightarrow \mathbb{Z}_3^3$, where

$$f_1 = x_1 + x_2,$$
$$f_2 = x_1 x_3 + 1,$$
$$f_3 = x_2 x_3. \qquad \triangledown$$

Exercise 3.2. Compute the fixed points of the PDS in Exercise 3.1 (1) algebraically. Compare your answers to what you found in the previous exercise. Hint: A fixed point will satisfy $x_1 = x_1 + x_2$ and $x_2 = x_1 x_2$. $\qquad \triangledown$

Exercise 3.3. In Exercise 3.1 (2), you saw that the PDS has no fixed points. If instead we consider the system over $\mathbb{Z}_5$, would you expect to see fixed points? Verify your answer using the web tool Analysis of Dynamic Algebraic Models (ADAM) at
dvd.vbi.vt.edu/adam.html. $\qquad \triangledown$

Exercise 3.4. Find three different PDSs over $\mathbb{Z}_3$ that have the wiring diagram as in Figure 3.5. $\qquad \triangledown$

Exercise 3.5. Let's reverse engineer! Find a PDS on three variables over $\mathbb{Z}_2$ that fits the data as in Figure 3.6. Is the PDS unique? What if the entire state space was given? $\qquad \triangledown$

Exercise 3.6. [14] Phosphorylation is the addition of a phosphate group to a protein or any other organic molecule. It turns many protein enzymes on and off, thereby altering their function and activity. Suppose proteins A and B interact. Let x_1 reflect the absence (0)/presence (1) of protein A, whereas x_2 indicates if the protein B has

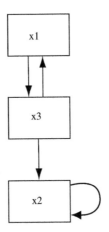

FIGURE 3.5

Wiring diagram for Exercise 3.5.

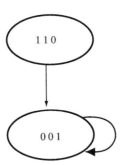

FIGURE 3.6

Partial state space for Exercise 3.6.

already been phosphorylated (1) or not (0) before the interaction. Furthermore, we want f_1 to indicate the absence/presence of protein A right after the interaction and f_2 to point out whether protein B has been phosphorylated or not.

1. Explain why f_1 should depend on x_1 only, while f_2 must depend on both x_1 and x_2.
2. Do you find the following Boolean model reasonable? Correct it if necessary.

$$f_1 = x_1,$$
$$f_2 = x_1 \ \text{OR} \ x_2.$$

3. Convert the above Boolean functions into polynomials over $\mathbb{Z}_2$.   ▽

Project 3.1.   Study the articles [15, 37] and construct the state spaces and wiring diagrams of the models.   ▽

Project 3.2.   Study the article [16] and, without computing the entire state space, find the cycle structure of the following Boolean monomial dynamical system: F : $\mathbb{Z}_2^7 \to \mathbb{Z}_2^7$, where

$$F(x_1, x_2, x_3, x_4, x_5, x_6, x_7) = (x_7, x_1 x_3, x_4, x_5, x_6, x_7, x_2).   ▽$$

Project 3.3.   In [17], the authors describe an equivalence among three discrete modeling classes, namely *logical models, bounded Petri nets*, and PDSs. What is the significance of such an equivalence? In addressing this question, consider the following:

1. In their article, the authors describe logical models and bounded Petri nets. How easy is it to describe or construct these types of models?
2. What algebraic structure do they have?
3. How do their description and algebraic structure compare to those of PDSs?
4. Do you think it is easier to develop an algorithm for PDSs than for logical models or Petri nets?   ▽

3.3 COMPUTATIONAL ALGEBRA PRELIMINARIES

In this section we introduce some basic concepts from computational algebra that we will need for the rest of the chapter. For a comprehensive introductory treatment of the subject some excellent sources are [19–21].

In linear algebra we learn how to solve systems of linear equations using Gaussian elimination. However, the algorithm cannot be easily adjusted for application to systems of polynomials of degree larger than one. Simultaneously solving systems of polynomial equations is needed in many areas but a systematic computational method was not developed until the 1960s when the so-called Gröbner bases were developed [22, 23]. The role of a Gröbner basis is similar to that of a basis for a vector space and the algorithm for finding it can be seen as a multivariate, nonlinear generalization of Gaussian elimination for linear systems.

An issue that arises when using multivariable polynomials is that polynomial division is generally not well defined. Polynomial division in one variable is always well defined in the sense that there is a unique remainder upon division of polynomials. This is because there is a unique monomial ordering (see Definition 3.4) and the division of polynomial f by polynomial g proceeds by dividing the highest power of the variable in g into the highest power in f. In other words, the one-variable monomials are ordered using degree ordering:

$$\cdots \succ x^{m+1} \succ x^m \succ \cdots \succ x^2 \succ x \succ 1.$$

With multivariate polynomials, however, there is more than one way of ordering their monomials (terms) and thus to carry out long division. So the remainder of long division is not uniquely determined in general.

Example 3.3. Consider the polynomials $f = x^2$ and $g = x^2 + xy^2$. The remainder of dividing f by g is $-xy^2$ or x^2, depending on whether we choose the initial monomial of g to be x^2 or xy^2, respectively.

As we will see below, multivariate polynomial division becomes well defined when using Gröbner bases.

Let k be a field. Here, we focus on the finite fields $\mathbb{Z}_p$ where p is prime; however, much of the presented results are valid for other fields, such as $\mathbb{Q}$. The linear combination of monomials of the form $x^\alpha = x_1^{\alpha_1} \cdots x_n^{\alpha_n}$ over k, where α is the n-tuple exponent $\alpha = (\alpha_1, \ldots, \alpha_n) \in \mathbb{Z}_{\geq 0}^n$ with nonnegative entries, forms a ring, called a *polynomial ring*, denoted $k[x_1, \ldots, x_n]$. For the purpose of polynomial division, it is necessary to be able to arrange the terms of a polynomial unambiguously in some order. This requires a *total* ordering on the monomials, i.e., for every pair of monomials x^α and x^β, exactly one of the following must be true:

$$x^\alpha \prec x^\beta, \qquad x^\alpha = x^\beta, \qquad x^\alpha \succ x^\beta.$$

Taking into account the properties of the polynomial sum and product operations, the following definition emerges.

Definition 3.4. A *monomial ordering* on $k[x_1, \ldots, x_n]$ is any relation $\succ$ on $\mathbb{Z}_{\geq 0}^n$ satisfying:

1. $\succ$ is a total ordering on $\mathbb{Z}_{\geq 0}^n$.
2. If $\alpha \succ \beta$ and $\gamma \in \mathbb{Z}_{\geq 0}^n$, then $\alpha + \gamma \succ \beta + \gamma$.
3. $\succ$ is a well ordering on $\mathbb{Z}_{\geq 0}^n$, i.e., every nonempty subset of $\mathbb{Z}_{\geq 0}^n$ has a smallest element under $\succ$.

To simplify the classification of monomial orderings, we will assume throughout that $x_1 \succ x_2 \succ \cdots \succ x_n$ (or $x \succ y \succ z$).

One of the most popular monomial orderings is the *lexicographic ordering (lex)* which is analogous to the ordering of words in dictionaries and under which $x^2 \succ_{lex} xy^2$. Another one is the *graded lexicographic ordering (grlex)* which orders by total degree first, then "breaks ties" using the lexicographic ordering. Under graded lexicographic ordering, $xy^2 \succ_{grlex} x^2$. Yet another very useful monomial ordering is the *graded reverse lexicographic ordering (grevlex)*. Like *grlex*, it uses the total degree first but to break ties it looks at the rightmost variable and favors the smaller power. For example, $y^2z \prec_{grlex} xz^2$ but $y^2z \succ_{grevlex} xz^2$. Algorithmic definitions are available at [19].

A monomial ordering can also be defined by a weight vector $\omega = (\omega_1, \ldots, \omega_n)$ in $\mathbb{Z}_{\geq 0}^n$. We require that ω have nonnegative coordinates in order for 1 to always be the smallest monomial. Fix a monomial ordering $\succ_\sigma$, such as $\succ_{lex}$. Then, for $\alpha, \beta \in \mathbb{Z}_{\geq 0}^n$, define $\alpha \succ_{\omega,\sigma} \beta$ if and only if

$$\omega \cdot \alpha \succ \omega \cdot \beta \quad \text{or} \quad (\omega \cdot \alpha = \omega \cdot \beta \quad \text{and} \quad \alpha \succ_\sigma \beta).$$

For example, with $x \succ y$, weight vector $\omega = (3, 1)$ orders the monomials of polynomial $g = x^2 + xy^2$ in the same way as the lexicographic order, while $\omega' = (1, 1)$ performs the same monomial ordering as the graded lexicographic ordering.

However, even after fixing the monomial ordering, there is still a problem with multivariate polynomial division: when dividing a given polynomial by more than one polynomial, the outcome may depend on the order in which the division is carried out. Let $f, h_1, \ldots, h_m \in k[x_1, \ldots, x_n]$. The famous *ideal membership problem* is to determine whether there are polynomials $h_1, \ldots, h_m \in k[x_1, \ldots, x_n]$ such that $f = \sum_{i=1}^m h_i f_i$. The following definition will help us state this in the language of abstract algebra and explains the name of the problem.

Definition 3.5.   Let k be a field. Then

$$I = \langle f_1, \ldots, f_m \rangle = \left\{ \sum h_i f_i | h_1, \ldots, h_m \in k[x_1, \ldots, x_n] \right\},$$

is called an *ideal* in $k[x_1, \ldots, x_n]$.

We know from abstract algebra that I is closed under addition and multiplication by any polynomial in $k[x_1, \ldots, x_n]$. This is analogous to a subspace of a vector space, where the "vectors" are polynomials.

We ask whether f is an element of I. In general, even under a fixed monomial ordering, the order in which f is divided by the generating polynomials f_i affects the remainder; perhaps even more surprisingly, a nonzero remainder of division does not imply $f \notin I$. Moreover, the generating set $\{f_1, \ldots, f_m\}$ of the ideal I is not unique but a special generating set $\mathcal{G} = \{g_1, \ldots, g_t\}$ can be selected so that the remainder of polynomial division of f by the polynomials in $\mathcal{G}$ performed in any order is zero if and only if f lies in I. A generating set with this property is called a *Gröbner basis*. A Gröbner basis exists for very ideal other than $\{0\}$ and for a fixed monomial ordering. Before we give a precise definition, we need to look at a special kind of ideal.

Definition 3.6.   The *initial ideal* of an ideal $I \neq \{0\}$ for a fixed monomial ordering $\succ$ is denoted $in_\succ(I)$ and is the ideal generated by the set of leading monomials (under the specified ordering) of the polynomials of I. That is, $in_\succ(I) = \langle in_\succ(f)|f \in I \rangle$, where $in_\succ(f)$ is the monomial of f that is ordered first under the specified monomial ordering, called a *leading* or *initial* monomial. The monomials which do not lie in $in_\succ(I)$ are called *standard monomials*.

Definition 3.7.   Fix a monomial ordering $\succ$. A finite subset $\mathcal{G}$ of an ideal I is a *Gröbner basis* if $in_\succ(I) = \langle in_\succ(g)|g \in \mathcal{G} \rangle$.

A Gröbner basis for an ideal is necessarily a generating set for the ideal but may not be unique even under a fixed monomial ordering. However, if we also require that for any two distinct elements $g, g' \in \mathcal{G}$, no term of g' is divisible by $in_\succ(g)$, such a Gröbner basis $\mathcal{G}$ is called *reduced* and is unique for an ideal and a monomial ordering $\succ$, provided the coefficient of $in_\succ(g)$ in g is 1 for each $g \in \mathcal{G}$.

Example 3.4.   Let $x \succ y$ and $f_1 = x^3 - 2xy$, $f_2 = x^2 y - 2y^2 + x$, and $I = \langle f_1, f_2 \rangle$ in $k[x, y]$ under *grlex* order. Notice that $\{f_1, f_2\}$ is not a Gröbner basis for I because $xf_2 - yf_1 = x^2 \in I$ and so $in_{grlex}(x^2) = x^2 \in in_{grlex}(I)$ but $x^2 \notin \langle in_{grlex}(f_1), in_{grlex}(f_2) \rangle = \langle x^3, x^2 y \rangle$. The Buchberger algorithm (or any of its improvements, such as the Buchberger-Möller algorithm [24]) generates the following *grlex* Gröbner basis: $\{x^3 - 2xy, x^2 y - 2y^2 + x, -x^2, -2xy, -2y^2 + x\}$.

Example 3.5.   In the previous example, we can see that the leading monomials (disregarding coefficients) of the Gröbner basis are $x^3, x^2 y, x^2, xy$, and y^2. So the initial ideal is generated by x^2, xy, and y^2: any other leading monomial can be written in terms of these generators. Furthermore, we see that the standard monomials associated to the Gröbner basis are 1, x, and y.

3.4 CONSTRUCTION OF THE MODEL SPACE: A REVERSE ENGINEERING ALGORITHM

In this section, we consider the problem of constructing PDSs from a given set of data collected from a biological system on n components, called *nodes*.

Let $\mathbb{F}$ be a finite field. Let $\mathbf{s}_1, \ldots, \mathbf{s}_m \in \mathbb{F}^n$ be a set of m input states of an n-node system, and $\mathbf{t}_1, \ldots, \mathbf{t}_m \in \mathbb{F}^n$ a set of corresponding output states, where $n, m \in \mathbb{N}$. Each of $\mathbf{s}_i$ and $\mathbf{t}_j$ has the form

$$\mathbf{s}_i = \left(s_{i1}, \ldots, s_{in} \right),$$
$$\mathbf{t}_j = \left(t_{j1}, \ldots, t_{jn} \right),$$

where states in bold are elements of $\mathbb{F}^n$ and italicized states are elements of $\mathbb{F}$. We aim to find all PDSs $F = (f_1, \ldots f_n)$ over $\mathbb{F}$ such that $F(\mathbf{s}_i) = \mathbf{t}_i$ for each $1 \leqslant i \leqslant m$ and where each $f_j \in \mathbb{F}[x_1, \ldots, x_n]$. We proceed by considering the subproblem of finding all transition polynomials f_j that satisfy

$$f_j(\mathbf{s}_1) = t_{1j}$$
$$f_j(\mathbf{s}_2) = t_{2j}$$
$$\vdots$$
$$f_j(\mathbf{s}_m) = t_{mj}.$$

Let p be the (prime) characteristic of $\mathbb{F}$. The identity $a^p = a$ for each $a \in \mathbb{F}$ imposes a degree restriction on polynomials when viewed as functions, namely $x_i^p = x_i$ for $1 \leqslant i \leqslant m$. This equation gives rise to the relation $x_i^p - x_i = 0$, for each $1 \leqslant i \leqslant m$. Since we are interested in polynomials that can be treated as functions, the local functions f_j are actually polynomials in the quotient ring $\mathbb{F}[x_1, \ldots, x_n] / \langle x_1^p - x_1, \ldots, x_n^p - x_n \rangle$. This ring contains polynomials that are the remainders of the elements of $\mathbb{F}[x_1, \ldots, x_n]$ upon division by the degree relations, that is, polynomials whose degree in any variable is less than p. In essence, the quotient ring is constructed by "modding out" by the elements of the "divisor" ideal. Thus we aim to find all polynomials $f_j \in \mathbb{F}[x_1, \ldots, x_n] / \langle x_1^p - x_1, \ldots, x_n^p - x_n \rangle$ that map each system state $\mathbf{s}_i$ to the node state t_{ij}.

Since the remaining discussion assumes that the coefficient field has positive characteristic, we can simplify the notation. We will denote the quotient polynomial ring simply as $\mathbb{F}\left[x_1, \ldots, x_n\right]$ and elements in the ring as polynomials with usual representation. That is, while an element of the quotient ring is of the form $f(x_1, \ldots, x_n) + \langle x_1^p - x_1, \ldots, x_n^p - x_n \rangle$, we will instead write it as $f(x_1, \ldots, x_n)$, understanding that all ring elements have been reduced modulo the ideal $\langle x_1^p - x_1, \ldots, x_n^p - x_n \rangle$.

Let F_j be the set of all polynomials that fit the data for node j as described above. We solve the problem similarly to solving a system of nonhomogeneous linear equations in that we construct a "particular" polynomial and the set of "homogeneous" polynomials. The "particular" polynomial is any that interpolates or fits the given data. There are numerous methods for constructing an interpolating polynomial function. Here we use a formula based on the ring-theoretic version of the Chinese Remainder Theorem:

$$f_j(\mathbf{x}) = \sum_{i=1}^{m} s_{i+1,j} r_i(\mathbf{x}),$$

where $r_i(\mathbf{x})$ is a polynomial that evaluates to 1 on $\mathbf{s}_i$ and 0 on any other input, and $\mathbf{x}$ represents $x_1, x_2, \ldots, x_n$. Specifically,

$$r_i(\mathbf{x}) = \prod_{k=1}^{m} (s_{i\ell} - s_{k\ell})^{p-2} (x_\ell - s_{k\ell}),$$

where ℓ is the first coordinate in which $\mathbf{s}_i \neq \mathbf{s}_k$ and $i \neq k$. For a detailed description of the construction, see [25]. For the set of "homogeneous" polynomials, we use the Ideal-Variety Correspondence from algebraic geometry [19]. We view the input data $\{\mathbf{s}_1, \ldots, \mathbf{s}_m\}$ as a *variety*, or a set of roots of a system of polynomials equations. The Ideal-Variety Correspondence states how to construct the ideal of polynomials that vanish on the data. To each input state $\mathbf{s}_i$, for $1 \leqslant i \leqslant m$, we associate the (maximal) ideal

$$\mathbf{I}(\mathbf{s}_i) = \langle x_1 - s_{i1}, \ldots, x_n - s_{in} \rangle .$$

Each ideal $\mathbf{I}(\mathbf{s}_i)$ is the set of polynomials that evaluate to 0 on the single data point $\mathbf{s}_i$. According to the correspondence, the intersection of these ideals contains all polynomials that vanish on the union of the input data. So the ideal of "homogeneous", or *vanishing*, polynomials is

$$\mathbf{I}(\{\mathbf{s}_1, \ldots, \mathbf{s}_m\}) = \bigcap_{i=1}^{m} \mathbf{I}(\mathbf{s}_i),$$

which we denote by I for simplicity. Therefore, the set F_j is described by $f_j + I :=$ $\{f_j + h | h \in I\}$.

Example 3.6.   Consider a 3-node system having the following states in $\mathbb{Z}_5$:

$$2\ 0\ 0$$
$$4\ 3\ 1$$
$$3\ 1\ 4$$
$$0\ 4\ 3$$

which we interpret as

$$s_1 = (2, 0, 0) \rightarrow t_1 = (4, 3, 1),$$
$$s_2 = (4, 3, 1) \rightarrow t_2 = (3, 1, 4),$$
$$s_3 = (3, 1, 4) \rightarrow t_3 = (0, 4, 3).$$

The corresponding ideals are

$$I(s_1) = \langle x_1 - 2, x_2, x_3 \rangle,$$
$$I(s_2) = \langle x_1 - 4, x_2 - 3, x_3 - 1 \rangle,$$
$$I(s_3) = \langle x_1 - 3, x_2 - 1, x_3 - 4 \rangle.$$

To compute the intersection I of the ideals, we can use the computer algebra system Macaulay 2 [26] with the following code:

```
R=ZZ/5[x1,x2,x3]
I1=ideal(x1-2,x2,x3)
I2=ideal(x1-4,x2-3,x3-1)
I3=ideal(x1-3,x2-1,x3-4)
I=intersect{I1,I2,I3}.
```

Then I is computed as

$$\Big(-x_1 + 2x_2 + x_3 + 2,\ 2x_2x_3 - 2x_3^2 + x_3,\ -x_2x_3 - x_3^2 + x_2 + x_3,$$
$$x_2^2 + x_2x_3 + 2x_2 + 2x_3,\ 2x_2^2 - 2x_2x_3 + x_2,\ x_1x_2 + x_1x_3 + x_2 + x_3,$$
$$-2x_1x_2 + 2x_1x_3 - x_1 - x_2 + x_3 + 2 \Big).$$

Note that negative coefficients can be written as positive numbers: $-1 \equiv 4 \bmod 5$, $-2 \equiv 3 \bmod 5$, etc.

To find the function $f_1(x_1, x_2, x_3)$ for node x_1 such that

$$f_1 (2, 0, 0) = 4,$$
$$f_1 (4, 3, 1) = 3, \qquad\qquad (3.1)$$
$$f_1 (3, 1, 4) = 0,$$

we compute the r polynomials:

$$r_1 (x_1, x_2, x_3) = (2 - 4)^3 (x_1 - 4) (2 - 3)^3 (x_1 - 3) = 3x_1^2 + 4x_1 + 1,$$
$$r_2 (x_1, x_2, x_3) = (4 - 2)^3 (x_1 - 2) (4 - 3)^3 (x_1 - 3) = 3x_1^2 + 3,$$
$$r_3 (x_1, x_2, x_3) = (3 - 2)^3 (x_1 - 2) (3 - 4)^3 (x_1 - 4) = 4x_1^2 + x_1 + 2.$$

We find that

$$f_1(x_1, x_2, x_3) = (4)\, r_1 + (3)\, r_2 + (0)\, r_3 = x_1^2 + x_1 + 3.$$

So the set F_1 of polynomials that fit the data for node x_1 is $F_1 = f_1 + I$. Using the above strategy, we can find similar sets $F_2 = f_2 + I$ and $F_3 = f_3 + I$ for x_2 and x_3, respectively, where $f_2(\mathbf{x}) = 3x_1^2 + x_1 + 4$ and $f_3(\mathbf{x}) = 2x_1^2 + 2x_1 + 4$.

The data set in the previous example is called a *time series*, since the data points are considered to be successive. We call $F_1 \times \cdots \times F_n$ the *model space* associated to the given data and any PDS $(f_1, \ldots, f_n) \in F_1 \times \cdots \times F_n$ is called a *model*. Note that the ideal I of vanishing polynomials is independent of the node or particular function.

For the following exercises, consider the time series over $\mathbb{Z}_3$, which contains five observations for a 3-node system.

$$
\begin{array}{ccc}
1 & 1 & 1 \\
2 & 0 & 1 \\
2 & 0 & 0 \\
0 & 2 & 2 \\
0 & 2 & 2 \\
\end{array}
$$

Exercise 3.7. Compute the ideal of the data points using Macaulay 2 and the *lex* monomial order with the variable order $x_1 \succ x_2 \succ x_3$. Use the previous example as a reference. In Macaulay 2, polynomial rings require an explicit choice of monomial ordering. For this exercise, you will use the following:

```
R=ZZ/3[x1,x2,x3,MonomialOrder=>Lex]
```

If none is given, then *grevlex* is used by default. Also the order of the variables in the polynomial ring indicates the variable order. Is the ideal different under the *grevlex* order?   ▽

Exercise 3.8. Let $D = \{\mathbf{c}_1, \ldots, \mathbf{c}_m\} \subseteq \mathbb{F}^n$ be a data set as in Example 3.6. A polynomial $f : \mathbb{F}^n \to \mathbb{F}$ that interpolates D can be constructed as

$$f(x_1, \ldots, x_n) = \sum_{i=1}^{m} f(c_{i1}, \ldots, c_{in}) \prod_{j=1}^{n} (1 - (x_j - c_{ij})^{p-1}).$$

Compute an interpolating polynomial for each node using this interpolation formula and the one provided in the text above. Do they differ on the data? Do they differ on other data points?   ▽

Project 3.4. An easy calculation shows that there are p^n states in $(\mathbb{Z}_p)^n$. From this we find that there are p^{p^n} polynomials in $\mathbb{Z}_p[x_1, \ldots, x_n]$. How many n-dimensional PDSs are there over $\mathbb{Z}_p[x_1, \ldots, x_n]$? Given a data set $D \subset (\mathbb{Z}_p)^n$, how many models are there in the corresponding model space?   ▽

3.5 MODEL SELECTION

As we saw in the previous section, the model space consists of all PDSs that fit a given data set. The problem is now to select models from the space.

There are numerous strategies for model selection. One naive approach is to simply choose the interpolating polynomial constructed above. The problem, however, is that predictions of the model are dependent on the form of the model. Specifically, the terms that appear in the polynomials will "predict" the interactions among the nodes as represented in the wiring diagram associated to the PDS. While it is possible that the chosen interpolation formula well characterizes the physical setting being modeled, it is more likely that the resulting model will have little to no predictive power. Therefore, strategies that incorporate appropriate priors are desirable (for example, see [27,28]).

It is believed that GRNs are sparse, in the sense that they have few (less than half of all possible) connections, and their edges follow a power-law distribution [10,11]. A model selection strategy that makes use of these assumptions is one based on Gröbner bases. Given a set $f_j + I$ of transition functions for a node, the method finds a polynomial in the set that contains no terms in I. This is done by computing the remainder of f_j upon division by the elements in I. As seen in Section 3.3, polynomial division in general is not well defined in the sense that remainders may not be unique, but is well defined given a Gröbner basis. The strategy requires that a Gröbner basis $\mathcal{G}$ for I is computed. Then the process of computing the remainder of f_j upon division by the elements of I results in a unique polynomial called the *normal form* of f_j with respect to $\mathcal{G}$ and is denoted $NF(f_j, \mathcal{G})$. Given a monomial ordering $\succ$ and a Gröbner basis $\mathcal{G}$ for I with respect to $\succ$, then $(NF(f_1, \mathcal{G}), \ldots, NF(f_n, \mathcal{G}))$ is a model that fits the given data and is reduced with respect to the data.

Since Gröbner bases depend on the choice of monomial ordering, so do resulting models.

Example 3.7. In Example 3.6, the graded reverse lexicographic ordering, *grevlex*, with $x_1 \succ x_2 \succ x_3$ will yield models involving the terms $\{1, x_1, x_2, x_3\}$, whereas the *lex* ordering with the same variable order results in models with the terms $\{1, x_3, x_3^2, x_3^3\}$. We can see this by doing the following in Macaulay 2 (comments are preceded by two dashes):

```
R=ZZ/5[x1,x2,x3] --grevlex order is assumed
I1=ideal(x1-2,x2,x3) --same ideals as above
I2=ideal(x1-4,x2-3,x3-1)
I3=ideal(x1-3,x2-1,x3-4)
I=intersect{I1,I2,I3}
f1=x1^2+x1+3
f2=3*x1^2+x1+4
f3=2*x1^2+2*x1+4
```

The next lines compute each f "modulo" I and is the same as $NF(\texttt{f}, \mathcal{G})$, where $\mathcal{G}$ is the reduced Gröbner basis for I with respect to the monomial order given in the description of the polynomial ring.

```
f1%I
f2%I
f3%I
```

We see that $NF(\texttt{f1}, \mathcal{G}) = -x_3 - 1$, $NF(\texttt{f2}, \mathcal{G}) = x_2 - 2$, and $NF(\texttt{f3}, \mathcal{G}) = -2x_3 + 1$, where $\mathcal{G}$ is the Gröbner basis for I with respect to *grevlex*.

Now let's change the order to *lex*. Since the f polynomials and I were defined as objects in the ring R, we need to "tell" Macaulay 2 to consider them as objects in a new ring with a different monomial order.

```
S=ZZ/5[x1,x2,x3, MonomialOrder=>Lex]
sub(f1,S)%sub(I,S)
sub(f2,S)%sub(I,S)
sub(f3,S)%sub(I,S)
```

This time, we see that while $NF(\texttt{f1}, \mathcal{G})$ and $NF(\texttt{f3}, \mathcal{G})$ are the same as the normal forms with the *grevlex* order, here we have that $NF(\texttt{f2}, \mathcal{G}) = 2x_3^2 + x_3 - 2$, where $\mathcal{G}$ is the Gröbner basis for I with respect to *lex*.

To see the explicit Gröbner basis that is being used, you can type

```
gens gb I
```

which returns the generators of the Gröbner basis as a matrix. To get the generators in list form, type

```
flatten entries gens gb I
```

Exercise 3.9. Based on your work in Exercise 3.7, which variables do you expect to be present in the normal forms? For example, do you expect x_1 to appear in any normal form?   ▽

Exercise 3.10. Use Macaulay 2 to compute the normal forms of the polynomials you found in Exercise 3.8 with respect to the ideals you found in Exercise 3.7. Are the resulting polynomials different when computed using *lex*, *grlex*, or *grevlex*?   ▽

Exercise 3.11. Compute the set of standard monomials for the monomial orders considered in Exercise 3.7. You can use the following command in Macaulay 2 to assist you:

```
leadTerm I
```

(This is the same as `leadTerm (gb I)` since a Gröbner basis is required.) What is the relationship between these monomials and the terms in the normal forms computed in Exercise 3.10?   ▽

3.5.1 Preprocessing: Minimal Sets Algorithm

We saw in the previous section that the choice of monomial ordering affects which model is selected from the model space. One way to minimize the effect of this choice is to minimize the number of variables in the ambient ring. We again restrict our attention to finding functions f_j for a node x_j. Consider the data set $D_j = \{(s_1, t_{1j}), \ldots, (s_m, t_{mj})\}$ as above and let F_j be the set of polynomials that fit the data. Next we describe the Minimal Sets Algorithm (MSA) which finds the smallest sets of variables such that a polynomial in F_j in those variables exists (see [18] for more details).

Definition 3.8. Let D_j and F_j be as above. Then $M = \{x_{i_1}, \ldots, x_{i_r}\} \subseteq \{x_1, \ldots, x_n\}$ is a *minimal set* if $\mathbb{F}[x_{i_1}, \ldots, x_{i_r}] \cap F_j \neq \emptyset$ and removal of any variable from M renders the intersection empty.

Minimal Sets Algorithm:

1. Input D_j.
2. Partition the input states $s_1, \ldots, s_m$ according to the output values $t_{1j}, \ldots, t_{mj}$. For each output value t_{ij}, let $X_{t_{ij}} = \{s_k | (s_k, t_{ij}) \in D_j\}$. Then the partition is given by $X = \{X_{t_{ij}} | 1 \leq i \leq m\}$.
3. Compute the square-free monomial ideal M generated by the monomials $m(s_k, s_\ell) = \prod_{s_{ki} \neq s_{\ell i}} x_i$ for all $s_k \in X_{t_{kj}}$ and $s_\ell \in X_{t_{\ell j}}$ in which $t_{kj} \neq t_{\ell j}$. The monomials $m(s_k, s_\ell)$ encode the coordinates in which the input states s_k and s_ℓ differ.
4. Compute the primary decomposition of M; that is, write M as the irredundant intersection $\bigcap_i Q_i$ of primary ideals.
5. Return the generating sets of the associated minimal prime ideals $\sqrt{Q_i}$.

The generating sets of the ideals $\sqrt{Q_i}$ are the sought after minimal sets (see Corollary 4 of [18]).

Theorem 3.1. *Let D_j be as above. The minimal sets for D_j are the generating sets of the minimal primes in the primary decomposition of the monomial ideal M as constructed above.*

Example 3.8.   Consider the data in Example 3.6. The minimal sets for x_1 are $\{x_1\}, \{x_2\},$ and $\{x_3\}$, meaning that there is a polynomial function in one variable that fits the data for x_1. In this case the minimal sets are the same for nodes 2 and 3.

Essentially what the MSA does is it computes the intersection of all possible "minimal" wiring diagrams. In the above example, the minimal sets for x_1 are $\{x_1\}$, $\{x_2\}$, and $\{x_3\}$ which indicates that ALL minimally described functions which fit the data for x_1 will be in terms of just one variable.

As the above example demonstrates, there are typically many minimal sets for a node. Next we introduce a method to score minimal sets, based on the sparseness assumption.

Let M_j be the set of minimal sets for a node x_j. Define Z_s to be the number of sets in M_j of cardinality s and $W_i(s)$ to be the number of sets X in M_j of cardinality

s such that $x_i \in X$. Then we score each variable x_i using the function

$$S(x_i) = \sum_{s=1}^{n} \frac{W_i(s)}{sZ_s}$$

and each set using

$$T(X) = \prod_{x_i \in X} S(x_i).$$

We can obtain a probability distribution on M_j by dividing the set scores by $\sum_{X \in M_j} T(X)$. The best performing minimal set is the one with the highest normalized set score.

Example 3.9.   Suppose we have the following minimal sets for a given node in a 6-node system: $X_1 = \{x_1\}$, $X_2 = \{x_2, x_3\}$, $X_3 = \{x_2, x_4\}$, $X_4 = \{x_3, x_5, x_6\}$. Then the scores are 1, 7/24, 1/8, and 7/108, respectively. In this case, the singleton set X_1 has the highest score. Even though X_2 and X_3 are the same size, X_2 has a higher set score than X_3 since the variables x_2 and x_3 have higher variable scores than x_4, namely, 1/2, 7/12, and 1/4, respectively.

For other scoring strategies, see [18]. Note that the strategy presented here corresponds to the (S_1, T_1) method in that paper.

Exercise 3.12.   Using the time series immediately preceding Exercise 3.7, compute the minimal sets for each node.   ▽

Exercise 3.13.   Why can the following not be minimal sets?

1. $\{\{x_1\}, \{x_1, x_2\}, \{x_3, x_4, x_5\}\}$
2. $\{\{x_1, x_2\}, \{x_2, x_3\}, \{x_1, x_2, x_3\}\}$
3. $\{\{x_1 x_2\}, \{x_3, x_4, x_5\}\}$
4. $\{\{x_2\}, \{x_3 + 1\}\}$ ▽

Exercise 3.14.   While sets of standard monomials depend on the choice of monomial order, minimal sets do not. What is the relationship between a minimal set and a set of standard monomials?   ▽

Project 3.5.   One can sometimes extract information about a data set from the structure (regularity or irregularity in the distribution) of a given collection of minimal sets. What can you infer from the following minimal sets for a 5-dimensional system? Each collection of minimal sets has a different corresponding data set. Look for patterns in occurrences of variables in the minimal sets and explore what these patterns say about the corresponding data set.

1. $\{\{x_1, x_4\}, \{x_2, x_4\}, \{x_3, x_4\}\}$. *Hint: What does it mean that x_4 is in every minimal set? That every minimal set has the same variable together with any other variable (except x_5)? That x_5 is not present?*
2. $\{\{x_1, x_2\}, \{x_2, x_3\}, \{x_1, x_3\}, \{x_5\}\}$ *Hint: What does it mean that x_5 appears in a minimal set by itself?*   ▽

3.5.2 Postprocessing: The Gröbner Fan Method

Recall from Section 3.5 that Gröbner bases depend on the choice of monomial ordering and so do resulting models. In order to avoid ambiguity in model selection, we would like to be able to (1) identify all monomial orderings that cause the Buchberger algorithm to produce different reduced Gröbner bases; (2) study the effect of these monomial orderings on the model dynamics; and (3) select the best model according to some criteria.

A combinatorial structure that contains information about all reduced Gröbner bases of an ideal, and thus fulfills item (1) from the above list, is the *Gröbner fan* of an ideal. It is a polyhedral complex of cones, each corresponding to an initial ideal, which is in a one-to-one correspondence with the *marked* reduced Gröbner bases of the ideal, where "marked" refers to the initial term of each generating polynomial being distinguished. Here we give a brief introduction to the concept of the Gröbner fan. For details see, for example, [29].

A fixed polynomial ideal has only a finite number of different reduced Gröbner bases. Informally, the reason is that most of the monomial orderings only differ in high degree and the Buchberger algorithm for Gröbner basis computation does not "see" the difference among them. However, they may vary greatly in the number of polynomials and shape. In order to classify them, we first present a convenient way to define monomial orderings using vectors. Again, we think of a polynomial in $k[x_1, \ldots, x_n]$ as a linear combination of monomials of the form $x^\alpha = x_1^{\alpha_1} \cdots x_n^{\alpha_n}$ over a field k, where α is the n-tuple exponent $\alpha = (\alpha_1, \ldots, \alpha_n) \in \mathbb{Z}_{\geq 0}^n$.

Definition 3.9.   Suppose that $\omega = (\omega_1, \ldots, \omega_n)$ is a vector with real coefficients. It can be used to define an ordering on the elements of $\mathbb{Z}_{\geq 0}^n$ by $\alpha \succ_\omega \beta$ if and only if $\alpha \cdot \omega > \beta \cdot \omega$.

We are now almost ready to give a definition of the Gröbner fan. We first define its building blocks, the Gröbner *cones*.

Definition 3.10.   Let $\mathcal{G} = \{g_1, \ldots, g_r\}$ be a marked reduced Gröbner basis for an ideal I. Write each polynomial of the basis as $g_i = x^{\alpha_i} + \sum_\beta c_{i,\beta} x^\beta$, where x^{α_i} is the initial term in g_i. The *cone* of $\mathcal{G}$ is $C_\mathcal{G} = \{\omega \in \mathbb{R}_{\geq 0}^n | \alpha_i \cdot \omega \geq \beta \cdot \omega$ for all i, j with $c_{i,\beta} \neq 0\}$.

The collection of all cones for a given ideal is the Gröbner fan of that ideal. The cones are in bijection with the marked reduced Gröbner bases of the ideal. Since reducing a polynomial modulo an ideal I, as the reverse engineering algorithm requires, can have at most as many outputs as the number of marked reduced Gröbner bases, it follows that the Gröbner fan contains information about all Gröbner bases (and thus all monomial orderings) that need to be considered in the process of model selection. For an example of utilizing the information encoded in the Gröbner fan of an ideal for reverse engineering of PDSs, see [30].

There are algorithms based on the Gröbner fan that enumerate all marked reduced Gröbner bases of a polynomial ideal. An excellent implementation of such an algorithm is the software package Gfan [31].

Example 3.10.   Consider the ideal

$$I = \left\langle z^2 - z, y^2 - y, xz + 1 - z - y + yz - x, xy - yz, x^2 - x \right\rangle \subseteq \mathbb{Z}_3[x, y, z].$$

This ideal has three distinct marked reduced Gröbner bases, $\mathcal{G}_1, \mathcal{G}_2, \mathcal{G}_3$ (below), that correspond to the given weight vectors (monomial orderings), ω_i. In other words, the Gröbner fan of I consists of three cones and each of the given weight vectors is an element of a different cone.

$$\mathcal{G}_1 = \{z^2 - z, y^2 - y, xz + yz - x - y - z + 1, xy - yz, x^2 - x\},$$
$$\omega_1 = \{2, 1, 1\},$$
$$\mathcal{G}_2 = \{z^2 - z, x^2 - x, yz + xz - y - x - z + 1,$$
$$xy + xz - y - x - z + 1, y^2 - y\}, \omega_2 = \{1, 2, 1\},$$
$$\mathcal{G}_3 = \{y^2 - y, x^2 - x, yz - xy, xz + xy - z - x - y + 1, z^2 - z\},$$
$$\omega_3 = \{1, 1, 2\}.$$

One can compute the first Gröbner basis, for instance, using the following Macaulay 2 code. Note that the ideal generators do not have to be a Gröbner basis.

```
R=ZZ/3[x,y,z, MonomialOrder=>{Weights=>{2,1,1}}]
I=ideal(z^2-z, y^2-y, x*z+y*z-x-y-z+1, x*y-y*z, x^2-x)
flatten entries gens gb I
```

The following are the normal forms of the polynomial $f = -xy + y + z \in \mathbb{Z}_3[x, y, z]$, calculated with respect to the Gröbner bases and monomial orderings above:

$$f_1 = -yz + y + z, \quad f_2 = xz - x + 1, \quad f_3 = -xy + z + y.$$

The software packages Gfan and others implement a special type of Gröbner basis conversion, known as the Gröbner walk [32]. Unless, however, the dimension of the fan is low or the number of its cones happens to be small, computing the entire fan is computationally expensive and some of its components do not have polynomial bounds. The following is an example that illustrates how infeasible computing the entire Gröbner fan could be.

Example 3.11.   For illustrative purposes, we will work over $\mathbb{Q}$ in this example. The Gröbner fan of the ideal

$$I = \left\langle x_1^5 - 1 + x_3^2 + x_3^3, x_2^3, x_2^2 - 1 + x_3 + x_1^2, x_3^3 - 1 + x_2^5 + x_1^6 \right\rangle \subseteq \mathbb{Q}[x_1, x_2, x_3]$$

has 360 full-dimensional cones. The intersection of the fan with the standard simplex in $\mathbb{R}^3$ is shown in Figure 3.2 [31].

Fortunately, in order to compute polynomial normal forms, the only informa-tion that we need to extract from the Gröbner cones of a fan is their corresponding reduced Gröbner bases and/or their *relative* volumes, where "relative" refers to the

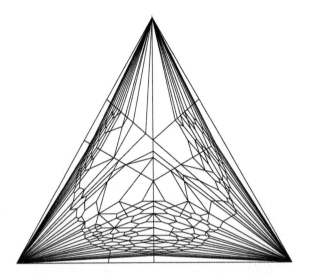

FIGURE 3.2

The Gröbner fan of the ideal in Example 3.10 intersected with the standard 2-simplex. The x_1-axis is on the right, the x_2-axis on the left, and the x_3-axis at the top.

cone volume when bounded by an n-sphere centered at the cone's vertex. While computing the exact relative cone volumes requires knowing the facets of the fan, that is, the fan itself, approximation of the relative volumes in many cases is sufficient [33]. A stochastic method for estimating the relative volumes of the Gröbner cones of a Gröbner fan without computing the actual fan, as well as a Macaulay 2 implementation for uniform sampling from the Gröbner fan, is presented in [34].

Exercise 3.15. In Example 3.10, the weight vector $\omega_1 = \{2, 1, 1\}$ generated Gröbner basis $\mathcal{G}_1 = \{z^2 - z, y^2 - y, xz + yz - x - y - z + 1, xy - yz, x^2 - x\}$. Can you find a different weight vector that produces the same Gröbner basis?   ▽

Exercise 3.16. Which one of the three Gröbner bases in Example 3.10 corresponds to the lexicographic monomial ordering?   ▽

Exercise 3.17. Let us construct a stochastic PDS. Let $F = (f_1, f_2)$ be a Boolean PDS that fits a data set. Let I be the ideal of functions that vanish on the data points. Suppose 20% of all monomial orderings generated the following normal form of F with respect to I:

$$f_1 = x_1 + x_2,$$
$$f_2 = x_1,$$

and 80% of the monomial orderings generated

$$f_1 = x_1,$$
$$f_2 = x_1 x_2.$$

Draw the state space and wiring diagram of this stochastic PDS labeling the edges with the corresponding probabilities. $\triangledown$

Project 3.6. Consider the four time series in Section 4.3 of [30]. Find the ideal of polynomials that vanish on the series and using the software package Gfan [31], compute its Gröbner fan. $\triangledown$

3.6 DISCRETIZATION

For reasons explained at the beginning of Section 3.2, we have been assuming that the experimental data we use for reverse engineering have already been discretized into a (small) finite number of states. Typically, however, experimental measurements come to us represented by computer floating point numbers and consequently data discretization is in fact part of the modeling process and can be viewed as a preprocessing step. We will use the definition of discretization presented in [35].

Definition 3.11. A *discretization* of a real-valued vector $\mathbf{v} = (v_1, \ldots, v_N)$ is an integer-valued vector $\mathbf{d} = (d_1, \ldots, d_N)$ with the following properties:

1. Each element of $\mathbf{d}$ is in the set $0, 1, \ldots, D - 1$ for some (usually small) positive integer D, called the degree of the discretization.
2. For all $1 \leq i, j \leq N$, we have $d_i \leq d_j$ if and only if $v_i \leq v_j$.

Without loss of generality, assume that $\mathbf{v}$ is sorted, i.e., for all $i < j, v_i \leq v_j$. *Spanning* discretizations of degree D satisfy the additional property that the smallest element of $\mathbf{d}$ equal to 0 and that the largest element of $\mathbf{d}$ is equal to $D - 1$.

There is no universal way for data discretization that works for all data sets and all purposes. Sometimes discretization is a straightforward process. For example, if a gene expression time series has a sigmoidal shape, e.g., $(0.1, 1.2, 2, 23.04, 26)$, it is reasonable to discretize it as $(0, 0, 0, 1, 1)$. More complicated expression profiles may be easy to discretize too and it is often true that the human eye is the best discretization "tool" whose abilities to discern patterns cannot be reproduced by any software. Large data sets, on the other hand, do require some level of automatization in the discretization process. Regardless of the particular situation, it is good practice to look at the data first and explore for any patterns that may help with the discretization before inputting the data into any discretization algorithm. Afterwards, the way you choose to discretize your data, which includes selecting the number of discrete states, should depend on the type and amount of data and the specific reason for discretization. Below we present several possible approaches which by no means comprise a complete list.

Binary discretizations are the simplest way of discretizing data, used, for instance, for the construction of Boolean network models for gene regulatory networks [36,37]. The expression data are discretized into only two qualitative states as either present or absent. An obvious drawback of binary discretization is that labeling real-valued data according to a present/absent scheme may cause the loss of large amounts of information.

Interval discretizations divide the interval $[v_1, v_N]$ into k equally sized bins, where k is user-defined. Another simple method is *quantile discretization* which places N/k (possibly duplicated) values in each bin [38]. Any method based on those two approaches would suffer from problems that make it inappropriate for some data sets. Interval discretizations are very sensitive to outliers and may produce a strongly skewed range [39]. In addition, some discretization levels may not be represented at all which may cause difficulties with their interpretation as part of the state space of a discrete model.

On the other hand, quantile discretizations depend only on the ordering of the observed values of **v** and not on the relative spacing values. Since distance between the data points is often the only information that comes with short time series, losing it is very undesirable. A shortcoming, common for both interval and quantile, as well as for most other discretization methods, is that they require the number of discrete states, k, to be user-provided.

A number of entropy-based discretization methods deserve attention. An example of those is Hartemink's *Information-Preserving Discretization* (IPD) [35]. It relies on minimizing the loss of pairwise mutual information between each two real-valued vectors (variables). The mutual information between two random variables X and Y with joint distribution $p(X, Y)$ and marginal distributions $p(x)$ and $p(y)$ is defined as

$$I(X; Y) = \sum_x \sum_y p(x, y) log \frac{p(x, y)}{p(x)p(y)}.$$

Note that if X and Y are independent, by definition of independence $p(x, y) = p(x)p(y)$, so $I(X; Y) = 0$. Unfortunately, when modeling regulatory networks and having as variables, for instance, mRNA, protein, and metabolite concentrations, the joint distribution function is rarely known and it is often hard to determine whether two variables are independent.

Another family of discretization techniques is based on clustering [40]. One of the most often used clustering algorithms is the k-means [41]. It is a non-hierarchical clustering procedure whose goal is to minimize dissimilarity in the elements within each cluster while maximizing this value between elements in different clusters. Many applications of the k-means clustering such as the MultiExperiment Viewer [42] start by taking a random partition of the elements into k clusters and computing their centroids. As a consequence, a different clustering may be obtained every time the algorithm is run. Another inconvenience is that the number k of clusters to be formed has to be specified in advance. Although there are methods for choosing "the best k" such as the one described in [43], they rely on some knowledge of the data properties that may not be available.

Another method is *single-link clustering* (SLC) with the Euclidean distance function. SLC is a divisive (top-down) hierarchical clustering that defines the distance between two clusters as the minimal distance of any two objects belonging to different clusters [40]. In the context of discretization, these objects will be the real-valued entries of the vector to be discretized, and the distance function that measures the distance between two vector entries v and w will be the one-dimensional Euclidean distance $|v - w|$. Top-down clustering algorithms start from the entire data set and iteratively split it until either the degree of similarity reaches a certain threshold or every group consists of one object only. For the purpose of data analysis, it is impractical to let the clustering algorithm produce clusters containing only one real value. The iteration at which the algorithm is terminated is crucial since it determines the degree of the discretization. SLC with the Euclidean distance function has one major advantage: very little starting information is needed (only distances between points). In addition, being a hierarchical clustering procedure it lends itself to adjustment in case that clusters need to be split or merged. It may result, however, in a discretization where most of the points are clustered into a single partition if they happen to be relatively close to one another. Another problem with SLC is that its direct implementation takes D, the desired number of discrete states, as an input. However, we would like to choose D as small as possible, without losing information about the system dynamics and the correlation between the variables, so that an essentially arbitrary choice is unsatisfactory.

In [44], a hybrid method for discretization of short time series of experimental data into a finite number of states was introduced. It is a modification of the SLC algorithm: it begins by discretizing a vector in the same way as SLC does but instead of providing D as part of the input, the algorithm contains termination criteria which determine the appropriate value of D. After that each discrete state is checked for information content and if it is determined that this content can be considerably increased by further discretization, then the state is separated into two states in a way that may not be consistent with SLC.

If more than one vector is to be discretized, the algorithm discretizes each vector independently. (For details on multiple vector discretization, see [44]). If the vector contains m distinct entries, a complete weighted graph on m vertices is constructed, where a vertex represents an entry and an edge weight is the Euclidean distance between its endpoints. The discretization process starts by deleting the edge(s) of highest weight until the graph gets disconnected. If there is more than one edge labeled with the current highest weight, then all of the edges with this weight are deleted. The order in which the edges are removed leads to components, in which the distance between any two vertices is smaller than the distance between any two components, a requirement of SLC. We define the distance between two components G and H to be $\text{dist}(G, H) = \min\{|g - h| \mid g \in G, h \in H\}$. The output of the algorithm is a discretization of the vector, in which each cluster corresponds to a discrete state and the vector entries that belong to one component are discretized into the same state.

Example 3.12. Suppose that vector $\mathbf{v} = (1, 2, 7, 9, 10, 11)$ is to be discretized. The diagram obtained by the SLC algorithm is given in Figure 3.3. The complete weighted graph in Figure 3.4 corresponds to iteration 0 of the diagram.

Having disconnected the graph, the next task is to determine if the obtained degree of discretization is sufficient; if not, the components need to be further disconnected in

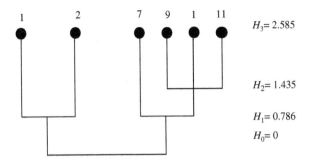

FIGURE 3.3

Diagram representing the SLC algorithm applied to the data of Example 3.12. The column on the right gives the corresponding Shannon's entropy increasing at each consecutive level.

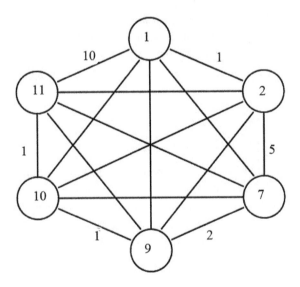

FIGURE 3.4

The complete weighted graph constructed from vector entries 1, 2, 7, 9, 10, 11. Only the edge weights of the outer edges are given.

a similar manner to obtain a finer discretization. A component is further disconnected if, and only if, both (1) and (2) below hold:

1. The minimum vertex degree of the component is less than the number of its vertices minus 1. The contrary implies that the component is a complete graph by itself, i.e., the distance between its minimum and maximum vertices is smaller than the distance between the component and any other component.

2. One of the following three conditions is satisfied ("disconnect further" criteria):

   a. The average edge weight of the component is greater than half the average edge weight of the complete graph.
   b. The distance between its smallest and largest vertices is greater than or equal to half this distance in the complete graph. For the complete graph, the distance is the graph's highest weight.
   c. Finally, if the above two conditions fail, a third one is applied: disconnect the component if it leads to a substantial increase in the information content carried by the discretized vector.

   The result of applying only the first two criteria is analogous to SLC clustering with the important property that the algorithm chooses the appropriate level to terminate. Applying the third condition, the information measure criterion may, however, result in a clustering which is inconsistent with any iteration of the SLC.

Exercise 3.18. Consider the following vector $v_1 = (1.3, 2.1, 7.2, 9.05, 10.5, 11.00)$. Plot the vector entries on the real number line and propose an appropriate discretization based on your intuition. How did you decide on the number of discrete states? Would your decision change if the entry 4.6 is added?   ▽

Exercise 3.19. Discretize vector v_1 from Problem 3.18 using the specified method.

1. Interval.
2. Quantile.
3. The hybrid method presented in this section.   ▽

Exercise 3.20. If you know that the experimental data is very noisy, would you be willing to use a smaller or a larger number of discrete states?   ▽

Exercise 3.21. Suppose a second vector, $v_2 = (0.8, 1.8, 3.1, 8.0, 9.5, 10.7)$, is to be discretized and it is known that v_1 and v_2 are strongly correlated (assume Spearman rank correlation). How would you discretize v_2 based on your discretization of v_1?   ▽

Project 3.7. Use the principle of relationship discretization that is behind Hartemink's information-preserving discretization method to discretize v_1 and v_2.   ▽

References

[1] Jacob F, Monod J. Genetic regulatory mechanisms in the synthesis of proteins. J Mol Biol 1961;3(3):318–56.
[2] Setty Y, Mayo A, Surette M, Alon U. Detailed map of a cis-regulatory input function. PANS 2003;100(13):7702–7.
[3] Mayo A, Setty Y, Shavit S, Zaslaver A, Alon U. Plasticity of the cis-regulatory input function of a gene. PLoS Biology 2006;4(4):0555–61.

[4] Stigler B, Veliz-Cuba A. Network topology as a driver of bistability in the lac operon. J Comput Biol 2008;18(6):783–94.

[5] de Jong H. Modeling and simulation of genetic regulatory systems: a literature review. J Comput Biol 2002;9(1):67–103.

[6] Bornholdt S. Less is more in modeling large genetic networks. Science 2005;310:449–50.

[7] Brazhnik P, de la Fuente A, Mendes P. Gene networks: how to put the function in genomics. Trends Biotechnol 2002;20(11):467–72.

[8] Gardner T, Faith J. Reverse-engineering transcription control networks. Phys Life Rev 2005;2:65–88.

[9] Bansal M, Belcastro V, Ambesi-Impiombato A, di Bernardo D. How to infer gene networks from expression profiles. Mol Sys Biol 2007;3(78):1–10.

[10] Burda Z, Krzywicki A, Martin OC, Zagorski M. Distribution of essential interactions in model gene regulatory networks under mutation-selection balance. Phys Rev E 2010;82(1).

[11] Yeh H-Y, Cheng S-W, Lin Y-C, Yeh C-Y, Lin S-F, Soo V-W. Identifying significant genetic regulatory networks in the prostate cancer from microarray data based on transcription factor analysis and conditional independency. BMC Med Genom 2009;2(70).

[12] Lidl R, Niederreiter H. Finite fields. Encyclopedia of mathematics and its applications, vol. 20. Cambridge University Press; 1997. p. 369.

[13] Strogatz S. Non-linear dynamics and chaos: with applications to physics, biology, chemistry and engineering. Perseus Books; 2000.

[14] Termanini A, Tieri P, Franceschi C. Encoding the states of interacting proteins to facilitate biological pathways reconstruction. Biol Direct 2010;5(52).

[15] MacLean D, Studholme DJ. A boolean model of the pseudomonas syringae hrp regulon predicts a tightly regulated system. PLoS ONE 2010;5(2).

[16] Colon-Reyes O, Laubenbacher R, Pareigis B. Boolean monomial dynamical systems. Ann Comb 2004;8:425–39.

[17] Veliz-Cuba A, Jarrah A, Laubenbacher R. Polynomial algebra of discrete models in systems biology. Bioinformatics 2010;26(13):1637–43.

[18] Jarrah A, Laubenbacher R, Stigler B, Stillman M. Reverse-engineering of polynomial dynamical systems. Adv Appl Math 2007;39(4):477–89.

[19] Cox D, Little J, O'Shea D. Ideals, varieties, and algorithms. New York: Springer-Verlag; 1997.

[20] Adams WW, Loustaunau P. An introduction to Gröbner bases. Graduate studies in Math, vol. III. American Mathematical Society; 1994.

[21] Eisenbud D. Introduction to commutative algebra with a view towards algebraic geometry. Graduate texts in mathematics. New York: Springer; 1995.

[22] Buchberger B. An algorithm for finding a basis for the residue class ring of a zero-dimensional polynomial ideal. PhD thesis, University of Innsbruck; 1965.

[23] Hironaka H. Resolution of singularities of an algebraic variety over a field of characteristic zero: I. Ann Math 1964;79(1):109–203.

[24] Buchberger B, Möller HM. The construction of multivariate polynomials with preassigned zeros. In: Calmet J, editor. Proceedings of the European computer algebra conference EURO CAM'82. LNCS, vol. 144. Springer; 1982. p. 24–31.

[25] Laubenbacher R, Stigler B. A computational algebra approach to the reverse engineering of gene regulatory networks. J Theor Biol 2004;229:523–37.

[26] Grayson D, Stillman M. Macaulay 2, a software system for research in algebraic geometry; 2006. <http://www.math.uiuc.edu/Macaulay2>.

[27] Gustafsson M, Hornquist M, Lombardi A. Constructing and analyzing a large-scale gene-to-gene regulatory network Lasso-constrained inference and biological validation. IEEE/ACM Trans Comput Biol Bioinform 2005;2(3):254–61.

[28] Thorsson V, Hornquist M, Siegel A, Hood L. Reverse engineering galactose regulation in yeast through model selection. Stat Appl Genet Mol Biol 2005;4(1): 1–22.

[29] Mora T, Robbiano L. Gröbner fan of an ideal. J Symb Comput 1988;6(2/3): 183–208.

[30] Dimitrova ES, Jarrah AS, Laubenbacher R, Stigler B. A Gröbner fan method for biochemical network modeling. In: Proceedings of the international symposium on symbolic and algebraic computation (ISSAC). ACM; 2007. p. 122–6.

[31] Fukuda K, Anders J, Thomas R. Computing Gröbner fans. Math Comput 2007;76(260):2189–212.

[32] Collart S, Kalkbrener M, Mall D. Converting bases with the Gröbner walk. J Symb Comput 1997;24:465–9.

[33] Dimitrova ES, Garcia-Puente LD, Hinkelmann F, Jarrah AS, Laubenbacher R, Stigler Brandilyn, et al. Parameter estimation for Boolean models of biological networks. J Theor Comput Sci 2011;412:2816–26.

[34] Dimitrova ES. Estimating the relative volumes of the cones in a Gröbner fan. Math Comput Sci 2010;3(4):457–66.

[35] Hartemink A. Principled computational methods for the validation and discovery of genetic regulatory networks. PhD dissertation, Massachusetts Institute of Technology; 2001.

[36] Kauffman SA. Metabolic stability and epigenesis in randomly constructed genetic nets. J Theor Biol 1969;22:437–67.

[37] Albert R, Othmer H. The topology of the regulatory interactions predict the expression pattern of the segment polarity genes in Drosophila melanogaster. J Theor Biol 2003;223:1–18.

[38] Dougherty J, Kohavi R, Sahami M. Supervised and unsupervised discretization of continuous features. In: Prieditis A, Russell S, editors. Machine learning: proceedings of the 12th International Conference; 1995.

[39] Catlett J. Megainduction: machine learning on very large databases. PhD dissertation, University of Sydney; 1991.

[40] Jain A, Dubes R. Algorithms for clustering data. Prentice Hall; 1988.

[41] MacQueen J. Some methods for classification and analysis of multivariate observations. In: Proceedings of the 5th Berkeley symposium of mathematical statistics and probability, vol. 1; 1967. p. 281–97.

[42] Saeed A, Sharov V, White J, Li J, Liang W, Bhagabati N, et al. TM4: a free, open-source system for microarray data management and analysis. BioTechniques 2003;34(2):374–8.

[43] Crescenzi M, Giuliani A. The main biological determinants of tumor line taxonomy elucidated by of principal component analysis of microarray data. FEBS Lett 2001;507:114–8.

[44] Dimitrova ES, McGee J, Laubenbacher R, Vera Licona P. Discretization of time series data. J Comput Biol 2010;17(6):853–68.

Global Dynamics Emerging from Local Interactions: Agent-Based Modeling for the Life Sciences

David Gammack*, Elsa Schaefer* and Holly Gaff[†]

*Department of Mathematics, Marymount University, Arlington, VA, USA
[†]Department of Biological Sciences, Old Dominion University, Norfolk, VA, USA

4.1 INTRODUCTION

4.1.1 Agent-Based Modeling and the Biology Mind Set

Agent-based (or individual-based) models (ABMs) are stochastic simulations built on interactions between individuals in a population (agents) or a set of populations and their environment. The modeler defines an ABM by identifying a set of autonomous agents, each defined by individual characteristics, who interact locally using adaptive behavior that is based on their characteristics, their neighbors' characteristics, and the local environment. The culmination of the model design ultimately leads to emergence: local interactions leading to global phenomena. Bonabeau writes that agent-based modeling is "a mind set more than a technology" [1]. Indeed, ABMs are particularly helpful because of the intuitive nature of their structure and approach. Biological research seeks to identify truths about biological systems at the most fundamental level. This often involves a great amount of detail that is not easily incorporated into population-level models. Biologists appreciate the ability of the ABM to include intricate details at multiple scales. The rules applied to agents within a model are intuitive analogies to biological interactions. Additionally, ABMs can be built with language that is easily understood, in contrast to modeling efforts that abstract populations and interactions to mathematical constructs.

As we will explore in more detail in Section 4.3, ABM frameworks also encourage repeated simulations mimicking the replication of a typical experimental setup. This parallel helps scientists understand how a model is simulating the experiment and how one can interpret the results. ABMs allow for agents to adapt and grow, which also mimics most biological systems. While error propagation is a concern with many calculations, careful model development and sensitivity analysis can help reduce the likelihood of spurious results from accumulating errors [2]. Overall, ABMs provide a natural translation of biological experiments into computer-simulated models that allow the exploration of research questions and hypotheses beyond what can be done in a lab or field setting because of cost, time, biological or ethical constraints.

Mathematical Concepts and Methods in Modern Biology. http://dx.doi.org/10.1016/B978-0-12-415780-4.00004-1

Susceptible
Humans

Infected
Humans

Recovered
Humans

FIGURE 4.1

We use blue, pink, and yellow balls to represent susceptible, infected, and recovered humans in our hands-on agent-based modeling activity.

4.1.2 A Brief Note About Platforms

Agent-based modeling in its most simple form can be accomplished using human agents performing hands-on experiments. A fun activity is to use colored marbles or beads to simulate a disease spreading through a classroom population (the full details are available in the article by Gaff et al. [3]). These experiments provide a light introduction to the concept of global phenomena that can result from individual behaviors.

4.1.2.1 ABM Modeling Exercise: Disease Spread

Let's simulate a disease moving through a population. Gather a bunch of beads, marbles, or balls in three different colors. In our explanation, we'll follow Gaff's model [3] and assume that blue balls represent people who have not caught this disease, the susceptible people. We will assume that pink balls represent people who are infected with the disease, and we'll assume that yellow balls represent people who have recovered from this disease, and are now immune from re-infection (see Figure 4.1).

Each person or pair in the class will conduct a simulation on a population of 20 balls that are hidden in an open container, such as a paper bag. How would the number of infected balls that we start with change the dynamics of how the disease moves through the population? We suggest dividing up as follows:

- Last names that start with A-I, place 19 blue balls and 1 pink ball in the bag.
- Last names that start with J-P, place 18 blue balls and 2 pink balls in the bag.
- Last names that start with Q-Z, place 16 blue balls and 4 pink balls in the bag.

We are going to simulate a disease spreading through a population by moving through a series of chance encounters. We will play a "game" in which each "turn" consists of removing and replacing two balls from our bag, and in this way we'll mimic a series of chance daily encounters between humans who may be infected with the disease.

For the first turn, without looking, randomly pull two balls out of the bag. If the two balls are the same color, then put them back in the bag. If one ball is blue and the other is pink, replace the blue ball with a pink ball, and put the two pink balls back into the bag.

Exercise 4.1.   What assumption are we making when we replace the blue ball with a pink ball? Do you think we should always replace the blue ball with a pink ball when we draw a pink and blue ball together? How else might we create rules for this turn?   ▽

Let's modify our rule for drawing a pink ball and a blue ball together so that some of the disease simulations will assume a lower probability of catching the disease through contact with an infected person.

- If your birthday is in January – June, then always replace the blue ball with a pink ball.
- If your birthday is in July – December, then flip a coin and only replace the blue ball with a pink ball if you flip a heads.

As we take repeated "turns" of drawing and replacing pairs of balls, we wish to simulate disease recovery by periodically replacing a single pink "infected" ball with a single yellow "recovered" ball.

Exercise 4.2.   What real-life assumption is mimicked by replacing a pink ball with a yellow ball every 10 turns?   ▽

We suggest dividing assumptions for length-to-recovery through the class as follows:

- First names that start with A-I: every 5th turn, before drawing your pair of balls, first replace one of the pink balls with a yellow ball.
- First names that start with J-P: every 10th turn, before drawing your pair of balls, first replace one of the pink balls with a yellow ball.
- First names that start with Q-Z: every 15th turn, before drawing your pair of balls, first replace one of the pink balls with a yellow ball.

At this point, we're ready to repeat the simulation steps through many "days," until there are no pink balls left in the cup. Note that if a yellow ball is chosen with a pink or blue ball, no replacements are to be made. At the end of each "turn," record the number of blue, pink, and yellow balls that are in the bag.

Exercise 4.3.   What real-life assumption is mimicked by not replacing the yellow ball with a pink ball when we draw a pink and yellow ball together?   ▽

Exercise 4.4.   Complete your simulation. How many "turns" did it take for the simulation to end?   ▽

Exercise 4.5.   What are the agents in this model? What characteristic(s) does each agent have? What are the rule(s) governing the system?   ▽

Exercise 4.6.   Class project: Compare the results of the simulations as a class. Which of the three choices that guided the individual simulations seemed to most impact the duration and severity of the disease spread, and what are the real-life implications of your observations?   ▽

While the activity above makes for a fun day of class, if we introduce repetition of the experiments to correct for chance outcomes or if we introduce new complications to our model and repeat, the activity becomes tedious, and we begin to hope for a piece of software that can easily play the game for us. ABMs can be coded using any object-oriented language, but a number of platforms including SWARM, RePast, or its simplified component ReLogo, and NetLogo, are available that make it easier to jump in and to visualize results without programming skills. These platforms simplify the processes of creating and observing movement through space, formation of patterns among individuals, effects of diffusion and mixing, and evolutionary processes. NetLogo, in addition to being freely distributed online [4], is well documented and easy to learn and use, and is a fine tool for even serious ABM initiatives with the exception of models that are particularly computationally intensive or for which the NetLogo conceptual style of short-term local interactions of agents on a discrete grid is not appropriate [5,6].

The NetLogo library has an extensive repository of biological and other user-contributed models that are useful both as an end-product for classroom simulation/experimentation and as starting-points for the creation of new models. Several examples are included in Figure 4.2, and we will explore several other models following.

The NetLogo environment is not difficult to use. The agents by default are called "turtles" and syntactically one typically begins a command "ask turtles" to perform

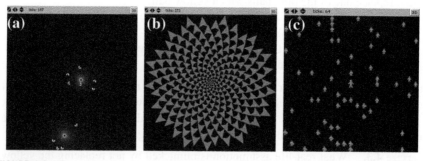

FIGURE 4.2

The NetLogo program includes multiple ready-to-use models. Pictured here are (a) a model which shows how a set of simple behavioral rules cause moths to circle a light [72], (b) a model exploring the interlocking spirals found in seeds and petals that occur naturally as the flower grows [73], (c) a model that mimics a disease moving through a population; here the user can control the movements of one individual [74].

various actions. Students can learn the language and environment through exploration of the NetLogo library models, but there is also an excellent text by Grimm and Railsback that thoroughly walks a new ABM explorer through model development and the NetLogo platform [7].

Exercise 4.7.   Google "NetLogo" to find its Northwestern University homepage, and download the free software onto your computer.   ▽

4.1.3 A Brief History of Agent-Based Modeling

Although ABMs became widely popular for computation in the 1990s, their first instantiations were modest. In the 1940s, Von Neumann created a simple grid on paper to emulate a machine whose sole purpose was to copy itself [8]. Each cell in this grid has a state, "on" or "off" (see Figure 4.3). The cell's neighborhood is defined with respect to each individual cell, e.g., any cell less than two cells away. With each time step, a new generation of each cell's state is defined using a function of the current state of the cell and the states of the cell's neighbors. Thus, the cell's current state influences and is influenced by its neighbors' states. Global patterns can be found based on the function used to create each new generation of cells. This process is now known as cellular automata.

The next innovation in ABMs is from Conway and his Game of Life [9]. Despite its grand moniker, the Game of Life is a simple zero player game, meaning that its only human input is in its initialization. A grid of Von Neumann's cells is encoded with some very basic rules concerning the state of the cell, i.e., "live" or "dead", and the generation function, which, as above, changes the state based on the current state and the state of its neighbors. Despite its simplicity, the Game of Life perfectly demonstrates emergent behavior: patterns and self-organization evolve from simulations of various initial conditions.

Another early pioneer in the development of agent-based modeling is Thomas Schelling, who in 1978 published what is now known as Schelling's Tipping Model. The model highlighted the process of observing large-scale societal ramifications for

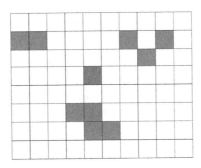

FIGURE 4.3

An illustration of a basic cellular automata grid.

the behavior choices made by individuals as he explored the causes of segregation in a society which he perceived to not express outright racism [10,11]. Schelling's model consists of a checkerboard, with each grid space representing a possible home. The agents can be thought of as red and green turtles that fit onto a checkerboard grid, and who are given an opportunity to move to a new grid space at each time step. The turtle will choose to move if some proportion of its neighbors are not the same color as itself. Schelling found that very low levels of personal preference can lead to the global phenomenon of segregation. Schelling won the 2005 Nobel Prize in Economic Sciences "for having enhanced our understanding of conflict and cooperation through game-theoretic analysis," an award which Schelling believes was at least in part due to this work [11].

4.1.3.1 *ABM Modeling Exercises: Segregation*

We can experiment with Schelling's ideas using the NetLogo Models Library which is located under the "File" directory in NetLogo. There, in the Social Sciences section, we find the model "Segregation," which is inspired by Schelling's work on housing patterns in cities [12]. The model consists of green and red turtles who prefer to live near their own kind. The model allows us to choose the number of turtles that we begin with, as well as the percentage of like-turtles that need to be present to keep a turtle from moving. The model reports the average percentage of same-color neighbors for each turtle, as well as the percent of turtles who are forced into an "unhappy" situation of too many turtle neighbors with differing colors. Open the Segregation model, and familiarize yourself with the three tabs "Interface," "Information," and "Procedures." The "Interface" tab is where we run our simulations for this multicolor turtle dilemma. The "Information" tab explains the model's assumptions and outputs. Finally, the "Procedures" tab shows the code that generates the model. Let's begin by exploring the model using the Interface tab.

Exercise 4.8.   Using the `number` slider, choose to start with 1000 turtles, and assume that turtles will be happy if at least 20% of their neighbors are their same color by setting the `%-similar-wanted` slider. Click the setup button, followed by the go button.

- **a.** If this process is repeated three times, will the output change for the percent of same-colored turtle neighbors?
- **b.** What percentage of turtles seem to be happy at the end of each simulation? (Note that sliders should be set as stated in the exercise.)
- **c.** Did the end-proportion of numbers of like-colored turtle neighbors mirror the minimum requirement for each turtle to be happy?   ▽

Exercise 4.9.   This exercise explores the effect of turtle density to the resulting segregation. How do the output measures change as the number of turtles increases and decreases on this fixed grid? Do not change the assumption of 20% as the `%-similar-wanted`.   ▽

Exercise 4.10.   This exercise considers the sensitivity of outputs to the choice of the preference-level of each turtle to have its neighbors be the same color. Return to the setting of 1000 turtles on the grid. By the way, it will be helpful here to notice that the simulation only stops if every turtle is happy. If you want to stop the simulation yourself, just click on the go button.

a. Is it possible to choose a `%-similar-wanted` level at which some turtles are not happy at the end? If so, about what value of `%-similar-wanted` seems to be the breaking-point between all-happy and some-unhappy turtles? Also, what is the corresponding end percent of like-colored turtle neighbors?
b. Without running any simulations, conjecture what you think would happen if the turtles were more crowded. How would the results above likely change if we fit 2000 turtles into the same space?
c. Repeat the simulations and analysis from part (a) for the problem with 2000 turtles. Did any of the results surprise you?   ▽

The remaining exercises provide an introduction to the syntax within the NetLogo code. Click on the "Procedures" tab. The commands following the phrase to setup in the code for the simulation tell NetLogo what to do when the user clicks the setup button. In this case, we see:

```
Ask n-of number patches
   [sprout 1
   [set color red]] Note that number has been defined
   with a specific numerical value elsewhere in the code.
```

Exercise 4.11.   Highlight `sprout` and click on the Help menu to look this up in the NetLogo Dictionary. What does the command `sprout 1` accomplish?   ▽

Exercise 4.12.   Look up the phrases `n-of` and `set color` similarly. What do the three lines of code above accomplish all together?   ▽

Exercise 4.13.   How does the code determine if a turtle is happy?   ▽

Exercise 4.14.   How far can a turtle move in a given step to try to find happiness?   ▽

Exercise 4.15.   Did the assumption of how far the turtles could move have a big influence on results? Would any of the outcomes of this simulation change if the turtle could move as far as 20 units each time? Make this change, and repeat a few of your experiments.   ▽

Exercise 4.16.   Change the code so that there are only a tenth as many green turtles as red turtles and repeat some of your experiments. Does this make a difference? ▽

Exercise 4.17.   What are the agents in this model? What characteristic(s) does each agent have? What are the rule(s) governing the system?   ▽

Exercise 4.18.   Project: Modify the model code so that there are three colors of turtles. Do any of the conclusions change in this situation?   ▽

Another significant contributor to the development of ABMs is Craig Reynolds, who, while he has not yet earned a Nobel prize, does have a 1998 Academy Award for "pioneering contributions to the development of three dimensional computer animation for motion picture production" [13]. In 1986, Reynolds created a computer animation that simulated flocking by having each of his animated birds make its choices for movement based only on the positions and velocities of its neighboring birds, providing a strong illustration of how global properties emerge from local choices. An interesting and unexpected property of his work is the unpredictability of a flock's motion past short time scales [14]. Reynolds maintains a nice movie and description of the flocking project on his personal webpage [14], where he also provides an extensive list of his and others work in ABM creation and other computational models of group motion.

4.1.3.2 *ABM Modeling Exercises: Flocking*

An example inspired by Reynold's work with flocking is also available as a sample model within NetLogo under the Biology models [15]. The model gives each bird the same set of instructions — there are no leaders with unique rules — and yet we observe an emergent flocking behavior as the model runs. There are three sets of rules: alignment, separation, and cohesion.

Exercise 4.19.   Open the model and click the setup button. Do you see any pattern in the alignment of the triangles?   ▽

Exercise 4.20.   In what way(s) do you observe a uniform flocking throughout the grid over time, and in what ways do you see individual variation affecting the global patterns?   ▽

Exercise 4.21.   What changes can you make to the sliders that increases the global organization of the birds?   ▽

Exercise 4.22.   Explore the "Procedures"[1] tab and give detailed, intuitive descriptions of what is meant by alignment, separation, and cohesion.   ▽

Exercise 4.23.   What are the agents in this model? What characteristic(s) does each agent have? What are the rule(s) governing the system?   ▽

Since the work of the pioneers, agent-based models have been used extensively to explore patterns in biology at various system granularities. The models allow us to consider how the nature of individual variation, including variability in space, demographic variability or life cycle details, phenotypic variation or behavior, experience and learning, and/or genetics and evolution combine to create population-level "emergent" patterns or characteristics. A great resource for many specific examples is SwarmWiki [16]. ABMs are used at the cellular level to understand intracellular signaling and metabolic pathways [17–20], cancer [21,22] or infectious diseases [23,24]. At the cellular level, ABMs may be used to model vascular biology [25,26]

---

[1]Instructor's note: While the code under "Procedure" is short, a careful answer to this exercise may take an hour or more to complete.

or acute inflammation [27]. At the organismal level, we may use ABMs to consider disease transmission vectors [28,29], and social and cultural effects and consequences for infectious diseases [30].

In the following three sections, we show how agent-based models may be appropriate for any level of research. In Section 4.2, we walk through the creation of a simple ABM that mimics some of the neuronal connections made in our brains. This is a model that was originally developed by an undergraduate student. We then present in Section 4.3 a more complex model that suggests how the disease cholera might spread through a small village. A more detailed version of this model was developed by a team of undergraduate students, and within Section 4.3 we dive into how we can use the Behavior Space function within NetLogo to replicate experiments. This will allow us to separate the chance events that occur in a single simulation from broader trends that accurately reflect the impact of modifications to the model or its parameters. In Section 4.4, we describe how ABMs are complex enough to support a full-fledged research effort, and we give a quick overview of the way in which such models should be presented.

4.2 AXON GUIDANCE

In this section we will develop a model of axon guidance. The motivation for this comes from an undergraduate honors thesis titled: "Agent-based Modeling of Commissural Axon Guidance along the Midline" by Worku [31].

4.2.1 Background

Healthy brain development in the embryo is, in part, determined by correct electrical and chemical pathways being established. These pathways are used by brain cells to communicate effectively with all the other parts of the brain and the body. The human central nervous system is bilateral; thus the transfer of information between the sides of the body must be conducted across the midline of the brain. Neurons (nerve cells) in the brain have axons that grow away from the cell body (Figure 4.4) and can cross the midline. These axons facilitate communication between cells. To accomplish this, the "head" of an axon (the growth cone) uses filopodia (spikes) that monitor the local environment and uses this information to guide the direction and development of the axon. To be affected by a given guidance cue a neuron must express a specific receptor: think of this as a lock (receptor) and key (guidance cue), the key is useless if it fits the wrong lock.

There are trillions of neurons in the brain, and the complexity of the connections between all the cells is incredible. Therefore, undergraduate research tends to study either a single guidance cue, or a neuron expressing receptors that are thought to respond to a specific guidance cue. The aim here is to investigate complex mechanisms involved in the developing brain, where multiple cues are released and guide a neuron expressing multiple receptors.

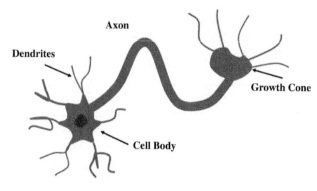

FIGURE 4.4

Neuron structure.

To develop an ABM of this (or any) specific system, the researcher must understand the system as a whole, break the system down into its salient features, and then develop an ABM which describes the whole system as the combination of its features. Therefore, one needs to understand the biological domain (axon biology), break the problem down, make assumptions, and then develop and implement a model within an ABM framework.

4.2.2 Understanding the Domain: Axon Biology

Neurons are found in the peripheral and central nervous system and are responsible for a functioning body and mind. They control everything, from the beating of the heart, to pain sensation, to the ability to communicate. The structure of a neuron is shown in Figure 4.4; following is a brief description of the axon components. The trillions of neurons communicate with one another via axons (long nerve fibers) that extend from the cell body to other parts of the brain [32]. The dendrites, finger-like protrusions around the cell body, receive signals from the local environment and pass that information to the cell body, and to other cells via the axon. Axons transmit signals via synapses protruding off the growth cone. The growth cone is a specialized structure that contains receptors enabling neurons to recognize and respond to environmental chemical cues. It continually monitors the environment via filopodia and this information causes it to grow in a particular direction [33].

Exercise 4.24.   Open the NetLogo model *Diffusion1.nlogo*. Answer the following questions (these are also listed in the Information tab for the model):

a. What are the agents in this model? What characteristic(s) does each agent have? What are the rule(s) governing the system?

b. How is diffusion being modeled?

c. What happens to the particles when they get to the edge of the environment?

d. The "Settings…" tab is located near the top of the Interface screen. Alter the "Settings…" so that particles bounce off the walls.

e. Click on the "Procedures" tab in NetLogo. You will see sections of the code with the following form:

```
;; COMMENT
;; what does this do?
NetLogo Code
;; END COMMENT
```

For each occurrence, delete the phrase `;; what does this do?` and replace it with a clear description of what that part of the NetLogo code does. (This is called *commenting* your code.)

f. Alter the code so that there is a constant source of 25 particles in the center of the environment (at the origin). Hint: use `create-particles`.   ▽

In the body, the growth cone responds via four main mechanisms: contact-mediated attraction or repulsion, chemotaxis (attraction up gradients of the cue) or chemorepulsion (repulsion down gradients in the cue) [32]. Contact-mediated responses depend on the amount of guidance cue in the environment and the gradient of the guidance cue. The guidance cues, governing the direction of axon growth, originate and then diffuse out from the midline [34].

There are various guidance molecules that have been identified and divided into different families, namely Ephrin, Semaphorin, Netrin, and Slit [35]. In an effort to create a manageable initial model, we will focus on two types of guidance cues (Netrin and Slit) and three types of neuronal receptors (DCC, UNC5, and Robo 1). Netrin both attracts and repulses axons, depending on the type of receptor on the axon. Axons expressing the DCC receptor are attracted to areas with higher levels of Netrin, which acts as a chemoattractant, whereas axons expressing the UNC-5 receptor are repelled from high concentrations of Netrin, which acts as a chemorepellent [32,34]. Axons expressing the Robo 1 receptor are repelled by Slit. For some axons, the Robo 1 receptor is upregulated once the axon has crossed the midline resulting in the axon being unable to return to the other side [36]. In short, one chemical (Netrin) acts as both a chemo-attractant and -repellant, the other (Slit) acts as a chemorepellant, with both of these being dependent on the receptor(s) expressed by the neuron. A schematic of the known responses of different growth-cone types is shown in Figure 4.5.

Certainly the description above represents a significant simplification. Axons usually have more than one kind of receptor, and there are multiple guidance cues. For example, one axon might express both DCC and Robo 1 receptors, and it will behave according to a combined response to the multiple cues.

Exercise 4.25.   Open the NetLogo model *Diffusion2.nlogo*. Answer the following questions after reading the code in the "Procedures" tab.

a. What is the difference between diffusion in this model and the model in *Diffusion1.nlogo*?

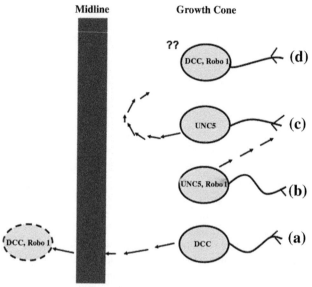

FIGURE 4.5

Diagram of the growth cones modeled. Netrin and Slit are produced at the midline and then diffuse into the rest of the environment. There are four growth cones that will be modeled, three of which have known responses: a - attracted over the midline (by Netrin) and then repelled away from the midline (by Slit); b - repelled from the midline (by Netrin and Slit); c - repelled from the midline (by Netrin, when concentration is large enough); d - unknown. (Adapted from: Evans and Bashaw [75])

b. Fill in all the sections requiring comments.
c. Use the `find-gradients` command to make the *agents* move up positive gradients of particles (chemotaxis).
d. Use the `find-gradients` command to make the *agents* move down negative gradients of particles (chemorepulsion).   ▽

4.2.3 Breaking Down the Problem

An essential part of research is being able to take complex phenomena and break them down into their most important attributes. ABMs require the hard coding of how *agents* interact with each other and their environment, which forces the developer to restrict the information to its most important features.

In the following exercise you will develop the assumptions that will help build the agent-based model to study the unknown response of a neuron to given guidance cues. The guidance cues diffuse out from the midline of the brain, therefore the rate of diffusion of these cues is an essential parameter of the system.

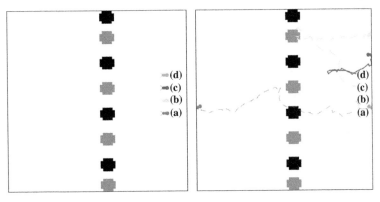

FIGURE 4.6

Basic view of the Agent Based Model. On the left is the initial setup with black and gray circles representing Netrin and Slit sources and bug-like shapes (*agents*) representing the four growth cone types a–d, labeled as in Fig. 4.5 and §4.2.4. The right hand picture shows the results of a single run of the code (see text for information).

Exercise 4.26. Make sure you refer to §4.2.2 to help answer these questions.

   a. What are the *environmental* features?
   b. Which part of the biology should be identified as the *agent*?
   c. How should the *agent* interact with the *environment*?   ▽

   These assumptions can now be transferred into an ABM.

4.2.4 Constructing a Model of Axon Development

Figure 4.6 shows output from the model of Worku, taken from her Honors Thesis "Agent-based Modeling of Commissural Axon Guidance along the Midline" [31]. The midline is represented by black and gray circles which indicate where Netrin and Slit are produced. The four bug shaped *agents* (a–d) represent the four types of growth cones discussed earlier, see Figure 4.5. A basic simulation, where a single axon of each receptor type is modeled, can be seen in Figure 4.6, and the results can be summarized as follows:

   a. growth cones which initially express only DCC receptors are able to cross the midline, whereupon the Robo 1 receptor is expressed, restricting the growth cone to the opposite side of the midline and repulsing it away.
   b. growth cones expressing both UNC5 and Robo 1 receptors are repelled from the midline and grow away from it.
   c. growth cones expressing only the UNC 5 receptor move randomly until they are close to the midline (where the Netrin levels are high) whereupon the guidance cue gradient repels the growth cone.

d. growth cone expressing both DCC and Robo 1 receptors initially moves toward the midline, then away.

Before you proceed with building a model on axon development, let us recap some of the assumptions we have made:

- the *agents* in the ABM mimic the behavior of the developing axons, see exercise 4.25;
- the diffusion of Netrin and Slit is modeled via the `diffuse` command in NetLogo, see exercise 4.25;
- the response to the chemicals is dependent on the type of axon, see exercise 4.26.

There are some important aspects of the biology that we have not yet worked out how to code in NetLogo, for example:

1. tracking the path of the axon (the *agent*): we need to visualize the path of the developing growth cone (the *agent*) because it needs to represent whole neuron (cell body, axon and growth cone, see Figure 4.4);
2. contact-mediated chemo-attraction/repulsion (this requires there to be "enough" chemical in the environment before the axon will respond to gradients);
3. in some cases, the axonal response changes once the axon crosses the midline (see axon a in Figure 4.5);
4. how can the simulation be stopped? (We do not want the simulation to carry on indefinitely.)

Exercise 4.27.   Tracking the path of the axons:

a. Which command traces the path of an agent? (Hint: search for *pen* in the Netlogo dictionary.)
b. Use the command from (a) in *Diffusion2.nlogo* to display the paths of the *agents*. (You should probably restrict the number of *agents* to 5.)   ▽

Exercise 4.28.   Contact-mediated movement:
   Adapt your program so that the agent only responds to chemical gradients when there are at least 10 particles the agents patch. (This mimics contact-mediated chemo-taxis.)   ▽

Exercise 4.29.

a. Adapt your program so that a single *agent* starting at (30,0) is attracted up gradients of the "chemical." However, when the *agent* crosses $x = 0$ the response changes and the chemical acts as a chemorepulsant.
b. Alter the code so that you can change how fast the *agent* moves.   ▽

Exercise 4.30.   Stopping the code:
   Look up the *stop* command and use it to stop the simulation after 500 ticks.   ▽

Exercise 4.31.   In this question, you will develop code that models two axons that (due to differences in their receptors) respond react differently to the effect two chemicals in the environment.
Use your previous work to create a program to the following specifications:

a. There are two (C1 and C2) chemicals, with sources at (0, 10) for chemical C1 and at (0, −10) for chemical C2, both of which have diffusion rates that are user-defined via a slider.
b. There are two *agents*:

The first *agent* starts at (30,10) and reacts to chemical C1 via chemotaxis and is unaffected by chemical C2.
The second *agent* starts at (30, −10) and initially acts in the same way as the first *agent*. However, when this *agent* crosses the middle of the environment ($y = 0$) the response switches and the *agent* no longer responds to chemical C1 and is chemorepulsed by chemical C2.   ▽

Exercise 4.32.   Project: Construct a model of axon guidance.
Important aspects that you will need to include are: A midline where there are two sources of chemical (C1 and C2) that can act as chemoattractants/chemorepulsants.

• C1 and C2 both diffuse through the environment.

Four types (A-D) of axon (*agent*) that act in the following ways:

• type A: attracted to C1, until it crosses the midline, then repulsed by C2
• type B: repulsed by C1 and C2, however the repulsion by C2 is contact mediated
• type C: repulsed by C2 (contact mediated)
• type D: attracted to C1, repulsed by C2

Make sure there are sliders for the diffusion rates. (A slider can be added using *Button* on the NetLogo Interface: an example can be found in the model *Diffusion1. nlogo*.)   ▽

4.3   AN AGENT-BASED MODEL FOR CHOLERA AND THE IMPORTANCE OF REPLICATION

Cholera is a diarrheal disease caused by human ingestion of the bacteria *Vibrio cholerae*, a bacteria that can live in environments that lack proper sanitation. An outbreak typically begins through contact with a contaminated water source, but becomes amplified by a fecal-oral spread among humans [37]. The rapid dehydration that accompanies a cholera illness can lead to death within a few hours if not treated. Cholera causes millions of illnesses each year, and leads to more than 100,000 annual deaths worldwide, primarily in countries which lack the infrastructure to provide clean water and adequate sanitation [38]. As an illustration of how an ABM might be used to implement the scientific method, we consider the setting of an outbreak of cholera in a small village, such as one in a developing African country, and the hypothesis that household cleaning can stop an outbreak from growing in a village even while the water supply remains contaminated [39]. In exploring the hypothesis, we first describe our physical space — we create a landscape with divisions, e.g., households, schools, markets, etc., needed to simulate a village as well as the water supply that is contaminated with the cholera-causing bacteria. We then create our "agents," the humans inhabiting the space, and define daily activities of these agents within the village according to household, gender, and age. We incorporate the ways in which the cholera-causing bacteria might spread through the community as a result of fecal-oral contamination, including rules for how humans become infected and how they in turn may contaminate their environment. From these relatively intuitive rules, a simulation is built from which global properties of the disease spread emerge. Once satisfied that our basic simulation matches what is known about the dynamics of cholera, we can explore options for decreasing the course of the disease. In this ABM model, we introduce of household sanitation. We can test whether this household change was able to affect the village-wide measure of numbers of people who become ill. Because of the probabilistic nature of each simulation, each model experiment is repeated many times to determine the "average" baseline behavior and the "average" improvement with the sanitation effort.

4.3.1 Model Description

In the online materials we have included the NetLogo file "cholera.nlogo" which is the result of the discussion herein. It may be helpful to open this file in NetLogo while working through the model description that follows.

For the purposes of this initial exploration, we consider a simplified setting of three extended households and one common space for our "village." We additionally make the simplification of ignoring gender- and age-specific activities or behavior. We will assume that the village shares a water source that is contaminated with cholera. The level of contamination can be varied in our simulations using the slider-set parameter `environmental-reservoir`. Each household visits the water regularly, and each household member is in contact with the contaminated water at "dinner." Each household has a sleeping area and a latrine area, and its residents are identified by a common color. The configuration described above is also pictured with a simulation screenshot in Figure 4.7.

Cholera is known to spread in several ways [37]. There is generally a small chance of contracting cholera from the local water supply. In fact, with the baseline choices of parameters chosen for our simulation, the model generally runs for many "days" before any inhabitants become infected. However, it is known that once one or two people in a village contract cholera, they become active spreaders of the bacteria as a result of shared contact with household objects and linens, and even by amplifying the environmental contamination of the water supply due to a lack of sanitary latrines [37]. Additionally, researchers have shown that the bacteria coming from freshly shed feces, waste less than 5 h old, may be 700 times more infectious than is the bacteria

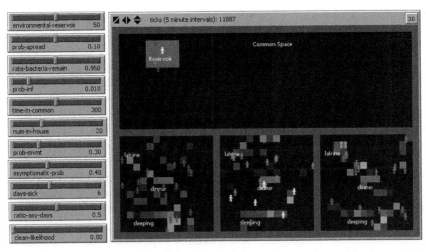

FIGURE 4.7

This figure shows a partial screenshot from the NetLogo simulation screen of the cholera model. The sliders on the left represent input parameters that can be altered by the user. The graphic shows three households sharing a common space. Each household obtains its water from an environmentally contaminated water source. Each household has a sleeping, eating, and latrine area and set activities that combine pre-determined placement with random movement. Patch colors show the current level of bacterial contamination; bright to dark patches show the concentration. In the color version seen during a simulation or in the online version of the text, pink shows highly infectious bacteria and green shows less-infectious bacteria.

in the water supply [40]. We refer to the bacteria in this 5-h window as being highly infectious.

Our model only considers the spread of bacteria through the environment as humans move about, and ignores the potential for amplification of the contamination of the water supply. The model watches humans moving through interactions with their environment in 5-min timesteps. We assume that humans who have cholera may leave behind highly infectious bacteria in the spot where they stand with probability `prob-spread`. Humans who encounter a spot that has been contaminated by the water source or by infected or contaminated humans have the same probability `prob-spread` of picking up the bacteria and moving it to another spot. Each time step, only `rate-bacteria-remain` of the bacteria survive, so gradually the contamination fades from the grid patches as well as from the humans in the grid. Each grid block's color shows the contamination levels: the high-infectious patches are white–pink when they hold the maximum-allowed concentration of bacteria, and they shift to dark-pink as the concentration fades. Likewise, the patches contaminated with less-infectious bacteria move from white-green to dark-green as the concentrations fade.

The patch colors showing contamination can be observed during a simulation or in the online color version of the text, and a grayscale version can be seen in Figure 4.7.

The model assumes that humans contract the disease through contact with a patch that has been contaminated with bacteria. The likelihood of contracting disease is found through the product of the contamination level of the patch and the slider-set parameter `prob-inf`. Due to the difference in infectivity of the freshly-shed bacteria, the probability of infection from contact with a patch that has only the low-infectious bacteria is found by dividing the product above by 700. The current value for that slider has been chosen with other initial slider values so that some, but not all, humans become infected in our baseline experiments. Humans are pictured as red in the simulation once they are infected. Recovered humans are immune, and are pictured as white. Since the disease is transmitted through shared contact with patches, we hypothesize that the amount of time the residents spend in the common space versus in their individual homes might contribute to the spread of disease, and this is regulated with the slider `time-in-common`. We additionally hypothesize that human density could be a significant factor. Since shrinking the NetLogo grid requires significant re-coding, to increase density in our experiments we must change the population size. The number of humans in each household is set by the slider `num-in-house`. Finally, we suspect that the amount of individual movement would also change the likelihood of cross-contamination of patches and thus the number of infections. This parameter is adjusted using the slider `prob-mvmt`.

There are several unknown parameters that may have great impact on the dynamics of a cholera outbreak. The parameters are often not known because the regions of the world affected by cholera are those with little infrastructure and few resources for the careful study and recording of outbreak data, but additionally the bacteria mutate and do not behave in the same way in all outbreaks [37]. Hence, a key motivation for an agent-based study of cholera is to gain insight into the significance of the choice of various ranges of parameter values, and to better understand how daily activities might contribute to the spread of cholera.

A parameter of debate is the proportion of people who may be infected with cholera and show no symptoms [37,41,42]. Those with asymptomatic infections are most likely ill for less time than those with symptoms; however, because they are not suffering, they are moving around more within their population and may have a greater opportunity to spread the bacteria. Our agent-based simulations will give us an opportunity to explore under what conditions having a high number of asymptomatic illnesses might increase or decrease the spread of disease. The model assumes that that humans who have symptomatic cholera are spending all of their time in the household, and most of that time near the latrine, as pictured in Figure 4.7, while those who have asymptomatic illnesses leave the house along with other villagers during common times. We will vary the parameter `asymptomatic-prob` to discern the effect of asymptomatic disease sufferers. Additionally, while there is varying information about the length of time a human will be sick with cholera — anywhere from 1–2 days to 1–2 weeks and set by our slider `days-sick`, most certainly, asymptomatic infections are significantly more brief. We may vary `ratio-asy-days` to see if a smaller length of illness for asymptomatic infected humans changes our conclusions.

Finally, the slider-set parameter `clean-likelihood` will allow us to test our main hypothesis which that some level of individual cleaning could control an outbreak of cholera even in the case that the local water remains contaminated. Some examples of individual household cleaning include frequent washing of hands, household objects, and linens, and the levels of `clean-likelihood` can reflect typical adherence to the rules for cleaning.

4.3.2   **ABM Modeling Exercises: Cholera and the NetLogo BehaviorSpace**

We begin with a series of exercises that will help us understand the behavior and sensitivities of our model as the parameters change value. We suggest that all students complete exercises 4.33, 4.34, 4.35, and for the remaining exercises, either divide the tasks among the class to report back to the groups as a whole, or allow students to refer to the online file "Master Summary of Simulations.pdf" to interpret the results of the simulations requested. These exercises are time-consuming if students are required to perform all simulations, but otherwise take only a minute or two to consider.

Exercise 4.33.   Download our *cholera.nlogo* model from the textbook's online materials. Observe the "Interface" tab. In addition to the slider-set parameter values we discussed previously, you will see eight gray boxes at the bottom of these screen. These are "Monitors" allow us to display the values of any expressions, such as the number of agents which have a certain characteristic. Click "setup" and then "go," and stop the model after about 100 days have passed. How much variation do you observe in the reported values in the gray monitor boxes as you run the simulation a few times?   ▽

In a single simulation, the results depend on the probabilities of where each agent moves, and the probabilities of each agent picking up, spreading, and/or becoming infected with bacteria. While we can observe general behaviors by observing a series of disease simulations, we cannot draw firm conclusions about specific quantitative measures of the spread of the disease (as measured by the numbers in the gray boxes) based on a single simulation, or even based on several simulations.

Fortunately, NetLogo has a terrific tool built in that allows us to repeat and record multiple simulations while we step away from the computer. We will use this tool, the BehaviorSpace, to carefully analyze our model.

Exercise 4.34.   From within the cholera model interface, click on "Tools" followed by "BehaviorSpace," and in the window that pops up you will see a number of experiments that we have already created. Click on the top experiment, baseline experiment, and click the "Edit" button (see Figure 4.8a). Here is a detailed list of the options we will use in the "Experiment" interface that pops up (see Figure 4.8b):

- In the box that follows the instructions "Vary variables as follows" we have already listed all values that we want to hold for our baseline experiment.

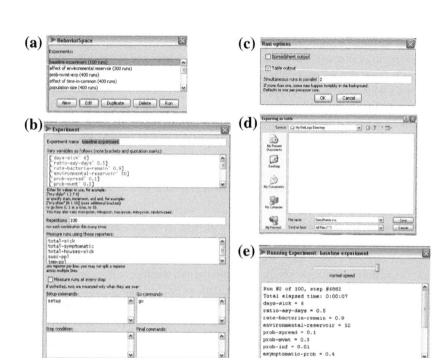

FIGURE 4.8

These screenshots illustrate the steps for using NetLogo's BehaviorSpace to run multiple simulations. We edit the baseline experiment (a); observe the input parameters, the number of repetitions, the output measures, and the time limit (b); run the experiment and select Table output (c); save the file with an easy name, a nice folder, and a `csv` extension (d); and select the fastest speed with no updates to allow the simulations to run (e).

- To the right of "Repetitions," we have entered 100, because you typically need at least 100 repetitions so that the average of the quantitative results do not reflect outlying individual simulations. However, it takes a long time to run 100 simulations on most normal computers, so you might want to change this to from 100 to 10 here and following since we are only exploring and learning, and not trying to produce scientific results.
- Under the heading "Measure runs using these reporters" we have listed the variables that we measure in the gray boxes on the interface screen. You will not need to modify these.
- We do not check the box for "Measure runs at every step."
- To the right of "Time Limit" we've entered 28800 which is how many 5-min intervals are in 100 days.

Click the "OK" button at the bottom of the "Experiment" interface, which returns you to the "BehaviorSpace" interface. Click the "Run" button while the baseline experiment is still highlighted. When you come to the "Run options" box (see Figure 4.8c), you should uncheck the "Spreadsheet output" box and check the "Table output" box. Ignore the third box, and click the "OK" button. This will take you to a box where you can save your file. We recommend giving the file a name and folder you will remember and be sure to add the extension .csv (see Figure 4.8d). Once you are finished saving the file, you finally make it to the interface "Running Experiment." Go ahead and move the speed slider to the right (fastest) and uncheck "Update view" and "Update plots and monitors" (see Figure 4.8e). At this point, you just wait for your simulations to complete.   ▽

Exercise 4.35.   After the simulation is complete, open the file you've created in a spreadsheet software such as Excel, OpenOffice Calc, or Google Spreadsheet. We suggest that you save the file within your spreadsheet software at this point so that you don't lose the work that follows. The spreadsheet has columns for all of the input and output variables that we set up in the BehaviorSpace. We want to calculate the averages of the output measures so that we have a basis for comparison in the exercises that follow. The following instructions assume that you ran 100 simulations, and you should modify them accordingly if you chose to run fewer. Click your mouse on the cell B110. Type = average (B8..B107). Obviously, you did not need to average this value because it is the same in all runs, but now we can copy that formula over to the columns where it is needed. Click and hold your mouse on the small dot that's in the bottom-right corner of cell B110, and, still holding the mouse button, drag your mouse over to cell U110. Release and you'll see that you have copied the formula across all rows. Save this file for future reference. Compare the results of these averages to the simulations that you ran in Exercise 4.33.   ▽

Exercise 4.36.   Redo the previous two exercises so that you have new averages from a new set of simulations (or compare the results of your simulations to our results in the outline materials accompanying our chapter). How much variation is there between the two sets of runs, and in the case where the input parameters differed, what kind of difference in outcome values would you need to see to be convinced that the end behavior had truly changed?   ▽

Exercise 4.37.   There are a number of experiments pre-programmed into the Behav- iorSpace. Because the individual repeated simulations are time-consuming, it is most helpful to divide these experiments throughout the class so that results can be com- pared as a whole. Each experiment should be run following the instructions above, and then averaged, using a modification of the spreadsheet instructions. You'll see that column A, [run number], often has its entries a little scrambled, and as a result, when multiple experimental values are used, the spreadsheet sometimes ends up with rows that need to be moved so that all like-values are listed together. Be mindful of this task before averaging the experimental results as you complete any exercises below.

a. Run the experiment `effect of environmental-reservoir`, and create a spreadsheet that contains only the header row, and the averaged output from each separate reservoir value, including the baseline averages. (The command paste special-values in spreadsheet is helpful here to copy the averaged values and not the formulas.)

What does this slider change, in real-life terms, in the simulation? (In answering this question, it may be helpful to both look at how `environmental-reservoir` is defined in the Procedures as well as to watch portions of several simulations in which differing values for `environmental-reservoir` are chosen.).

What effect, if any, did changing the reservoir value have? What would happen if the reservoir value was set to 0?

b. Run the experiment `prob-mvmt-exp`, and create a spreadsheet that contains only the header row, and the five averaged output rows for each separate movement value. What does this slider change, in real-life terms, in the simulation?

What effect, if any, did changing the movement slider have?

How can you make intuitive meaning of this perhaps surprising result?

c. Run the experiment `population size`, and create a spreadsheet that contains only the header row, and the five averaged output rows for each separate movement value. What slider does this experiment change?

What conclusions can you draw from this experiment?

d. Run the experiment `effect of time-in-common`, and create a spreadsheet that contains only the header row, and the five averaged output rows for each separate movement value. What does this slider change, in real-life terms, in the simulation?

What effect, if any, did changing the time-in-common slider have?

Do you think the effect is significant?

e. Let's see if the results are the same above if the village is more crowded. Click the mouse on the experiment `effect of time-in-common`, but before you run the experiment, edit it and change the value of `num-in-house` to 40. Run this new experiment. Of course, more will be sick than before because there are more people, but is it a greater percentage? Also, does the slider `effect of time-in-common` have any more impact now? Any thoughts?

f. Run the experiment `rate-bacteria-remain`, and create a spreadsheet that contains only the header row, and the five averaged output rows for each separate movement value. What does this slider change, in real-life terms, in the simulation?

What conclusions can you draw from this experiment?

g. Run the experiment `prob-inf`, and create a spreadsheet that contains only the header row, and the five averaged output rows for each separate movement value. What does this slider change, in real-life terms, in the simulation?

What conclusions can you draw from this experiment?

h. Where might there be some gaps to the numbers suggested for `prob-inf`. Modify the experiment to include a few needed values (and do not repeat the values for which you have already run simulations). How did this improve your understanding of the choice of values for this parameter?

i. Run the experiment `asymptomatic-prob`, and create a spreadsheet that contains only the header row, and the five averaged output rows for each separate movement value.

Why might increasing the probability of asymptomatic cases cause the spread of disease to slow?

Why might increasing the probability of asymptomatic cases cause the spread of disease to increase?

What is the actual result of changes to the probability of asymptomatic infections?

j. Run the experiment `prob-spread`, and create a spreadsheet that contains only the header row, and the five averaged output rows for each separate movement value. What does this slider change, in real-life terms, in the simulation? What conclusions can you draw from this experiment?

k. Run the experiment `days-sick`, and create a spreadsheet that contains only the header row, and the five averaged output rows for each separate movement value. What does this slider change, in real-life terms, in the simulation? What conclusions can you draw from this experiment?

l. Create an experiment by choosing `Duplicate` with the experiment `baseline` experiment. Set `random-mvmt` to be true. Run the experiment. What does this change in real-life terms, and what do the results suggest about the household and common space boundaries and individual behaviors?

m. Create an experiment by choosing `Duplicate` with the experiment `baseline` experiment. Set `dinner` to be true. Run the experiment. What does this change in real-life terms, and what do the results suggest about the importance of creating the dinner space as it is?   ▽

Let's now turn to the main hypothesis. If individuals clean in their homes, could that slow or end an outbreak?

Exercise 4.38.   Run the experiment `clean-likelihood`, and create a spreadsheet that contains only the header row, and the five averaged output rows for each separate movement value. What exactly does this slider change, in real-life terms, in the simulation?

What conclusions can you draw from this experiment?   ▽

Exercise 4.39.   Modify the experiment `clean-likelihood` to set `Purple-Cleaning-Only` to false. Run the experiment, adding the value originally covered by the baseline experiment.

Does one clean household prevent illnesses in the other households?

Can one clean household have the same impact as three clean households? Explain.   ▽

Finally, there are numerous changes that could be made to the simple model for cholera, with or without consideration of the question of cleaning.

Exercise 4.40.   Save the NetLogo cholera file with a new name, and modify the code so that individuals with symptomatic cholera are only confined to the home, and not constantly heading for the latrine. Rerun the baseline experiment with the new behavior. Does the in-home behavior of symptomatic individuals affect the outcome measures?   ▽

Exercise 4.41.   Save the NetLogo cholera file with a new name, and modify the code so that the Purple household always begins with one turtle who is infected and symptomatic. Run the baseline experiment twice: once with the environmental-reservoir at 50 and once with the reservoir at 0. Compare to the original baseline, and discuss the results.   ▽

Exercise 4.42.   Modify the assumptions for cleaning so that individuals clean wherever they roam, and not just inside the home. Does this change make any difference to the conclusions in Exercises 4.38 and 4.39?   ▽

There are a number of complications you could add to the model for a much more in-depth project, such as increasing the number of households or including genders and ages within the households, with appropriate activities such as markets and schools. These would be appropriate to consider as part of a long-time-frame project that compares results with those already discovered within the exercises.

4.4   USE AND DESCRIPTION OF ABM IN RESEARCH: TICK-BORNE DISEASE AGENT-BASED MODELS

ABMs provide a powerful tool for research at all levels including undergraduates. As part of a research project focused on understanding the dynamics of ehrlichiosis, a human tick-borne diseases, an ABM was created and published [29]. This model was used as the basis for building a more complex ABM to explore another tick-borne disease, Rocky Mountain spotted fever, by an undergraduate student as part of a summer research program working with Gaff [43]. Here we provide a basic overview of these two models as an example of the use of ABM in a research-focused project as well as how ABM can be reported for publication.

Vector-borne diseases are spread through ticks, mosquitoes, and other arthropods. In the US, tick-borne diseases are the most common arthropod-borne disease and have been for many years [44,45]. Risk of tick-borne disease, e.g., Lyme disease (*Borrelia burgdorferi*), Rocky Mountain spotted fever (*Rickettsia rickettsii*), Tidewater spotted fever (*Rickettsia parkeri*), and ehrlichiosis (*Ehrlichia chaffeensis*), is a combination of expected frequency of encounter with a competent vector and the prevalence of disease within that vector population. Predicting either of these factors is a task well suited for mathematical modeling and simulation.

In order to understand many of these diseases, we must come to understand the underlying dynamics of the tick populations themselves [46]. Unlike other arthropod vectors, ticks have a unique life history that significantly impacts the prediction and control of tick-borne diseases. The majority of tick species in the United States are classified as hard-bodied ticks (Ixodid). These ticks have four distinct life stages: egg, larva, nymph, and adult. For the three latter stages, ticks will quest, i.e., seek a single vertebrate host for a bloodmeal that will last several days. If successful in obtaining a complete bloodmeal as a larva or nymph, these ticks will then remain dormant for months to years before emerging as the next life stage to quest again. All adult female ticks and some adult male ticks require a bloodmeal before reproduction and death. Every tick species has a different set of preferred hosts for each life stage, different strategies to find hosts, different lengths of time between life stages and different competence for transmissible pathogens. For example, *Ixodes scapularis*, the black-legged tick, prefers rodents for a larval meal, medium-size mammals for a nymphal meal, and deer for the adult meal. This life history pattern creates unique temporal scale patterns that are also dependent on off-host survival factors such as weather and predation.

A number of types of models have been used to look at tick-borne diseases including differential equation-based models [47–49], age-structured difference equation models [50–57], GIS models [58–60], and others [61,62]. While each of these models are helpful to investigate the dynamics of tick and tick-borne diseases, they all use a population approach with generalized mass-action interactions between ticks and their hosts, ignoring many of the unique features of the tick life history. An agent-based model provides the ideal tool to test the validity of population-level models for ticks. The two preliminary models, TICKSIM and RMSF-SIM, presented here are described in more detail in Gaff [29] and Tillinghast and Gaff [43], respectively. TICKSIM models ehrlichiosis while RMSF-SIM models Rocky Mountain spotted fever.

4.4.1 The Model

One challenge of ABM is the lack of standardization in published descriptions of ABM, but Grimm and coauthors have attempted to solve this with a careful set of instructions first established in 2006 [63,64] and revised in 2010 [65]. This provides a standard framework for reporting described as the "Overview, Design concepts, and Details (ODD)" for a given model. This example follows the ODD protocol which consists of seven elements. The first three elements provide an overview, the fourth element explains general concepts underlying the model's design, and the remaining three elements provide details. In the *overview*, one states the purpose of the exploration and the basic spatial and agent-behavior assumptions. The *design concepts* explore complexities within a model, including basic assumptions underlying the design, emergence of global traits from individual behaviors, adaptive traits that have been assigned, and changes in states that occur by sensing surrounding locations or agents. The *details* are primarily parameter and initial values for the ABM. While many of the details may seem trivial, these details along with the rest of the ODD protocol provide the recipe for reproducibility of results.

4.4.2 Purpose

The main purpose of both models is to explore disease dynamics in a model that includes the explicit life history characteristics of ticks and hosts at an individual level. The results demonstrate the potential for spatially and temporally varying risk of disease. TICKSIM is a very simple model used to demonstrate the ability of an agent-based model to simulate ticks and tick-borne diseases. RMSF-SIM builds on TICKSIM to add host dynamics including predator–prey interactions and environmental sensing.

4.4.3 State Variables and Scales

In TICKSIM, two populations are considered: ticks and hosts which are loosely modeled after the lone star tick, *Amblyomma americanum*, and the white-tailed deer, *Odocoileus virginianus*. The environment is a series of patches of equal quality with wrapping boundaries, i.e., agents moving off one edge will reappear instantly on the opposite edge. In the refined model RMSF-SIM, three populations are considered: ticks, predator hosts, and prey hosts. The parameter values of the tick populations are based on those of the American dog tick *Dermacentor variabilis* while the predator and prey agents are generic animals. The tick agents feed on the prey animals, an assumed smaller animal, for a bloodmeal as a larvae and nymphs and feed on the predator animals as an adult. The environment is a series of patches with mix of forest, a safe home for all host agents, and grassland, the primary food source for prey host agents, with reflecting boundaries. Reflecting boundaries are more biologically realistic as they represent physical boundaries that prevent movement out of or into the system.

For both models, each tick agent has a unique identification number, sex, age, location, life stage (egg, larva, nymph, adult), time in current life stage, infection status (susceptible, infectious), current activity (resting, questing, feeding, laying eggs (females) and hosts (list of all hosts used for blood meals). For TICKSIM, host agents have a unique identification number, age, list of all ticks current on the host, list of all ticks ever on the host and infection status (susceptible, infectious, immune). For RMSF-SIM, host agents also have an energy level and safety level.

4.4.4 Process Overview and Scheduling

The model follows the same steps every day of the simulation as shown in Figure 4.9. The mortality of ticks is based on the time of the year with higher probabilities of death in the winter and summer. If a host dies, all ticks on that host are also assumed to die. The host is also immediately replaced to maintain a constant host population.

The RMSF model follows the same steps of every parameter execution and resulting day-to-day cycle (see Figure 4.9). The mortality of ticks is based on the time of the year with the highest probabilities of death in the winter and summer. Unless otherwise stated within the table of parameters, all parameters are kept static to ensure that the destabilization or success of the disease is due to inherent

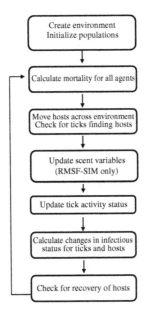

FIGURE 4.9

Flow diagram for TICKSIM and RMSF-SIM models.

characteristics of itself and not the ecosystem being unable to supply due to its own inherent breakdown or unrealistic overabundance of a resource.

4.4.5 Design Concepts

The tick agents in both models can sense hosts to use for blood meals but still only have a small probability of actually successfully attaching to that host. The density dependence for ticks is only indirectly implemented by giving a maximum number of ticks per host as assigned for each host species.

In TICKSIM, the hosts simply move randomly around the uniform grid of patches. In RMSF-SIM, the forest patches serve as home for host agents but has no food. The grassland has a variable amount of food available depending on the time in the year and the consumption by prey agents. The system for both models is closed.

In RMSF-SIM, there is an additional layer of interactions for a predator–prey relationship between the two host species. The ABM tracks three variables that govern each host's activities: age, energy, safety. The probability of death from other environmental causes increases as a function of age for both predators and prey. Both predators and prey must maintain a set energy level to prevent starvation. Energy is lost over time and through movement, and is gained through food. Safety is a measure of time spent away from the home area in the forest. All host agents have to navigate back to their home area on a regular basis. Host agents are kept at a constant population through replacement of a new agent for every death. Finally, in RMSF-SIM, predator and prey agents emits a "scent" as they move through the environment. This allows predators to navigate toward a prey while prey will move away from predators.

The tick agents are the primary vector of the disease. TICKSIM is based on the dynamics of *Ehrlichia chaffeensis*. For both models, there is no direct host-to-host transmission of the disease so all disease in hosts is from the ticks. RMSF-SIM models *Rickettsia rickettsii*, which is able to be transmitted from an infected female tick to the eggs she lays. Every successive generation that is born carrying the disease has their respective egg-laying capability cut to one-third of total capacity for three generations. For more information on disease models, see Chapter 6.

4.4.6 Input

For TICKSIM, each simulation is initiated in a uniform 25x25 patch grid with an initial 100 hosts randomly spread across the grid. The probability that a given host will start the simulation infected is 0.1. Additionally, 1000 ticks are randomly spread across the grid. The probability that a given tick will be infected is also 0.1. The simulations are assumed to begin on June 1, at which time larval stages would not be present, and thus the initial ticks are split into adults and nymphs with approximately 100 adults and 900 nymphs. All other parameters are given in Gaff [29].

For RMSF-SIM, each simulation is initiated in a divided 40X30 patch grid of one fourth forest and three fourths grassland with an initial 75 predators and 500 prey spread randomly across the grid. Additionally 15,250 ticks (10,000 eggs, 5000 larvae, 250 nymphs) are randomly spread across the grid. The simulations are assumed to begin on January 1, at which time adult stages will not be present.

4.4.7 Simulation Experiments

An example experiment of TICKSIM was used to explore the possibility of the establishment of a new tick population into a completely naive area. From this simulation, it can be shown that as few as two nymphs dropping off in a given area can establish a new population approximately 33% of the time. Simulations with RMSF-SIM show that while the transovarially transmitted disease always remains in a population initially, the mortality and reduced fecundity precludes the disease from remaining for more than 20 years. The combined results of these models show that a tick-borne disease that cannot be transmitted from a female to her eggs has a limited chance of establishment initially. When the females lay infected eggs, the disease always remains in the system initially.

TICKSIM was also modified and run with various combination of disease dynamics to explore the likelihood of a new disease entering the system with the new tick. Four scenarios for tick dynamics were used: (1) no transovarial transmission and no reduced fecundity from infection; (2) no transovarial transmission but with reduced fecundity from infection; (3) transovarial transmission at 10% and no reduced fecundity from infection; and (4) transovarial transmission at 10% and reduced fecundity from infection at 92%. Each scenario was initialized with 10 ticks of which on average 5 were infected, and the simulation was run 25 times each for 10 years. For Scenario 1 and 2, the tick population was established, but the disease died out within two years for all simulation runs. For Scenario 3, both the tick population and the disease remained in the system at the end of all but one simulation. For Scenario 4, even with a reduction in fecundity of 92% for infected females, both the tick population and the disease remained in the system at the end of all simulations. This demonstrates the significant impact of transovarial transmission on the establishment of a tick-borne disease in a new area.

Exercise 4.43.   Using the ODD Template of Grimm et al. [65] provided here and demonstrated in this section, return to your favorite model in this chapter and complete the ODD Template for that model.

1. Purpose
   Question: What is the purpose of the model?
2. Entities, state variables, and scales
   Questions: What kinds of entities are in the model? By what state variables, or attributes, are these entities characterized? What are the temporal and spatial resolutions and extents of the model? Most ABMs include the following types of entities:

   - Agents/individuals
   - Spatial units (e.g., grid cells)
   - Environment
   - Collectives

3. Process overview and scheduling
   Questions: Who (i.e., what entity) does what, and in what order? When are state variables updated? How is time modeled, as discrete steps or as a continuum over which both continuous processes and discrete events can occur? Except for very simple schedules, one should use pseudo-code to describe the schedule in every detail, so that the model can be re-implemented from this code. Ideally, the pseudo-code corresponds fully to the actual code used in the program implementing the ABM.
4. Design concepts
   Questions: There are eleven design concepts. Most of these were discussed extensively by Railsback (2001) and Grimm and Railsback (2005; Chapter. 5), and are summarized here via the following questions:

   a. Basic principles. Which general concepts, theories, hypotheses, or modeling approaches are underlying the model's design?
   b. Emergence. What key results or outputs of the model are modeled as emerging from the adaptive traits, or behaviors, of individuals?
   c. Adaptation. What adaptive traits do the individuals have? What rules do they have for making decisions or changing behavior in response to changes in themselves or their environment?

    d. Objectives. If adaptive traits explicitly act to increase some measure of the individual's success at meeting some objective, what exactly is that objective and how is it measured?

    e. Learning. Many individuals or agents (but also organizations and institutions) change their adaptive traits over time as a consequence of their experience? If so, how?

    f. Prediction. Prediction is fundamental to successful decision-making; if an agent's adaptive traits or learning procedures are based on estimating future consequences of decisions, how do agents predict the future conditions (either environmental or internal) they will experience?

    g. Sensing. What internal and environmental state variables are individuals assumed to sense and consider in their decisions?

    h. Interaction. What kinds of interactions among agents are assumed?

    i. Stochasticity. What processes are modeled by assuming they are random or partly random?

    j. Collectives. Do the individuals form or belong to aggregations that affect, and are affected by, the individuals?

    k. Observation. What data are collected from the ABM for testing, understanding, and analyzing it, and how and when are they collected?

5. Initialization

Questions: What is the initial state of the model world, i.e., at time $t = 0$ of a simulation run? In detail, how many entities of what type are there initially, and what are the exact values of their state variables (or how were they set stochastically)? Is initialization always the same, or is it allowed to vary among simulations? Are the initial values chosen arbitrarily or based on data? References to those data should be provided.

6. Input data

Question: Does the model use input from external sources such as data files or other models to represent processes that change over time?

7. Submodels

Questions: What, in detail, are the submodels that represent the processes listed in "Process overview and scheduling"? What are the model parameters, their dimensions, and reference values? How were submodels designed or chosen, and how were they parameterized and then tested?   ▽

4.5 COMMENTS FOR INSTRUCTORS

While ABMs have provided rich insights into multiple research areas for many decades, they have not traditionally been involved in the classroom. As a conclusion to this chapter, we take a few moments to explore some of ways in which explorations in agent-based modeling might benefit students.

    ABMs contain easily explained concepts, and provide a framework to allow students to explore new tools and subjects. Through ABM creation and exploration, our students gain computational and modeling expertise as well as the ability to work within a team to break down a large, unfamiliar problem into smaller, achievable steps. The ABM framework can provide an excellent exercise in teamwork as well as independent creation that ends with the satisfaction of a visually and intuitively appealing product.

A certain benefit of the ABM development experience to the student is the experience of using unfamiliar software packages. Once outside of academia, students will be required to explore and take advantage of software available to them. Instructors may choose an additional challenge of asking students to evaluate and choose between a number of available platforms, such as Swarm, RePast, NetLogo, AgentSheet, and Ascape. Each of these tools have various strengths and weaknesses, including the built-in operations and libraries for the ABM, coding requirements, GUIs, and observer flexibility. Worku, an undergraduate who developed the axon ABM, actually first evaluated all choices listed above, and then chose to use NetLogo for the following reasons: "I have chosen to use NetLogo since it is the most user friendly and does not need developers to have a significant background in coding. However, NetLogo enables the coder to code complex environments and gives the observer flexibility to change these environments." A student with different skills may have chosen a different tool; the process of evaluating the software is a valuable experience for students.

Another great aspect of using ABM development as a teaching tool is the ease of model refinement and expansion as it is essential for students to learn how to assess and analyze their model, assumptions, limitations, and the future direction of their work. In the case of her thesis, Worku identified several limitations. Her model assumes the number of receptors on each axon is the same and that they are evenly distributed. However, it has been documented that the number of receptors found on each growth cone, and their distribution, alters the efficacy of the growth cone response to the guidance cues [66]. She also assumed that all growth cones have the same shape and size, which remains constant. However, it has been documented that growth cones can change their size while they are moving through the environment [67]. Finally, it is assumed that the growth cone itself does not produce any guidance cue. Another theory states that the growth cone can produce chemoattractant which leads to grouping of axons, which is then undone as axons approach the midline [68]. These are examples that Worku identified as limitations in her model showing that ABMs are complex enough to stretch a student's technical abilities, but tractable enough for them to understand its limitations.

Another part of the appeal of ABM in education is its local-level viewpoint that encourages gathering and integration of data that is both technical and social in nature [69]. As observed in Section 4.3, for example, a disease model may incorporate relatively technical information about spatial configurations, population densities, causes, and likelihood of disease transmission, waning rates, etc., but the model may also encourage intuitive explorations, such as how individual human behavior might affect disease spread.

Working with agent-based models can engage students at all levels of Bloom's Taxonomy [70]. Experiments using pre-programmed ABM simulations offer the opportunity to inform students in a classroom setting, helping them to understand and remember basic mechanisms that contribute to the global dynamics [71]. For example, students could adjust parameters in pre-created epidemic models, running repeated simulations to understand the impact of various disease properties or health policies on the spread of disease. Students could explore a simulation of wound healing to gain an intuitive understanding of the factors that positively and negatively influence healing. They could explore a model that explains how ants might queue themselves into our kitchens. Through a careful evaluation of pre-existing models, students have an opportunity to develop a deep and memorable understanding of underlying behaviors. In addition, students are able to observe that a single simulation may have dramatically different results from what is most commonly observed in other simulations, thus reinforcing valuable lessons about stochasticity and the importance of repetition.

Students who take the time to develop and code individual simulations have the most opportunity to learn. As we mentioned earlier, through the creation of an ABM and its subsequent analysis, students have a platform which allows them to postulate and experiment with hypotheses. The process of model creation requires students to consider all factors that are contributing to the agent-environment interactions and evolution. Students are forced to explicitly describe all relevant assumptions, and the creation process will most likely alert students to gaps in their understanding, encouraging a search for new sources of information or perhaps a re-reading of existing research papers with a new mind set. Complex models are best approached with a divide-and-conquer technique, and teamwork becomes rather meaningful and inspiring as individual group members contribute code for the behaviors they choose to describe. Now group work, rather than feeling burdensome, allows the team to come together and see a whole that is far better than the sum of its parts, and students begin to appreciate the collaborative nature of science. Through the culmination of the model creation and analysis process, students often gain a first experience as true scientific researchers.

Agent-based models provide a powerful tool for research and exploration in many subjects including biology. The intuitive nature of the ABM framework makes the exploration of existing ABMs an inviting and informative tool for biologists regardless of their prior computational experiences. Software platforms such as NetLogo make model development a realistic goal for scientists from all backgrounds.

Acknowledgments

HDG would like to thank Daniel Drake Tillinghast, David Gauthier, Dan Sonenshine, and Wayne Hynes at Old Dominion University and Colleen Burgess for her help in an initial version of TICKSIM. HDG also thanks the ODU Honors College for support of Research for Undergraduates in Math and Science. ES and DG thank the DISCOVER program at Marymount University and its support of undergraduate research. DG would like to thank axon development collaborators: undergraduate students Alelena Hilary and Bezawit Worku, and Dr. Amanda Wright at Marymount University. ES and DG thank ABM cholera model collaborators: undergraduate students Mike Bokosha, Stacey Cole, Eric Kamta, Shazia Khattak, Hannah Korbach, and Samantha Rimkus. HDG was supported by Grant No. K25AI067791 from the National Institute Of Allergy And Infectious Diseases. The content is solely the responsibility of the authors and does not necessarily represent the official views of the National Institute Of Allergy And Infectious Diseases or the National Institutes of Health. HDG and ES were supported by the National Science Foundation DMS-0813563.

4.6 SUPPLEMENTARY MATERIALS

All supplementary files and/or computer code associated with this article can be found from the volume's website http://booksite.elsevier.com/9780124157804

References

[1] Eric Bonabeau. Agent-based modeling: Methods and techniques for simulating human systems. PNAS 2002;99:7280–7.

[2] Simeone Marino. Hogue Ian B, Ray Christian J, Kirschner Denise E. A methodology for performing global uncertainty and sensitivity analysis in systems biology. J Theor Biol 2008;254:178–96.

[3] Gaff H, Lyons M, Watson G. Classroom manipulative to engage students in mathematical modeling of disease spread: 1+1 = Achoo!. Math Model Nat Phenom 2011;6:215–26.

[4] Wilensky U. Netlogo http://ccl.northwestern.edu/netlogo, Center for Connected Learning and Computer-Based Modeling, Northwestern University, Evanston, IL; 1999.

[5] Lytinen SL, Railsback SF. The evolution of agent-based simulation platforms: a review of NetLogo 5.0 and ReLogo. In: Proceedings of the Fourth International Symposium on Agent-based Modeling (ABModSim-4), Vienna, Austria; 2012.

[6] Railsback Steven F, Lytinen Steven L, Jackson Stephen K. Agent-based simulation platforms: review and development recommendations. Simlation 2006;82:609–23.

[7] Railsback SF, Grimm V. Agent-based and individual-based modeling: a practical introduction. Princeton University Press; 2012.

[8] Burks AW, Von Neumann J. Theory of self-reproducing automata. University of Illinois Press; 1966.

[9] Gardner M. Mathematical games: the fantastic combinations of John Conway's new solitaire game Life. Sci Am 1970;223:120–3.

[10] Axelrod RM. The complexity of cooperation: agent-based models of competition and collaboration. Princeton University Press; 1997.

[11] Schelling Thomas C. Micromotives and macrobehavior, revised ed. W.W. Norton & Company; 2006.

[12] Wilensky U. Netlogo Segregation Model http://ccl.northwestern.edu/netlogo/models/Segregation, Center for Connected Learning and Computer-Based Modeling, Northwestern University, Evanston, IL; 1997.

[13] IMDb. Awards for Craig Reynolds http://www.imdb.com/name/nm0721662/awards, [accessed 03/30/2012].

[14] Reynolds Craig. Boids: Background and Update http://www.red3d.com/cwr/boidsurlwww.red3d.com/cwr/boids [accessed 03/30/2012].

[15] Wilensky U. Netlogo Flocking Model http://ccl.northwestern.edu/netlogo/models/Flocking, Center for Connected Learning and Computer-Based Modeling, Northwestern University, Evanston, IL; 1998.

[16] SwarmWiki. AgentBased Models in Biology and Medicine http://www.swarm.org/wiki/AgentBasedModelsinBiologyandMedicine.

[17] Cannata N, Corradini F, Merelli E, Omicini A, Ricci A. An agent-oriented conceptual framework for systems biology. Trans Comput Syst Biol 2005;III: 105–22.

[18] Gonzalez PP, Cardenas M, Camacho D, Franyuti A, Rosas O, Lagunez-Otero J, et al. Cellulat: an agent-based intracellular signalling model. BioSystems 2003;68:171–85.

[19] Pogson M, Smallwood R, Qwarnstrom E, Holcombe M. Formal agent-based modelling of intracellular chemical interactions. Biosystems 2006;85:37–45.

[20] Walker DC, Southgate J, Hill G, et al. The epitheliome: agent-based modelling of the social behavior of cells. Biosystems 2004;76:89–100.

[21] Lollini PL, Motta S, Pappaiardo F. Discovery of cancer vaccination protocols with a genetic algorithm driving an agent based simulator, BMC. Bioinformatics 2006;7.

[22] Zhang L, Wang Z, Sagotsky JA, Deisboeck TS. Multiscale agent-based cancer modeling. J Math Biol 2009;58:545–59.

[23] Castiglione F, Pappalardo F, Bernaschi M, Motta S. Optimization of HAART with genetic algorithms and agent-based models of HIV infection. Bioinformatics 2007;23:3350.

[24] Segovia-Juarez JL, Ganguli S, Kirschner D. Identifying control mechanisms of granuloma formation during M. tuberculosis infection using an agent-based model. J Theor Biol 2004;231:357–76.

[25] Bailey Alexander M., Thorne Bryan C., Peirce Shayn M. Multi-cell agent-based simulation of the microvasculature to study the dynamics of circulating inflammatory cell trafficking. Ann Biomed Eng 2007;35:916–36.

[26] Katie Bentley, Holger Gerhardt. Bates Paul A. Agent-based simulation of notch-mediated tip cell selection in angiogenic sprout initialisation. J Theor Biol 2008;250:25–36.

[27] Dong X, Foteinou PT, Calvano SE, Lowry SF, Androulakis IP. Agent-Based modeling of endotoxin-induced acute inflammatory response in human blood leukocytes. PLoS ONE 2010;5:e9249.

[28] Barrett Chris L., Eubank Stephen G., Smith James P. If smallpox strikes Portland. Sci Am 2005.

[29] Gaff H. Preliminary analysis of an agent-based model for a tick-borne disease. Math Biosci Eng 2011;8:463–73.

[30] Epstein Joshua M., Parker Jon, Cummings Derek, Hammond Ross A. Coupled Contagion Dynamics of Fear and Disease: Mathematical and computational explorations. PLoS ONE 2008;3:e3955.

[31] Worku B. Agent-based modeling of commissural axon guidance along the midline honors Thesis; 2011.

[32] Goodman CS. Mechanisms and molecules that control growth cone guidance. Annu Rev Neuro 1996;19:341–377.

[33] Grunwald IC, Klein R. Axon Guidance: receptor complexes and signaling mechanisms. Curr Opin Neurobiol 2002;12:250–59.

[34] Tessier-Levigne M, Goodman CS. The molecular biology of axon guidance. Science 1996;274:1123–33.

[35] Mortimer D, Fothergill T, Pujic Z, Richards LJ, Goodhill GJ. Growth cone chemotaxis. Trends Neurosci 2008;31:90–8.

[36] Brose K, Tessier-Levigne M. Slit proteins: key regulators of axon guidance, axonal branching, and cell migration. Curr Opin Neurosci 2000;10:95–102.

[37] Nelson Eric J, Harris Jason B, Glenn Morris J, Calderwood Stephen B, Camilli Andrew. Cholera transmission: the host, pathogen and bacteriophage dynamic Nature. Rev Microbiol 2009;7:693–702.

[38] Organization World Health. Cholera Fact Sheet $N°$ 107, August 2011 http://www.who.int/mediacentre/factsheets/fs107/en/index.html

[39] Gammack David, Schaefer Elsa, Korbach Hannah, Cole Stacey, Rimkus Samantha. An ABM exploration of cholera dynamics in a village. (in preparation).

[40] Merrell DS, Butler SM, Qadri F, et al. Host-induced epidemic spread of the cholera bacterium. Nature 2002;417:642–5.

[41] King AA, Ionides EL, Pascual M, Bouma MJ. Inapparent infections and cholera dynamics. Nature 2008;454:877–80.

[42] Miller Neilan Rachael L, Schaefer Elsa, Gaff Holly, Fister K. Renee, Lenhart Suzanne. Modeling optimal intervention strategies for cholera. Bull Math Biol 2010;72:2004–18.

[43] Tillinghast DD, Gaff HD. An agent-based model of the dynamics of a tick-borne disease. In: Proceedings of VMASC Student Capstone 2012; to appear.

[44] Barbour AG, Fish D. The biological and social phenomenon of Lyme disease. Science 1993;260:1610–6.

[45] Centers for Disease Control and Prevention. Lyme Disease Data http://www.cdc.gov/lyme/stats/index.html

[46] Sonenshine DE, Mather TN. Ecological dynamics of tick-borne zoonoses. Oxford University Press; 1994.

[47] Gaff H, Gross LJ. Analysis of a tick-borne disease model with varying population sizes in various habitats. Bull Math Biol 2007;69:265–88.

[48] Gaff H, Schaefer E. Metapopulation models in tick-borne disease transmission modelling. In: Michael E, Spear R, (Eds.), Modelling parasitic disease transmission: biology to control. Austin, TX, USA: Landes Bioscience/Eurekah; 2008.

[49] Gaff LJ. Gross, Schaefer E. Results from a mathematical model for human monocytic ehrlichiosis. Clin Microbiol Infect 2008;15:1–2.

[50] Awerbuch TE, Sandberg S. Trends and oscillations in tick population dynamics. J Theor Biol 1995;175:511–6.

[51] Haile DG, Mount GA. Computer simulation of population dynamics of the lone star tick, *Amblyomma americanum* (Acari: Ixodidae). J Med Entomol 1987;24:356–69.

[52] Mount GA, Haile DG. Computer simulation of population dynamics of the American dog tick (Acari: Ixodidae). J Med Entomol 1989;26:60–76.

[53] Mount GA, Haile DG, Davey RB, Cooksey LM. Computer simulation of Boophilus cattle tick (Acari: Ixodidae) population dynamics. J Med Entomol 1991;28:223–40.

[54] Mount GA, Haile DG, Barnard DR, Daniels E. New version of LSTSIM for computer simulation of *Amblyomma americanum* (Acari: Ixodidae) population dynamics. J Med Entomol 1993;30:843–57.

[55] Mount GA, Haile DG, Daniels E. Simulation of blacklegged tick (Acari: Ixodidae) population dynamics and transmission of *Borrelia burgdorferi*. J Med Entomol 1997;34:461–84.

[56] Mount GA, Haile DG, Daniels E. Simulation of management strategies for the Blacklegged tick (Acari: Ixodidae) and the Lyme disease spirochete, *Borrelia burgdorferi*. J Med Entomol 1997;90:672–83.

[57] Sandberg S, Awerbuch TE. Spielman A. A comprehensive multiple matrix model representing the life cycle of the tick that transmits the age of Lyme disease. J Theor Biol 1992;157:203–20.

[58] Bunnell JE, Price SD, Das A, Shields TM, Glass GE. Geographic information systems and spatial analysis of adult ixodes scapularis (Acari: Ixodidae) in the middle atlantic region of the U.S.A. J Med Entomol 2003;40:570–6.

[59] Das A, Lele SR, Glass GE, Shields T, Petz J. Modelling a discrete spatial response using generalized linear mixed models: application to Lyme disease vectors. Int J Geogr Inf Sci 2002;16:151–66.

[60] Randolph S. Epidemiological uses of a population model for the tick Rhipicephalus appendiculatus. Trop Med Int Health 1999;4:A34–42.

[61] Ding W. Optimal control on hybrid ODE systems with application to a tick disease model. Math Biosci Eng 2007;4:633–59.

[62] Ghosh M, Pugliese A. Seasonal population dynamics of ticks, and its influence on infection transmission: a semi-discrete approach. Bull Math Biol 2004;66:1659–84.

[63] Grimm V, Berger U, Bastiansen F, et al. A standard protocol for describing individual-based and agent-based models. Ecol Model 2006;198:115–26.

[64] Grimm V, Railsbeck SF. Individual-based modeling and ecology. Princeton University Press; 2005.

[65] Volker Grimm, Uta Berger, DeAngelis Donald L, Gary Polhill J, Giske Jarl, Railsback Steven F, et al. The ODD protocol: a review and first update. Ecol Model 2010;221:2760–8.

[66] Goodhill GJ. Mathematical guidance for axons. Trends Neurosci 1998;21: 226–31.

[67] Krottie J, Ooyen AA. Mathematical framework for modeling axon guidance. Bull Math Biol 2007;69:3–31.

[68] Graham BP, van Ooyen A. Mathematical modeling and numerical simulation of the morphological development of neurons. BMC Neurosci 2006;7:1–12.

[69] Friesen MR, Laskowksi M, Demianyk B, McLeod RD. Works in progress—developing educaitonal opportunities in agent-based modeling in Frontiers in Education Conferiece (FIE); 2010 IEEE:T2F-1—T2F-3; 2010.

[70] Bloom BS (Ed.), Taxonomy of educational objectives Book 1: cognitive domain, 2 ed. Addison Wesley; 1984.

[71] Huang C-Y, Tsai Y-S. Wen T-H. Simulations for epidemiology and public health education. Journal of Simulation 2010;4.

[72] Wilensky U. Netlogo Moths Model   http://ccl.northwestern.edu/netlogo/models/Moths, Center for connected learning and computer-based modeling, Northwestern University, Evanston, IL; 2005.

[73] Wilensky U. Netlogo Sunflower Model   http://ccl.northwestern.edu/netlogo/models/Sunflower, Center for connected learning and computer-based modeling, Northwestern University, Evanston, IL; 2003.

[74] Wilensky U. Netlogo Disease Solo Model http://ccl.northwestern.edu/netlogo/models/DiseaseSolo, Center for Connected Learning and Computer-Based Modeling, Northwestern University, Evanston, IL; 2005.

[75] Evans TA, Bashaw GJ. Axon guidance at the midline: Of mice and flies. Curr Opin NeuroSci 2010;20:79–85.

Agent-Based Models and Optimal Control in Biology: A Discrete Approach

Reinhard Laubenbacher*, Franziska Hinkelmann† and Matt Oremland*

*Virginia Bioinformatics Institute, Virginia Tech, Blacksburg, VA, USA
†Mathematical Biosciences Institute, The Ohio State University, Columbus, OH, USA

5.1 INTRODUCTION

The need to control complex systems, both engineered and natural, pervades our lives. From the thermostat that controls the temperature in our homes to the software that controls flight characteristics of the space shuttle during landing, the vast majority of engineered systems have built-in control mechanisms. Being able to control certain biological systems is no less important. For instance, we control ecosystems for agriculture and wildlife management; we control different parts of the human body to treat and cure diseases such as hypertension, cancer, or heart disease. And we control microbes for the efficient production of a vast array of biomaterials. With control comes the requirement to carry it out in a fashion that is optimal with respect to a given objective. For instance, we want to devise a schedule for administering radiation therapy to cancer patients in a way that maximizes the number of cancer cells killed while minimizing side effects. We apply pesticides to fields in a way that minimizes environmental damage. And we aim to control the metabolism of engineered strains of microbes so they produce the maximal amount of biofuel. Thus, the need for *optimal control* is a problem we face everywhere. This chapter will focus on optimal control of biological systems.

The most common approach to optimal control is through the use of mathematical models, often consisting of one or more (ordinary or partial) differential equations. These equations model key features of the system to be controlled and include one or more variables that represent control inputs. The following example illustrates this approach; it is taken from [1], where more details can be found. The problem we want to focus on is the optimization of cancer chemotherapy, taking into account certain immunological activity. The two relevant variables are x, which represents the volume of the tumor to be treated, and y, which quantifies the density of so-called immunocompetent cells, capturing various types of T-cells activated during the immune reaction to the tumor. These two variables are governed by the system of ordinary differential equations,

$$dx/dt = \mu_C F(x) - \gamma xy,$$
$$dy/dt = \mu_I (x - \beta x^2)y - \delta y + \alpha.$$

Mathematical Concepts and Methods in Modern Biology. http://dx.doi.org/10.1016/B978-0-12-415780-4.00005-3

Here, F is a function that represents the carrying capacity of the tumor, a concept adapted from ecology, in that it represents that ability of the environment to sustain the tumor. The Greek letters are constant parameters of the model. For instance, δ represents the rate of natural death of T-cells. For our purposes the exact model structure is not important, and details can be found in [1]. For instance, the term $-\beta x^2$ represents the observation that tumors suppress the activity of the immune system.

The control objective here is to reduce the tumor volume through the use of a chemotherapeutic agent. We will refer to this as the "treatment." We implement this control with a term $-xu$, where $u = u(t)$ represents the control input, namely the amount of the chemotherapeutic agent administered. The factor x takes account of the assumption that the effect of chemotherapy on the tumor is proportional to the tumor volume. Thus, the first equation becomes

$$dx/dt = \mu_C F(x) - \gamma xy - \kappa xu.$$

Here, $F(x)$ represents the carrying capacity of the tumor, as above. The parameter μ_C captures the growth rate of the tumor, with dimension *cells/day*. The parameter γ denotes the rate at which cancer cells are eliminated as a result of activity by T-cells, so that the term γxy in the equation models the beneficial effect of the immune reaction on the tumor volume. Finally, we assume that the elimination terms are proportional to the tumor volume. We therefore subtract a term κxu in the x-equation.

What kind of drug regime is appropriate for a given cancer patient depends in part on the particular state of this patient's disease. Our goal now is to optimize the influence of the control variable u, that is, the treatment, with respect to several factors. On the one hand we want to shrink the tumor through the administration of a maximal dose of the agent, and on the other hand we need to take into account the toxic side effects of the medication. Also, the treatment should be as short as possible. All this can be combined into the cost function

$$J(u) = ax(T) - by(T) + \int_0^T (cu(t) + d)dt.$$

This equation represents the cost of applying treatment schedule u for a length of time T. The optimal control problem now is to find a control u that minimizes this cost function. Solving such control problems is an active area of research, and algorithms have been developed to find u [2]. See also [3] for optimal control problems in systems biology.

Modeling biological systems through differential equations has its limitations, however. In many cases, the processes involved might be fundamentally discrete rather than continuous. For instance, in the case of a predator–prey relationship between two species inside an ecosystem, both populations are comprised of discrete individuals that engage in typically binary discrete interactions. Thus, it is not immediately clear whether one can apply methods such as differential equations, which assume that the quantities modeled vary continuously. In molecular biology, when we study regulatory relationships between genes inside a cell, these relationships are based on the interactions of discrete molecules. Modeling such systems using differential equations is based on two assumptions: first, there are many individuals involved, so that we can view them collectively as a continuous quantity; second, that we are able to describe the individual interactions in a "global" manner, as a term in a set of differential equations, usually involving one or more global parameters. Sometimes, both of these assumptions are justified, such as for large populations of bacteria or large quantities of chemicals in a fermenter. But at other times, one or the other, or both, of these assumptions fail. For example, in a cell, there might only be two or three molecules of a particular protein present at any given time, so that its role in a regulatory network becomes discrete and stochastic, and cannot be accurately modeled with continuous functions. And, as another example, in an ecosystem with several species, continuous models sometimes cannot accurately capture extinction events when one species reaches a very low count.

For these reasons, and others, several other frameworks have been used to model biological systems. One of these consists of so-called "agent-based," also called "individual-based" models. These types of model have a long history, going back to the 1940s and the work of John von Neumann [4].

Von Neumann was interested in the process of self-replication and aimed to construct a machine that could faithfully replicate itself. A theoretical instantiation of such a machine turned into the concept of a *cellular automaton* as a computational model. A very well-known example of a cellular automaton is John Conway's *Game of Life* [5]. Since it includes many of the basic concepts of agent-based models, we describe it briefly.

The *Game of Life* takes place on a chess-board-like 2-dimensional grid. This grid can either be finite or extend infinitely in all directions, thereby yielding two different versions of the game. Each square, or *cell*, on the grid can be either black or white. Since the *Game of Life* is intended to simulate life, a cell is instead referred to as either alive (black) or dead (white). Each cell away from the boundary of the grid has eight neighbors on the grid that physically touch it, four with which it shares an edge, and four that touch only on the corners. Cells on the boundary have fewer neighbors. To make all cells uniform, one can make the assumption that a cell at an edge of the grid, but away from the corner, has as additional neighbors the corresponding cells on the opposite edge of the grid. Thus, we effectively "glue" opposite edges together, so that the grid is situated on a torus rather than in the plane. Now that all cells have eight neighbors, we establish a rule that determines the evolution of the cells on the grid by determining what the status of a given cell is, depending on the status of its eight neighbors. The rule is quite simple.

- Any live cell with fewer than two live neighbors dies;
- Any live cell with two or three live neighbors stays alive;
- Any live cell with more than three live neighbors dies;
- Any dead cell with three live neighbors comes alive.

Thus, the rules are reminiscent of a population whose survival is affected by under- and overpopulation. If we now initialize this "Game" by assigning a "live" or "dead" status to each of the cells, then we can use this rule to evolve life on this grid by updating the status of all the cells in discrete time steps. The result is a vast array of different dynamics that can be observed, but is largely unpredictable from a particular initialization. There is a rich literature on this topic and many websites with *Game of Life* simulators. Before continuing, the reader is encouraged to try some of these and explore the question of predicting dynamic behavior from particular initializations.

General agent-based models have many features similar to this set-up. There is a collection of agents that are distributed across a spatial landscape. In the *Game of Life*, the spatial landscape is the grid, and there is one agent per cell. Each agent can be in one of a (typically finite) number of states, such as "dead" or "alive," and has a set of rules attached to it that it uses to determine its state, based on the state of those other agents it interacts with at any given time. Beyond the Game of Life, in general agent-based models, agents are able to move around in the spatial landscape, and there are rules that determine the agents' movement patterns. Typically, the rules are stochastic, rather than deterministic, and are governed by a collection of probabilities. For instance, agents might be predisposed to follow a certain gradient, representing, e.g., nutrient availability, but there is some chance that they might move in a different direction. While there are many other variations and features in particular agent-based models, these few basic features characterize the agent-based modeling framework.

These features are also sufficient to explain the basic differences between agent-based models and equation-based models, such as ordinary differential equations. Agent-based models lend themselves very well to a description of dynamical systems that arise from local interactions of many parts/agents, based only on local rules rather than on the configuration of the entire system at any given time. Also, it is very easy to represent a rich heterogeneous spatial environment that the agents navigate. Thus, the dynamics of the entire system, or its so-called global dynamics, "emerge" from these local interactions by applying the local rules repeatedly. In contrast, a system of differential equations, for instance, explicitly describes the global dynamics of the system up front. Furthermore, all the specifications for an agent-based model are intuitive, in the sense that they are direct computational representations of recognizable features in the actual system. This leads to models that are more faithful to the system to be modeled and that are more accessible to domain experts. With existing software for model building, they can even be built by domain experts directly, without the intervention of a modeler or mathematician. But, as always, there are no free lunches. These advantages come with some significant costs attached.

The reason that agent-based modeling became widespread only in the last 15 years or so is due to the fact that larger models with more features require extensive computational resources that were not available until recently. It is only now possible, barely, to build and analyze models that might have hundreds of millions of agents and tens of millions of spatial locations, with agents being very complex in terms of the states they can take on and the rules they follow. Even moderately complex models require high performance computing facilities for their analysis, which makes it difficult for individual researchers to use them. High computational complexity is compounded by the fact that there are very few mathematical tools available for the analysis of agent-based models. In essence, the only approach to model analysis is model simulation. That is, from given initializations of agent states and environmental parameters, one observes the time evolution of the system and hopes to observe patterns that might help one draw conclusions about, e.g., steady states of the model. Through choosing many initializations and doing many simulation runs from a given initialization in the case of a stochastic model, one aims to obtain global dynamic information about the model. Little else can be done because, in essence, the model consists of a computer program rather than a set of mathematical equations, and there are few things one can do with a computer program other than run it.

The lack of mathematical analysis tools extends in particular to the arena of optimal control. Existing approaches are heuristic in nature, based on domain knowledge. The goal of this chapter is to describe some of these approaches and to outline steps one can take to expand the repertoire of available tools to include techniques based on mathematics. The way to do this is to translate an agent-based model into a mathematical object, such as a system of equations of some sort, that makes it amenable to mathematical analysis tools. There are several possible ways in which to do this, and research is only in its early stages. Thus, the reader should see this chapter as a snapshot of an exciting research area that is evolving rapidly and providing rich opportunities for contributions at all levels. This chapter showcases one possible approach and the steps that have been accomplished on the road to developing mathematical tools for optimal control of agent-based models.

5.2 A FIRST EXAMPLE

Go to http://ccl.northwestern.edu/netlogo/models/RabbitsGrassWeeds. There you will find an agent-based model of rabbits in a field of grass and weeds. At each time step (or "tick") the rabbits move in a random direction (they lose energy by moving). If there is grass at their location, they eat it and gain energy. If their energy level climbs above a certain threshold, they give birth (in this model, a new rabbit is spontaneously created at the location of the parent). Upon birth, the parents' energy is halved, and the offspring is created with this halved energy level. Upon each tick, empty squares (or "patches") have a certain fixed probability of grass re-generating. Weeds can also be introduced in order to further complicate the dynamics; their behavior is similar to that of the grass.

Near the top of the page you will find an option that allows you to run this model in your web browser. Spend some time reading through the description in order to get a feel for how this model works. Click setup and then go to run the simulation (press go again to stop the simulation). Note that you can speed up or slow down the model by using the slider at the top of the interface. Each time you wish to re-initialize the model and start over, you must click setup again.

In this model, each of the grid squares (henceforth referred to as *patches*) are agents, and each rabbit is an agent as well. Notice that the rabbits move around randomly, eating grass as they encounter it. Note too that the patches are colored green when grass is present, and black when it is absent. Notice the sliders on the left-hand side: these values can be changed as the simulation proceeds, or prior to starting a simulation. What do you expect to happen to the rabbit population if the grass growth is set to a very low rate? What if the grass grows quickly? What effect would altering the birth threshold have on the amount of grass present at each time step? Do you think it is possible for the rabbit population and grass level to stabilize over time? Play around with the sliders to determine if your predictions are correct. Note that you may need to let the simulation run for several hundred time steps (or "ticks" in Netlogo) in order to observe consistent dynamics. Do the populations stabilize? Do they oscillate?

5.3 NETLOGO: AN INTRODUCTION

Hopefully, the example in Section 5.2 has given you an idea of what an agent-based model is: a computer simulation wherein agents interact with their local environment (possibly including other agents) based on a set of rules. In this section we guide you through several exercises involving agent-based models using Netlogo [6], a software tool developed to work with agent-based models. Netlogo can be downloaded for free at http://ccl.northwestern.edu/netlogo/. There you will find detailed instructions on how to install it. Netlogo is its own programming language, so named because it is a variation of the Logo language. While many key features are unique to Netlogo, users familiar with Logo will likely note similarities. Note too that Netlogo is continually updated and newer versions released. As of the time of this writing, the latest version is Netlogo 5.0. All files and exercises associated with this chapter will be conducted using this version; some adjustment may be necessary for newer (or older) versions. While there are many other software platforms available and in use for agent-based modeling, for the remainder of this chapter we will use Netlogo to introduce and examine agent-based models. It has the convenient advantage of providing the user with an intuitive graphical interface, which we will use to aid our understanding of the models we will examine. In addition, the standard installation comes with a library of models, all of which are open source. Thus we may build models from scratch or we may choose to alter an existing model. You may wish to go through the tutorials found at http://ccl.northwestern.edu/netlogo/docs/. A complete dictionary of programming terms can be found at http://ccl.northwestern.edu/netlogo/docs/dictionary.html. All exercises in this section refer to the "Rabbits Grass Weeds" model introduced in Section 5.2. It should be noted that for Exercises 5.1 and 5.2, the web-based version of the model can be used (via http://ccl.northwestern.edu/netlogo/models/RabbitsGrass Weeds), but Exercises 5.3 and 5.4 require installation of the Netlogo software.

Exercise 5.1. Note the two sliders `weeds-grow-rate` and `weed-energy`. By setting `weeds-grow-rate` to be any nonzero value, you will notice that the landscape of the model has been altered, as there are now patches containing weeds interspersed with the grass (represented as purple patches). What effect does this have on the rabbit population? How about on the grass population? What happens when you increase `weed-energy`? Set `grass-grow-rate` and `weeds-grow-rate` to be equal, and set `grass-energy` and `weed-energy` to be equal as well. What do you think will happen to the population levels now? Determine slider values so that the following situations occur:

a. The grass and weeds stay at (roughly) the same level (as shown in the plot window).
b. The weed levels (as shown in the plot window) are higher than the rabbit levels and the rabbits die out.
c. The weed levels (as shown in the plot window) are higher than the rabbit levels and the rabbits do not die out.   ▽

Exercise 5.2. Set `weed-grow-rate` and `weed-energy-level` to 0, and `grass-grow-rate` and `grass-energy` to 5 if they are not already at those levels. Set `birth-threshold` to 15, and press `setup` and `go` to restart the simulation. As the simulation runs, gradually decrease `birth-threshold` to 10, and then to 5. You will notice in the plot window that the rabbit population oscillates at a higher amplitude as you decrease `birth-threshold`. Why does this happen? As you further decrease `birth-threshold` to 2 or 1, the rabbits die out. Why does this happen?   ▽

Exercise 5.3. Right-click on the graphical interface and select `Edit...`. Notice that the boxes labeled 'World wraps horizontally' and 'World wraps vertically' are checked. This means that when a rabbit moves left from a leftmost patch, it will reappear on the corresponding right-most patch (and similarly if the rabbit moves vertically from a top-most patch). If you uncheck those boxes, the rabbits will be bound by the edges of the map (so we can think of them as being "fenced in"). What effect does checking these boxes have on the rabbits? What effect does it have on the grass?   ▽

Exercise 5.4. By altering the code in the "Code" tab, change the patches in the model so that there is a river three patches wide that separates the field into two halves (this can be done by altering the code in the `grow-grass-and-weeds` function). Ensure that grass and weeds cannot grow in the river. Alter the `setup` function so that rabbits cannot begin in the river, and alter the `death` function so that a rabbit dies if it goes into the river. What effect does this have on the rabbit population over time?   ▽

5.4 AN INTRODUCTION TO AGENT-BASED MODELS

Mathematical modeling is a method of encoding relationships and interactions in a natural or engineered system into a formalized system; the models can then be studied and analyzed using a variety of mathematical approaches. Models allow researchers from a wide variety of disciplines to examine systems and their emergent behavior. For example, a model may be used in order to make predictions about the future behavior of a system, or it may be used to solve a complicated problem explicitly.

In developing the best drug to administer to treat a particular disease, researchers in the past have relied on trial-and-error methods via repeated experimentation on live subjects. With an accurate model of the subject, however, such experiments can often be faithfully reproduced *in silico*; that is, they can be run on a computer and analyzed with far less preparation and expense. This highlights another benefit to mathematical modeling: given an accurate model of the system in question, a year's worth of real time can be simulated in several minutes.

Experimentation remains part of the process - a model will always have a limited ability to predict and must be correlated against empirical data in order to ensure that the models are indeed faithful simulations of the actual physical system. Thus, models are best used to inform *in vivo* experimentation, which in turn produces results that can be used to calibrate the model further.

Agent-based models are a class of computer models in which entities (referred to as agents) interact with each other and or their local environment. Formally:

Definition 5.1 (Agent-based model). A computer model that consists of a collection of agents/variables that can take on a typically finite collection of states. The state of an agent at a given point in time is determined through a collection of rules that describe the agent's interaction with other agents. These rules may be deterministic or stochastic. The agent's state depends on the agent's previous state and the state of a collection of other agents with whom it interacts. [7]

Systems (such as the human immune system) are increasingly being implemented in the form of agent-based models (with individual cells as the agents, of which there may be many types) as more and more research involves the use of *in silico* simulation to study the properties of this and other similarly complex systems.

Agent-based models have the advantage of being well suited for modeling many different types of systems. They have been used to study social interactions among individuals, the spread of disease through populations, scheduling and efficiency of factory processes, how cells react to drug treatments, and many other systems. It is perhaps worth noting that many of the current issues in the scientific community are interdisciplinary. Finding a cure for cancer will involve geneticists, biologists, mathematicians, chemists, and perhaps many other specialists. Agent-based models have an intuitive formulation and can often be examined via a graphical interface. Thus they are a natural tool for promoting interdisciplinary research, as the mathematics underlying the models is hidden in the programming; in other words, it is possible for biologists, chemists, and other researchers to make use of agent-based models without a full background in the mathematics that are involved in creation and analysis of the model. At the same time, the mathematical structure remains and can be explored concurrently by mathematicians.

Even within the scope of mathematical modeling, there are several noteworthy advantages to agent-based modeling. One such advantage is that they are effective for modeling systems wherein many agents follow the same set of rules (e.g., rabbit populations or blood cell types). In particular, models containing spatial heterogeneity can be effectively represented via agent-based models while perhaps not so easily using, for example, ordinary or partial differential equations (ODEs). Altering the spatial dynamics of an agent-based model consists of changing several lines of code or less, while such changes can be difficult (and perhaps even impossible) to implement in an ODE model.

While agent-based models provide a convenient and natural setting for studying complex systems, there are several issues within the research community that are currently unresolved. A 2006 paper [8] written by a large group of researchers identified two main obstacles: the first is the lack of standardization of the description of agent-based models, and the second is a lack of rigorous mathematical formulation of the system itself. Descriptions of agent-based models can vary in different settings, and as of this writing there is no standard definition that is universally agreed upon. Some models are developed to simulate physical processes, and others are developed in the framework of graph theory. In some cases models are developed in order to study only a certain aspect of the given system; thus the model may have more variables pertaining to certain processes than others. A research article may begin with the statement of an objective, or it may begin with a description of the model itself. In some cases, the rules that govern the updating of the agents' state variables are deterministic, and in other cases they are stochastic. All of these issues would be resolved if there was a standard protocol for describing agent-based models (indeed, one such protocol is proposed in [8]). The need for an agreed-upon structure to be followed is perhaps most clearly felt in the model's presentation. In particular, the layout of the model presentation ought to be standardized so that a reader immediately knows where to look in the description in order to learn what the model describes and what its rules are. In the current literature models are presented in myriad forms, and the descriptions of the agents, environments, and rules come in no particular order. Thus a reader is required to scour the paper for pertinent details that might otherwise be presented in some standard way.

The second major issue concerns the lack of rigor in the formulation of the model. In many cases, the description of a model is given in several paragraphs, describing in some imprecise manner what the agents are and how they interact with each other. In fact, in order for agent-based models to be implemented on computer software, there are intricate rules embedded in the programming of the models. These rules and equations are often glossed over if not entirely omitted; if they are given, it is typically through a verbal description rather than a strict mathematical formulation. Looking back on the example in Section 5.2, the precise way in which the rabbits move is not described. For example, it is unclear from a simple description whether the rabbits can move diagonally, or whether the distance they move at each time step is variable or constant. In fact, even running the model does not immediately answer this question—it is only made explicitly clear by thorough examination of the computer code. In order to describe such a model rigorously, it is necessary that this information (and other similarly imprecise descriptions) be presented clearly and unambiguously.

In addition to these issues, an author may spend several paragraphs explaining how the model was formulated that could just have easily been given in one or two equations. Short and precise definitions can save time for the reader and also make the author's intentions explicit. One proposition to overcome this lack of formalization is presented in [9]; this work proposes a rigorous mathematical representation for agent-based models as polynomial dynamical systems. A further description of such systems can be found in [10] and is described in Section 5.8.

5.5 OPTIMIZATION AND OPTIMAL CONTROL

Agent-based models are typically created to simulate some real-world process in order to aid investigation. Once a model has been created, a natural next step is to ask: what are we to do with it? In this section we give a brief overview of optimal control and optimization, and introduce these ideas as they apply to agent-based models.

Optimization is the process of finding the best solution with respect to a particular goal. For example, suppose we have a model of the immune system battling a bacterial infection, and we wish to study the effects of certain drugs on this battle. We may wish to find out which drug does the least amount of tissue damage while curing the infection—this is an optimization problem, because we are searching for the best drug with respect to the stated goal. On the other hand, perhaps our goal is to cure the infection in the shortest time possible, regardless of the tissue damage caused. This would be another optimization problem. It is likely, then, that the solution to the optimization problem depends on the optimization goal. In this scenario, the drug which cures the infection most quickly may consequently do more tissue damage than other drugs (though not necessarily); on the other hand, the drug that causes the least tissue damage might not cure the infection in the shortest possible time. In this example, optimization is a process of minimization: in one case we wish to minimize tissue damage and in the other we wish to minimize the healing time. However, it is also possible that the solution of an optimization problem is a process of maximization: given a model of an immune system fighting a fatal illness, what treatment will enable the patient to survive the longest? In fact, it may be that the goal of the optimization problem is to minimize one value while maximizing another. For example, our optimization goal may be to minimize tissue damage and maximize the expected lifespan of the patient, given that we can only administer a particular drug a fixed number of times. In general, however, optimization is the process of finding the best solution, depending on the objective.

Typically, optimization is a complicated process, and becomes even more so when realistic constraints are put in place. Through the use of agent-based models it may be possible to obtain a solution to an optimization problem that is not feasible in actuality: for example, the solution may exceed monetary limitations, may require actions that are not permitted by health care regulations, or may require interaction with the patient in an impractical way (e.g., if the solution calls for treatment every hour for 100 consecutive hours, or for 100% of the population to receive vaccination). Nevertheless, optimization remains a natural means of applying mathematics to solve real-world problems. Such problems are often framed as questions of optimization: what is the best outcome that can be produced based on the properties of the model?

Optimal control is a slightly different notion. In an optimal control problem, we ask the following question: given a particular state that we hope to reach, what is the most efficient way of reaching that state? In other words, we know the state of the system that we hope to reach, and the control problem is to find the solution that steers the system to that state in the best manner. Using a model of the immune system fighting an infection as an example, we may formulate an optimal control problem as follows: given that we wish to eliminate the number of infected cells in one month, what is the best drug treatment schedule we can devise? To summarize: in an optimal control problem we know the state we hope to reach and we search for the best way to reach it, whereas in an optimization problem, a goal is stated and we compare solutions to see which solution maximizes or minimizes the stated goal of the problem.

We now present an example explaining several terms related to optimization and optimal control that will be used in Sections 5.6 and 5.7. We then conclude this section with definitions of said terms. We do this in order to standardize and clarify the terminology within this chapter. Note that while many of these terms are also used in optimal control for continuous systems, there might be subtle differences in their meanings when applied to agent-based models. Furthermore, as this chapter is meant to provide an introduction to the topic, more formal definitions are outside the scope of this text. For a more formal treatment see [11].

Example 5.1.   Suppose that we are modeling lung cancer and we wish to study the effect of a certain drug. We formulate our optimal control problem as follows: which treatment schedule should we choose in order to reduce the number of cancer cells so that it remains below some fixed threshold over the course of one year, given that we wish to minimize the number of times the drug is administered and maximize the number of healthy cells? Note here that we use the term "treatment schedule" instead of "treatment" because the drug we are administering is fixed—we want to determine which days the drug should be administered in order to obtain optimal results.

In this case, there will be many variables: the number of healthy cells, the number of cancer cells, the rate at which cancer cells grow, the rate at which healthy cells regenerate, the expected lifespan of the patient, the frequency with which the drug is administered, the type of drug, and so on. Some of these variables will have fixed values: for example, the rate at which healthy cells regenerate (during intervals when no treatment is administered) can be determined through experimental measurements, and in fact this rate helps to define the model itself. Such variables are referred to as *model parameters*—they are a part of the specification of the model. The repeated interactions of the entities in the model, such as immune cells or rabbits, are referred to as the *model dynamics*. Note that we will only have direct control over some of the variables, and others we will simply measure by observation. For example, during each day of the simulated treatment we can decide whether or not to administer the drug; thus we have direct control over the value of this variable. We refer to these as *control variables*, because we have direct control over their values and they exercise control over certain aspects of the model. On the other hand, we might not be able to control over other aspects, such as the number of white blood cells present at the site of an infection. We refer to variables of this type as *state variables* because they help to describe the state of the system at any given time.

In this example, our only control variable is a binary decision whether or not to administer the drug on a given day, thus there are only two possible values for this control variable at each time step. For any given day, we may represent that the drug was administered that day by assigning that variable the value 1, and represent that the drug was not administered by assigning that variable the value 0. In this situation, then, a treatment schedule is a vector of length 365, of which each entry is either 0 or 1. Thus there are a total of 2^{365} possible treatment schedules. However, it is likely that not all possible treatment schedules will reduce the number of infected cells below the fixed level. The treatment schedules that *do* achieve this are *solutions* to the optimal control problem. The *solution space* is the set of all such solutions. Note that this space is entirely separate from the *state space*, which consists of all possible states a patient may be in throughout any simulation of the model.

Recall that we are trying to find the best solution among many candidates. In order to do this, we must have some way of ranking individual solutions. We do this by introducing a *cost function*. We are attempting to steer the system to a particular state, and each solution achieves that goal at a certain cost. We wish to determine the best solution (i.e., the optimal control); thus we wish to find the solution that minimizes the cost function. Earlier we noted that in the immune system example the best solution was the solution that involved the least number of treatment days throughout the treatment schedule; thus, in this example a reasonable cost function might involve the sum of the entries in the treatment schedule (i.e., the number of days on which the drug was administered). For the purposes of this chapter, the goal of the optimization and the optimal control problems will always be to **minimize** the associated cost function.

In our example, we wish to minimize T, the total number of days the drug is taken, and maximize I, the number of healthy cells. Then a reasonable cost function for a treatment schedule S which contains S_t days on which the drug is administered and results in an expected value of S_i healthy cells might be $c(S) = S_t - S_i$. Here we want to maximize S_i over all treatment schedules S; this is the same as minimizing $-S_i$ (in this way we can always formulate our problem so that the goal is to minimize the cost function). However, it is perhaps more realistic that one of our optimization goals has a higher priority than the other. For example, we may suppose that maximizing the number of healthy cells is more important than minimizing the number of treatment days. In that case, we introduce weighting coefficients to alter the cost function. For example, it may become $c(S) = S_t - 5 S_i$. This has the effect of increasing the relative importance of the healthy cell count when evaluating the associated cost of a particular treatment schedule.

In order to use our model to obtain results, we simulate the model a number of times and make observations on the results. We might, for example, administer a treatment every day. This particular solution is represented as a string of 365 consecutive 1's; we implement this in the agent-based model and count how many healthy cells are there after one year of simulated time. We then evaluate the cost function based on this information. Given that agent-based models are often stochastic in nature (meaning that while the rules are fixed, the model dynamics depend on many random variables), results obtained from one simulation are typically not reliable. We eliminate such variation by simulating a given control schedule many times and using averages for evaluation in the cost function.

Of all the possible 2^{365} solutions, there must be one that is better (or at least no worse) than every other solution. Strictly speaking, such a solution is referred to as an optimal solution. However, in many cases we cannot simulate all possible solutions because of the monumental amount of computing time involved. Thus, for the purposes of this chapter, we refer to the best solution *we are able to find* as the *optimal solution*, with the caveat that there may indeed be a better solution elsewhere in the solution space.

We now define the following terms, each of which have been illustrated in this example.

Definition 5.2 (Terms).

- *Model parameters:* quantities that are part of the model specification. They have fixed values.
- *Model dynamics:* the relationships between the state variables in a model. In general, the model dynamics will be affected by the model parameters and the rules that govern the interaction of the variables.
- *Control variable:* a variable whose value can be specified by the user. Altering the value of a control variable will (in general) have some effect on the resulting model dynamics.
- *State variable:* a variable whose value is observed but cannot be directly specified by the user (i.e., not a control variable). State variable values affect the model dynamics; they are affected by the value of other state variables, model parameters, and control variable values.
- *Solution:* a sequence of inputs to the control variables. A full solution assigns a value to each control variable at each time step; a partial solution is a sequence wherein values are either only assigned to some of the control variables, or are assigned to the control variables at only certain time steps.
- *Solution space:* the set of all possible solutions. If $p_1, p_2, \ldots, p_n$ are the control variables and each parameter p_i (for $1 \leqslant i \leqslant n$) can take on a_i possible values, and there are a total of t time steps, then the solution space will consist of

$$\prod_{i=1}^{n} a_i^t = a_1^t \cdot a_2^t \cdot \ldots \cdot a_n^t = (a_1 a_2 \cdots a_n)^t$$

  solutions (thus each solution is a vector of length t, representing the sequence of inputs to the control variables).
- *Population:* a subset of the solution space. The population may be the entire solution space or a proper subset.

- *Cost function:* a function that assigns a value to each possible solution. In general, the cost function will depend upon a combination of state variables and other quantifiable aspects of the model dynamics.
- *Optimal solution:* in a general optimization problem the optimal solution is a solution that achieves the stated optimization goal in the best way. In an optimal control problem, it is the solution that is associated with the minimum value of the cost function.

Use the modified rabbit and grass model file RabbitsGrass.nlogo at http://admg.vbi.vt.edu/software/rabbitsgrass-netlogo-files/ to answer the questions in this section. Once you have opened the file, you will find a description of the interface settings under the Info tab.

Exercise 5.5.   On the left-hand side of the model, you will notice sliders for the following values: initial-rabbits, initial-grass, birth-threshold, grass-grow-rate, grass-energy, move-cost, world-size, poison-strength. Are these model parameters or state variables? Explain.   ▽

Exercise 5.6.   List three state variables and one control variable for this model. ▽

Exercise 5.7.   Suppose the control objective is to minimize the number of rabbits while also minimizing the number of days on which poison is used. What is the difference between a solution and the *optimal* solution?   ▽

Exercise 5.8.   Click restore-defaults, set poison-strength to 0.5, and then click setup. Now, click poison repeatedly until you have run through the entire simulation time. Note the cost. Now click degrade-poison so that it is in the On position. Click setup and then repeatedly click poison again. What do you notice about the cost this time? Is it higher or lower? How can you explain the difference?   ▽

Exercise 5.9.   Click restore-defaults, set poison-strength to 0.5, and click setup. Using the buttons poison and don't poison as you wish, run through the simulation several times, taking note of the cost for each solution. What is the lowest cost you are able to achieve? If you now reduce poison-cost to 5 and increase rabbit-cost to 15, you will notice that the cost when using the same schedule is now different. Is it lower or higher? Why do you think this is? Try to find a schedule that reduces the cost to less than 10,000. How about less than 5000? Turn degrade-poison? to the On position and try to achieve similarly low costs. Do you notice any patterns in the solutions in each case?   ▽

Exercise 5.10.   Run the same schedule twice (for 1 run each time). You should notice that the costs are different each time. Why does this happen? Explain why this is an important issue, and suggest a method for dealing with this issue.   ▽

Exercise 5.11.   Click restore-defaults, then setup. Change num-runs to 100, then click go. Note the average cost over these 100 runs. Now change the number of runs to 25, then 3 (note that unchecking the view updates box toward the top of the model interface will increase run time significantly). What do you notice about the cost? Is it stabilizing? What is the least number of runs required in order to achieve a reliable cost? How would you justify your answer mathematically?   ▽

Exercise 5.12.   What effect does altering `poison-cost` have on the cost of a fixed schedule? What effect does `rabbit-cost` have? What do you think are reasonable choices for these values if this were an actual field containing rabbits and grass? Justify your answer.   ▽

5.6 SCALING AND AGGREGATION

Our investigation of agent-based models and how to formulate control problems for them motivates this section and the next. Searching for an optimal solution in such models often requires running many thousands of simulations, thus performing such simulations as quickly as possible is a primary concern. In this section, we discuss means of reducing the run time and complexity of agent-based models via scaling and aggregation.

Scaling is a method of shrinking the size of an agent-based model in order to improve run time. In Exercise 5.1, we explored the "Rabbits Grass Weeds" model. The dimensions of this model are 43×45 patches for a total of 1935 patches. The default number of rabbits is 150. In an attempt to scale this model, we may reduce the dimensions (and correspondingly, the initial number of rabbits). If we change the dimensions to 25×25 for a total of 625 patches, we may choose the initial number of rabbits to be $150 \times \frac{625}{1935} \approx 48$. Our hope, then, is that the pertinent model dynamics remain the same at this reduced size, since in that case we can run all subsequent trials at this reduced size for a substantial decrease in run time.

The first question we need to answer is exactly what it means for the model dynamics to remain the same, and how we can verify that this is indeed the case? Since we are using the model with a specific control objective in mind, and the value of a solution relies on the associated cost function, this cost function will help us determine how to quantify whether our scaled model retains the pertinent dynamics. In Section 5.5, we used the "Rabbits and Grass" model to determine control schedules for poisoning the rabbits. In that case, our cost function relied on two parameters: the rabbit population, and the number of days in which poison was used. In particular, the day-by-day grass levels are not relevant to the cost function, nor are the energy levels of the rabbits; thus these parameters need not be preserved in the scaled model. In the scaled model, the number of days in which poison is used is unchanged, so this variable is also unaffected by scaling. Then the only parameter we need to preserve is the rabbit population. To be more specific, we need to answer the following question: what are the smallest dimensions we can use in the model (by scaling down the number of rabbits accordingly) so that the average day-by-day rabbit population dynamics in the scaled model follow the same pattern as those dynamics in the original model? We answer this question by simulating many runs at each size and keeping track of the population levels for each. We can then calculate the correlation using any of a number of statistical methods; one such method is now presented.

5.6.1 Correlating Data Sets

Suppose we obtain two sets of data and wish to know how closely related they are. The absolute numbers may not tell the whole story, since the patterns in the data may remain similar even if the magnitude of the values change. Here we describe how to calculate *Pearson's sample correlation coefficient*, a real number $r \in [-1, 1]$, which estimates how closely correlated two data sets are. At $r = 1$, there is perfect correlation, such as between the ordered data sets $\{1, 2, 3\}$ and $\{2, 4, 6\}$. At $r = -1$, there is perfect negative correlation, meaning that as the data increase in one set, they decrease by precisely the same proportional amount in the second data set (this value is obtained for data sets $\{1, 2, 3\}$ and $\{-8, -9, -10\}$, for example). At $r = 0$, there is no connection between the two data sets at all (this value is obtained for data sets $\{1, 2, 3\}$ and $\{1, 2, 1\}$). Naturally, the larger the data sets the more informative the correlation coefficient. We now describe how to calculate r in general and then provide an example.

Definition 5.3 (Pearson's sample correlation coefficient). Let x, y be data sets consisting of n points (labeled sequentially as $x_1, \ldots, x_n$ and $y_1, \ldots, y_n$), and let $\bar{x}$ and $\bar{y}$ be the mean value of x and y respectively. Then Pearson's sample correlation coefficient r is defined as

$$r = \frac{\sum_{i=1}^{n} (x_i - \bar{x})(y_i - \bar{y})}{\sqrt{\sum_{i=1}^{n} (x_i - \bar{x})^2 \sum_{i=1}^{n} (y_i - \bar{y})^2}}.$$

Example 5.2. Let $x = \{150, 30, 40, 54, 72\}$ and $y = \{72, 18, 10, 30, 40\}$. Then $\bar{x} = 69.2$ and $\bar{y} = 34$. Then

$$
\begin{aligned}
r &= \frac{\sum_{i=1}^{5} (x_i - \bar{x})(y_i - \bar{y})}{\sqrt{\sum_{i=1}^{5} (x_i - \bar{x})^2 \sum_{i=1}^{5} (y_i - \bar{y})^2}} \\
&= \frac{(80.8 \cdot 38) + (-39.2 \cdot -16) + (-29.2 \cdot -24) + (-15.2 \cdot -4) + (2.8 \cdot 6)}{\sqrt{(80.8^2 + (-39.2)^2 + (-29.2)^2 + (-15.2)^2 + 2.8^2)(38^2 + (-16)^2 + (-24)^2 + (-4)^2 + 6^2)}} \\
&= \frac{4476}{\sqrt{9156.8 \cdot 2328}} \approx 0.969.
\end{aligned}
$$

Thus we see that these two sets of data are in fact very closely positively correlated, in keeping with the observation that the values in y are approximately one half of the corresponding values in x.

Since agent-based models are generally stochastic in nature, the data obtained will seldom present a perfect description of the system, because an infinite number of simulations would be required. Thus sample correlation coefficients of 1 or -1 are very highly unlikely. We may choose our desired correlation coefficient r, and when scaling the model we simply select the smallest dimensions that produce data whose correlation coefficient is at or above this level.

5.6.2 Cost Function Analysis When Scaling

Once we have determined how small we can safely scale our model, we must also be cautious about numerical results obtained from this model, as indicated in the following example.

Example 5.3.   Let r_i be the number of rabbits alive on day i, and let u_i be the control decision on day i (thus if we use poison then $u_i = 1$ and if not, $u_i = 0$). Let $\vec{u} = [u_1 u_2 \ldots u_n]$, where n is the total number of simulated days. Suppose our cost function is $c(\vec{u}) = a \cdot \sum_{i=1}^{n} r_i + b \cdot \sum_{i=1}^{n} u_i$, where $a, b \in \mathbb{R}$ are constants. Once we have scaled our model, we attempt to use it to determine the control schedule which minimizes this cost function. When doing so, we must be careful to scale the constants a, b as well, since otherwise, we may obtain meaningless or misleading results, as the following results suggest.

Example 5.4.   Let $a = 100, b = 2000, n = 5, \vec{u} = [01101], \vec{v} = [10000]$. Suppose the average rabbit numbers for $\vec{u}$ for the 5 simulated days are $\{150, 30, 40, 54, 72\}$ and by scaling, we are able to reduce the model size and achieve average rabbit numbers $\{75, 15, 20, 27, 36\}$. Using another control schedule $\vec{v}$ we obtain population levels $\{150, 200, 40, 8, 10\}$ (and the corresponding scaled population values are $\{75, 100, 20, 4, 5\}$).
Comparing the cost of the two schedules at the original size, we have:

$$c(\vec{u}) = 100(150 + 30 + 40 + 54 + 72) + 2000(0 + 1 + 1 + 0 + 1) = 40600,$$
$$c(\vec{v}) = 100(150 + 200 + 40 + 8 + 10) + 2000(1 + 0 + 0 + 0 + 0) = 42800.$$

Without scaling coefficients a, b, we would obtain the following costs using the scaled model:

$$c(\vec{u}) = 100(75 + 15 + 20 + 27 + 36) + 2000(0 + 1 + 1 + 0 + 1) = 23300,$$
$$c(\vec{v}) = 100(75 + 100 + 20 + 4 + 5) + 2000(1 + 0 + 0 + 0 + 0) = 22400.$$

From the original model, we conclude that $\vec{u}$ is a better solution than $\vec{v}$ (since the associated cost is lower), but from the reduced model we conclude that in fact $\vec{v}$ is better than $\vec{u}$. This is due to the fact that when scaling the population values, even though the dynamics correlated perfectly, the coefficients now give less weight to the rabbit values, since these numbers are smaller in magnitude. In order to compensate for this, we must account for the rabbit numbers being halved by *doubling a*. At the same time, we need not scale b because the number of days remains constant at all model sizes. Scaling a and b accordingly gives the following revised costs for the scaled model:

$$c(\vec{u}) = 200(75 + 15 + 20 + 27 + 36) + 2000(0 + 1 + 1 + 0 + 1) = 40600,$$
$$c(\vec{v}) = 200(75 + 100 + 20 + 4 + 5) + 2000(1 + 0 + 0 + 0 + 0) = 42800.$$

Here we obtain essentially the same cost as the original model, and thus eliminate the discrepancy in cost.

The "Rabbits and Grass" model lends itself nicely to scaling because the initial distribution and location of rabbits and grass is random. Thus, when reducing the dimensions, we may still choose which patches have grass at random, and likewise may randomly choose where to place our rabbits at the start of the simulation. But what about models that are not so spatially homogeneous? Suppose the landscape included a river, or a rocky area in the upper right-hand corner of the map? In such cases, we cannot simply reduce dimensions because the spatial layout of the model may be critical to the dynamics. Thus in general, a more sophisticated approach is required for spatially heterogeneous models.

Suppose that in the "Rabbits and Grass" model, the field consists of a hill whose peak is at the center of the field. Going out from the peak of the hill, the altitude decreases; thus at the periphery of the map, the land is flat. Suppose further that we now distinguish between various levels of grass: each patch may have little or no grass, some grass, or a lot of grass. Finally, suppose the grass grows more abundantly at higher altitudes: thus at the peak of the hill there is a lot of grass, and at the periphery there is less. If we wish to model rabbits on such a landscape, it is important to maintain these characteristics as we scale the field.

The first step of our approach is to create a matrix that represents the physical landscape. We may do this by using the values 0, 1, 2, 3 (for example) to represent how much grass a given patch contains (with 3 being the most abundant and 0 representing little or no grass). Suppose our original model is 10×10 and has a layout as shown in Figure 5.1.

In order to scale the model, we reduce the landscape using the **nearest neighbor** algorithm. First, we decide what dimensions we would like our reduced model to have; suppose it is $n \times n$. We then overlay an evenly spaced grid of $n \times n$ points over the original landscape (see Figure 5.2). Finally, we select values for each point by choosing the value of the neighbor nearest to that point (see Figure 5.3).

Note that this is only one algorithm that can be used for scaling a spatially heterogeneous model. Other methods include bilinear interpolation and bicubic interpolation; the reader is urged to explore the details of these algorithms on their own as this chapter provides only an introductory look at scaling.

Aggregation is another method of reducing computation and run time by simplification of the agent-based model. Rather than physically scaling the entire model, we may aggregate certain agents into groups and view each group as an agent. Thus there are fewer entities to keep track of, and fewer decisions that need to be made. This strategy is particularly helpful in models consisting of many agents of the same

```
0  0  0  0  0  0  0  0  0  0
0  0  1  1  1  1  1  1  0  0
0  1  1  1  2  2  1  1  1  0
0  1  1  2  3  3  2  1  1  0
0  1  2  3  3  3  3  2  1  0
0  1  2  3  3  3  3  2  1  0
0  1  1  2  3  3  2  2  1  0
0  1  1  1  2  2  2  1  1  0
0  0  1  1  1  1  1  1  0  0
0  0  0  0  0  0  0  0  0  0
```

FIGURE 5.1

Original 10 × 10 grid representing topological landscape.

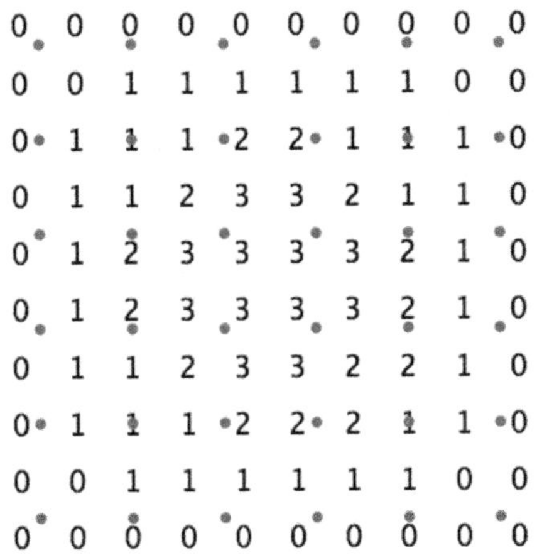

```
0  0  0  0  0  0  0  0  0  0
0  0  1  1  1  1  1  1  0  0
0  1  1  1  2  2  1  1  1  0
0  1  1  2  3  3  2  1  1  0
0  1  2  3  3  3  3  2  1  0
0  1  2  3  3  3  3  2  1  0
0  1  1  2  3  3  2  2  1  0
0  1  1  1  2  2  2  1  1  0
0  0  1  1  1  1  1  1  0  0
0  0  0  0  0  0  0  0  0  0
```

FIGURE 5.2

Original 10 × 10 grid with 6 × 6 overlaying points.

type, or many agents that follow the same set of rules. In modeling seasonal animal migration, for example, we may choose to aggregate a herd of antelope into one agent: the location of this agent would thus represent the average location of each individual in the herd. We can even represent certain antelope dying off and others being born by altering the size of the agent (e.g., as the antelope herd interacts with an aggregated prey agent such as cheetahs or lions, the size of each may expand or

$$
\begin{array}{cccccc}
0 & 0 & 0 & 0 & 0 & 0 \\
0 & 1 & 2 & 2 & 1 & 0 \\
0 & 2 & 3 & 3 & 2 & 0 \\
0 & 2 & 3 & 3 & 2 & 0 \\
0 & 1 & 2 & 2 & 1 & 0 \\
0 & 0 & 0 & 0 & 0 & 0
\end{array}
$$

FIGURE 5.3

Resulting 6 × 6 reduced grid.

contract accordingly). Of course, such methods may not be possible depending on the aim of the model, and certain agents may not be amenable to aggregation. In particular, this strategy tends to be more difficult to implement in models with a high level of spatial heterogeneity or a high level of specification at the agent level.

Whereas scaling requires the dimensions of the model to be reduced (and other model parameters accordingly scaled), aggregation generally requires reconstruction of the model, since the scope of the model is altered as the agent structure is reformulated. In that sense, aggregation can be more involved than scaling. On the other hand, the long-term goal of both techniques is to improve run time and reduce computation, so the extra steps may be well worth the effort. A combination of both techniques can provide substantial decreases in run time and simplification without loss of pertinent model dynamics or detail.

The following exercises refer to the modified Rabbits and Grass model, `Rabbits Grass.nlogo`, available at http://admg.vbi.vt.edu/software/rabbitsgrass-netlogo -files/. Please read the `Info` tab to familiarize yourself with the details of this model.

Exercise 5.13.   Run the model with the default values for `world-size` 44,11, 5. In each case, after the simulation ceases, take a snapshot and examine the population plots. What differences do you notice? What is the benefit of decreasing the world size? How would you determine the smallest world size that you could use to obtain reliable results?   ▽

Exercise 5.14.   Note in the cost function that we multiply the rabbit cost by $\frac{150}{\text{initial-rabbits}}$. Why is this? Why do we not multiply the poison cost by the same factor?   ▽

Exercise 5.15.   Turn `scale-rabbits?` off and run the model several times at various sizes. What differences do you notice in the population graphs? What is the effect of this option? Are results at smaller sizes more or less reliable when `scale-rabbits?` is off?   ▽

Exercise 5.16.   When `scaling-rabbits` is on and the world size is reduced, the rabbits become larger (graphically). Is reducing the number of rabbits an example of scaling or of aggregation? Justify your answer.   ▽

Exercise 5.17.   What other methods besides those mentioned in this section could be used to improve the run time of this model (i.e., to make it more computationally efficient)? $\triangledown$

5.7 A HEURISTIC APPROACH

In previous sections we have discussed optimal control as it relates to agent-based models, and techniques for improving the run time of such models. In theory, the process of finding the optimal solution requires only that we evaluate the entire solution space and choose the solution that minimizes the cost function. In practice, however, this remains unfeasible. For one thing, it is quite possible that the solution space is infinite; and even if it is finite, it is often the case that there are far too many solutions to possibly implement them all (in terms of realistic run times), even in models that have been subjected to scaling and aggregation techniques.

For example, a discrete model of the immune system might involve many millions of cells. Each will have its own rules describing its interaction with other cells, and may also have many variables attributed to it. Analytic methods fail in these cases, and exhaustive enumeration by computer is unfeasible due to the limitations of computer processing speeds. In fact, as the processing power of computers grows, so too does the possibility for creating models that increase in complexity, so that it is impossible to be certain that computers will ever be able to "catch up" to the complexity of the models.

For this reason, most of the optimization and optimal control methods currently used in discrete modeling (and with agent-based models in particular) are in the form of *heuristic algorithms*. An algorithm is a formal set of rules used in order to obtain a solution to a problem. A heuristic algorithm is a specific type of algorithm that conducts a "smart" search of the solution space. It may make use of certain aspects of the problem to avoid having to search through every possible solution, or it may update its search method based on the input it receives from previously found solutions. These algorithms begin with a choice of control input values (i.e., full or partial solutions) and use various methods to refine the values so as to decrease the value of the associated cost function until no better solution can be found. While these methods do not guarantee an optimal solution, they do provide an opportunity for analysis. In particular, if constructed and executed correctly, they are a vast improvement upon random searches of the solution space. One advantage to the heuristic approach is that it can be used with virtually any model. Since heuristic algorithms rely on results from simulation, it is not necessary to transform the model in any way in order to implement them. They provide a "brute force" technique that, while not always optimal, can always be used when other methods may fail. In particular, heuristic algorithms provide a baseline for results obtained via other, more rigorous methods. While a mathematician may be concerned with developing an algorithmic theory for explicitly finding a true optimal solution, in practice scientists and other researchers are often satisfied with a solution that is better than any previously known. As such, the heuristic approach to optimal control is a valuable option.

Of course, every heuristic algorithm has associated risks and advantages, and the particular algorithm that is best suited for a given model or problem will depend on these. As an introduction to the heuristic approach, we describe a so-called *genetic algorithm* in some detail, in the context of the "Rabbits and Grass" model. We conclude this section with a list of other heuristic algorithms.

5.7.1 Genetic Algorithms

Genetic algorithms (sometimes referred to as evolutionary algorithms) were originally developed as a means of studying the natural evolution process, and have been adapted for use in optimization since the 1960s [12–14]. Each solution is viewed as a chromosome. The chromosome is a string of values, or genes, each of which represents the value of one of the parameters. Each chromosome, then, represents an individual solution that can be implemented and whose associated cost can then be evaluated. A typical genetic algorithm functions in the following way: several chromosomes are selected to form a population. The cost of each is then evaluated. The chromosomes are ordered, beginning with the chromosome that produces the minimum cost. The best half (i.e., those with lower associated cost) of the chromosomes are then selected to carry on to the next generation; that is, they will be kept in the list to be evaluated later. Then the process of breeding comes into play: two random chromosomes from those that have been carried over are selected to serve as parents. A "child" chromosome (i.e., a new solution) is then bred from these parents by some method of crossover (*uniform crossover* is a typical method in which there is an equally likely chance that the child will take each particular gene from either parent). The next step is *mutation*: each gene has the possibility of being mutated at random, changing from its current value to any admissible value of that gene. The child chromosome produced from such breeding is added to the next generation of solutions. This process is repeated until the remaining half of the new population has been re-populated with new child chromosomes. The entire process is then repeated: the chromosomes are evaluated and the best are set aside, new chromosomes are bred from them, and so on.

The control objective with the "Rabbits and Grass" model was to determine a control schedule that minimized the number of rabbits while also minimizing the number of days on which poison was used. Suppose we wish to achieve this goal over the course of 10 days. Then a solution is a vector of length 10, each entry of which is either 0 or 1. Suppose we begin with two randomly chosen solutions, $p_1 = [0101010101]$, $p_2 = [1110110010]$. We create a "child" solution by going through each gene (i.e., each entry in the parent solutions) and randomly selecting one of the values. See Figure 5.4 for an example of such a "child" solution. The child solution is then subjected to mutation; see Figure 5.5 for an example.

The algorithm continues for a pre-determined number of steps, or until a certain condition is met. For example, we may choose to run the algorithm for 50 generations.

$$p_1 = [\,\mathbf{0}\,1\,0\,1\,0\,1\,\mathbf{0}\,1\,0\,1\,]$$
$$p_2 = [1\,1\,\mathbf{1}\,0\,1\,1\,0\,0\,\mathbf{1}\,0]$$
$$p_{new} = [0\,1\,1\,1\,1\,1\,0\,1\,1\,1]$$

FIGURE 5.4

Uniform crossover to "breed" a new solution.

$$[0\,1\,\mathbf{1}\,1\,1\,1\,0\,\mathbf{1}\,1\,\mathbf{1}] \quad \rightarrow \quad [0\,0\,1\,1\,1\,1\,0\,1\,1\,0]$$

FIGURE 5.5

Bold values are subjected to mutation.

Another method is to repeat the process until no better solution has been found for, say, ten consecutive generations. When the algorithm terminates, we choose the best current solution as our candidate for an optimal solution (note that there is no guarantee that the best solution found by the algorithm is actually the optimal solution; recall that we use *optimal solution* to mean the best solution we are able to find).

As with other heuristic algorithms, there are many variations and the one presented here is provided as a standard procedure. We may modify the algorithm, for example, so that the initial chromosomes are selected in a certain way: perhaps we wish to choose them at random, or perhaps we wish to choose chromosomes that are very different from one another (i.e., solutions that come from different regions of the solution space). We may modify the crossover process so that one parent is favored over another, or we may forego the mutation step altogether. The likelihood of mutation is another area where user input is important: a high level of mutation will result in more variation of child chromosomes, and thus will not incorporate the relative fitness of the parent chromosomes as much. On the other hand, if the mutation step is not included then we run a greater risk of our solution converging to some solution that is only locally minimal; that is, it may be better than all of the solutions that are similar to it, but not necessarily better than solutions that are radically different (we may think of these solutions as being farther away in the solution space). The advantage of using a genetic algorithm is that there is inherently some level of stochasticity; that is, there is always the possibility of mutating to a better solution (as long as the mutation rate is nonzero).

For the following exercises, use Netlogo to open the file entitled `Rabbits-Grass-GA.nlogo`, available at http://admg.vbi.vt.edu/software/rabbitsgrass-netlogo-files/. This file runs a genetic algorithm to attempt to determine the best poison schedule based on the cost function given in the model (the schedule which minimizes the number of rabbits and the days on which poison is used). Read the "Info" tab for information on each of the options. In addition, you may need to examine the code in this model in order to complete all exercises.

Exercise 5.18. Explain the difference between roulette and tournament selection. How would roulette selection be different if we were interested in schedules with the *highest* cost? ▽

Exercise 5.19. Explain the difference between uniform and 1-point crossover. Describe a control problem for which uniform crossover is more appropriate, and another for which 1-point crossover is more appropriate. ▽

Exercise 5.20. Explain the mutation methods `invert` and `neighbor swap`. Explain the advantages and disadvantages of each. ▽

Exercise 5.21. Devise and explain methods for selection, crossover, and mutation other than those available in the model. Describe advantages of each. ▽

Exercise 5.22. The `retention` slider determines how many of the top solutions should be carried over to the next generation. Describe a potential pitfall of setting this value to 0. Describe a potential pitfall of setting this value too high. ▽

Exercise 5.23. Notice that the default number of runs is 100. What benefit is there to setting this value lower? What is the risk? What are the advantages and disadvantages of setting this value higher? ▽

Exercise 5.24. Click `restore-defaults`, then click `setup`. Toward the top of the interface, make sure `view updates` is unchecked. Run the algorithm (this may take a while). Make note of the best schedule found. Now run it again, only changing the poison so that it degrades. Again, take note of the best schedule found (note that this data should be saved in a `.csv` file in the folder where the model is saved). What differences do you notice? Why do you think this is? ▽

Exercise 5.25. Make sure that `poison-degrade` is set to On and that `poison -strength` is set to 0.75. Choose values for the various sliders and methods on the right-hand side of the interface tab (i.e., the algorithm settings) in an attempt to minimize the cost. Use your intuition as a starting point, then revise the values based on the output (make sure `write-to-file?` is turned on). What is the minimum cost you are able to achieve? What pattern (if any) do you notice in the best schedule found? ▽

Exercise 5.26. Implement your ideas from Exercise 5.21 into the code and run the genetic algorithm using them. Do they perform better or worse than the original methods?[1] ▽

Exercise 5.27. Suppose the objective is simplified so that the goal is to simply minimize the number of rabbits, without regard to the poison cost. What do you think the best solution would be in this case? Alter the cost function in the code accordingly, and run the GA. Does the outcome reflect your intuition? ▽

Exercise 5.28. Determine a new objective and alter the cost function accordingly. Run the GA. Discuss the pattern of the best schedule found, and state whether or not this seems to make sense in light of your cost function. ▽

---

[1] This exercise requires knowledge of Netlogo programming.

5.7.2 **Other Heuristic Algorithms**

Genetic algorithms represent only one class of heuristic algorithms. There are many others, and as mentioned earlier, each have their own advantages and disadvantages. The purpose of this section is to provide you with an overview of how heuristic algorithms can be used as a brute force approach to optimal control of agent-based models. Depending on the nature of the model and the control objective, the reader may also be interested in simulated annealing [15, 16], tabu search [17–19], ant colony optimization [20, 21], and "squeaky wheel" optimization [22, 23] (to name just a few). The combinations of these algorithms and the fine-tuning therein provide a large framework from which to apply heuristic algorithms to agent-based models, and new algorithms are constantly being developed. While more rigorous mathematical theory is currently being developed, this approach is sometimes the best available analytical tool.

5.8 MATHEMATICAL FRAMEWORK FOR REPRESENTING AGENT-BASED MODELS

As addressed earlier, many agent-based models are too complex to find global optimal control purely by simulation. One option one might pursue in this case is to translate agent-based models into a different type of mathematical object for which more mathematical tools are available. This is similar to the approach pioneered by Descartes and his introduction of a coordinate system. In the plane, for instance, a Cartesian coordinate system allows us to represent lines by linear equations, circles by quadratic equations, etc. If we now want to find out whether two lines in the plane intersect we can either carry out an inspection of the two lines as geometric objects or, alternatively, we can determine whether the resulting system of linear equations has a solution. Translating agent-based models into a formal mathematical framework then opens up the possibilities to use theory from mathematics to analyze them from a different angle rather than just with simulations. One framework that has been studied is that of so-called polynomial dynamical systems (PDSs).

As in our running examples of rabbits and grass, agents in the models we consider here take on a finite number of states, such as a rabbit's position on the grid or the amount of grass on a patch. This finite set of states can be viewed as the analog of the points that make up a line in the plane, to continue with the analogy of the Cartesian coordinate system. Applying the rules of the model in order to update the states of all the agents can be thought of as a function that maps the collection of discrete states to itself. Since the state sets do not carry any kind of structure we are simply dealing with a function from a finite set to itself, about which we can say little. However, let us now introduce the analog of a Cartesian coordinate system into this set in a way that provides extra tools for us, analogous to the ability to use algebra to decide whether two lines in the plane intersect. We do this by way of an example.

Suppose that our set of states equals $\{low, medium, high\}$, which we can represent as $\{0, 1, 2\}$. We can easily turn this set into a number system by using so-called arithmetic modulo 3. That is, we can add and multiply the elements of this set according to rules that are identical to those used to multiply real numbers. Notice that the numbers in the set are all the remainders one can obtain under division by 3. Now we add and multiply "modulo 3," that is, $2 + 2 = 1$ and $2 \cdot 2 = 1$ as well, that is the remainder of $2 + 2 = 4$ under division by 3 is 1. The number system we obtain in this way is an example of a "finite field," which we will use for the remainder of the chapter. We denote the finite field with three elements by $\mathbb{F}_3$, or more generally, the finite field with p elements, where p is prime, by $\mathbb{F}_p$.

Given rules from an ABM that describe how the agents or variables change their states, we can find for every agent a polynomial that describes the given rule using elements of the finite field such as $\{0, 1, 2\}$ rather than $\{low, medium, high\}$. By constructing a polynomial for every single agent, we obtain a set of polynomials that describes the same behavior as the original ABM. The set is called a *polynomial dynamical system*.

Definition 5.4 (Polynomial dynamical system).   Let k be a finite field and $f_1, \ldots,$ $f_n \in k[x_1, \ldots x_n]$ be polynomials. Then $F = (f_1, \ldots, f_n) : k^n \to k^n$ is an n-dimensional polynomial dynamical system over k [24].

Here, k^n denotes the Cartesian product $k \times k \times \cdots \times k$ with n copies of k. $k[x_1, \ldots x_n]$ denotes the ring of polynomials in variables $x_1, \ldots, x_n$ and coefficients in k. For $(a_1, \ldots, a_n) \in k^n$ the function F is evaluated by

$$F(a_1, \ldots, a_n) = (f_1(a_1, \ldots, a_n), \ldots, f_n(a_1, \ldots, a_n)).$$

Example 5.5.   Let $k = \mathbb{F}_3 = \{0, 1, 2\}$ and $F = (f_1, f_2, f_3)$, where

$$f_1 = x_1 + x_2$$
$$f_2 = x_1 x_2 + x_3 + 1$$
$$f_3 = x_1 + x_2 + x_3^2.$$

The PDS F maps points in $\mathbb{F}_3^3$ to other points in $\mathbb{F}_3^3$, and iteration of F results in a dynamical system. F has one fixed point, or steady state, namely $(1, 0, 2)$ because $F(1, 0, 2) = (f_1(1, 0, 2), f_2(1, 0, 2), f_3(1, 0, 2)) = (1, 0, 2)$, and note that all calculation are done "modulo 3" because F is over $\mathbb{F}_3$. Furthermore, F has a limit cycle or oscillation of length five: applying F repeatedly to $(0, 2, 1)$ yields

$$(0, 2, 1) \to (2, 2, 0) \to (1, 2, 1) \to (0, 1, 1)$$
$$\to (1, 2, 2) \to (0, 2, 1) \to (2, 2, 0) \to \cdots$$

With software packages such as ADAM [25], one can compute the dynamics of a PDS and visualize it.

Most agent-based models can be translated into a polynomial dynamical system of this type. Each variable x_i represents an agent, a patch, or some other entity in the model. Each element in the finite field k represents a different state or condition of the variables. The corresponding polynomials f_i encode the behavior of the agent or patch as described in the model. In the rabbit and grass model, one could assign a variable x_i for each patch. The values of the finite field $k = \{0, 1, 2, \ldots, p - 1\}$ then represent how many rabbits and how much grass is currently on the patch. The polynomials f_i determine the number of rabbits and amount of grass on patch i at the next iteration, given the current number of rabbits and grass. We will explain in more detail how to derive the polynomials for an agent-based model in Section 5.9, however their existence is assured by the following result.

Theorem 5.5 ([26]).   *Let $f : k^r \to k$ be any function on a finite field k. Then there exists a unique polynomial $g : k^r \to k$, such that $\forall x \in k^r$, $f(x) = g(x)$.*

Any such mapping over a finite field can be described by a unique polynomial. Using *Lagrange interpolation*, we can easily determine the polynomial. Let $f : k^r \to k$ be any function on k. Then

$$g(x) = \sum_{(c_{i1}, \ldots, c_{ir}) \in k^r} f(c_{i1}, \ldots, c_{ir}) \prod_{j=1}^{r} \left(1 - (x_j - c_{ij})^{p-1}\right) \tag{5.1}$$

is the unique polynomial that defines the same mapping as f.

Example 5.6.   Suppose $k = \mathbb{F}_3, r = 2$, and the mapping f is defined on $\mathbb{F}_3^2 = \{0, 1, 2\} \times \{0, 1, 2\}$ as follows:

$$f(0, 0) = 0,$$
$$f(0, 1) = 1,$$
$$f(0, 2) = 2,$$
$$f(1, 0) = 1,$$
$$f(1, 1) = 2,$$
$$f(1, 2) = 0,$$
$$f(2, 0) = 2,$$
$$f(2, 1) = 0,$$
$$f(2, 2) = 1.$$

Then the polynomial g that defines the same mapping as f is constructed as follows:

$$g(x, y) = 0 +$$
$$1\left((1 - x^2)(1 - (y - 1)^2)\right) +$$
$$2\left((1 - x^2)(1 - (y - 2)^2)\right) +$$
$$1\left((1 - (x - 1)^2)(1 - y^2)\right) +$$
$$2\left((1 - (x - 1)^2)(1 - (y - 1)^2)\right) +$$
$$0 +$$

$$2\left((1-(x-2)^2)(1-y^2)\right) +$$
$$0 +$$
$$1\left((1-(x-2)^2)(1-(y-2)^2)\right)$$
$$= x + y.$$

Converting an agent-based model into a polynomial dynamical system provides us with a conceptual advantage, since rather than being limited to working with a computer simulation as our only means of analysis, methods and theory from abstract algebra and algebraic geometry can be used. For example, one might be interested in the steady states of a model, i.e., all the configurations of the system that do not change over time. Written as a polynomial dynamical system $F : k^n \to k^n$, these states are exactly the solutions to $F(x) = x$, the n-dimensional system of equations $f_1(x) = x_1, \ldots, f_n(x) = x_n$ in $k[x_1, \ldots, x_n]$ [25]. Equivalently, the solutions are the points in the variety $\mathcal{V}$ of the ideal generated by the polynomials $f_1(x) - x_1, \ldots, f_n(x) - x_n$. For an introduction to varieties, we recommend [27].

When the polynomials describing the biological system have a special structure, other analysis methods are available. For example, for conjunctive networks, i.e., a PDS over $\mathbb{F}_2$ where each polynomials is a monomial, the dynamics can be completely inferred by looking at the dependency graph or wiring diagram of the PDS [28]. We can determine the fixed point and limit cycle structure of this PDS without using any simulation.

We have mentioned *Conway's Game of Life*, a cellular automaton as a special case of agent-based models earlier in this chapter. Cells are agents that die or come to life based on a rule including their eight neighbors [5]. The Game of Life can be translated into a polynomial dynamical system. Each variable x_i represents a cell on the grid. Each polynomial $f_i : \mathbb{F}_2^9 \to \mathbb{F}_2$ depends on x_i's eight neighbors and itself and describes whether x_i will be dead (0) or alive (1) at the next iteration given the values of its neighboring cells. The fixed points of this system correspond to *still lives* such as blocks and beehives, two cycles to *oscillators*, e.g., blinkers and beacons. Working with the mathematical representation of the game, one can for example study still lives by using concepts from invariant theory using the symmetry of the rules as group actions.

Exercise 5.29.   Using Lagrange's interpolation formula (5.1), construct the polynomial $f_1 : \mathbb{F}_2^9 \to \mathbb{F}_2$ describing the behavior of cell x_1 with neighbors $x_2, \ldots, x_9$, i.e., the polynomial $f_1(x_1, x_2, \ldots, x_9)$ should evaluate to 0 or 1 (dead or alive) according to the rules of the Game of Life given the state of the cell x_1 and its eight neighbors. $\triangledown$

In the next section, we provide some polynomials for common rules in agent-based models. By providing such rules, we hope to simplify the process of translating an ABM to a PDS. The polynomials can be used as given or as a starting point to construct functions that represent more complex behavior.

5.9   TRANSLATING AGENT-BASED MODELS INTO POLYNOMIAL DYNAMICAL SYSTEMS

This section is authored jointly by Franziska Hinkelmann, Matt Oremland, Hussein Al-Asadi, Atsya Kumano, Laurel Ohm, Alice Toms, Reinhard Laubenbacher

Discrete models, including agent-based models, are important tools for modeling biological systems, but model complexity may hinder complete analysis. Representation as a PDS provides a framework for efficient analysis using theory from abstract algebra. In this section, we provide polynomials that describe common agent interactions. In the previous section we described how to interpolate agent behavior and to generate the appropriate polynomial. However, for a variable with many different states, this method of interpolation results in long and complex polynomials that are difficult to expand, simplify, or alter. Thus we provide some general "shortcut rules" for constructing polynomials that describe key agent and patch interactions present in many ABMs. Each of the following polynomials exists in the finite field $\mathbb{F}_p$.

Since we are particularly interested in ABMs describing complex biological systems, we use the term concentration to describe the states of a patch variable (for example, concentration of white blood cells on a patch). In this chapter we describe several polynomials that describe both basic movement, and movement according to the state of the neighboring patches.

5.9.1   Basic Movement Function

One can construct various polynomials to describe the movement of an agent on an n-by-n grid where the x- and y-coordinates of patches are numbered 0 to $n - 1$ from left to right with torus topology, i.e., there are no boundaries to the grid, so that if an agent on the left most patch moves to the left, he appears on the right side of the grid, and similarly for top and bottom. By moving forward we mean moving to the right, and potentially "wrapping" around to the left edge of the grid, and by moving backward we mean moving to the left.

For an agent moving forward one patch per time step, we want an agent on patch x to move to patch $x + 1$ unless the agent is on patch $x = n - 1$, in which case the agent will move to patch $x = 0$ on the next step due to the torus topology of the grid (see Table 5.1). The polynomial describing this movement is given in Eq. (5.2).

One can construct a similar polynomial for an agent moving backward one step per time step (see Eq. (5.3)), and using this polynomial along with the forward movement one can create a polynomial to describe movement of several steps along the x-axis as specified by a series of elements in $\{1, -1, 0\}$ (representing forward, backward, or no movement), (Eq. 5.4). Furthermore, one can generalize these polynomials to describe movement of a fixed step length m, (Eq. 5.5) and (Eq. 5.6).

Table 5.1 Movement of an agent on a grid with n patches.	
At Time t	**At Time $(t+1)$**
Patch 0	Patch 1
Patch 1	Patch 2
Patch 2	Patch 3
$\vdots$	$\vdots$
Patch i	Patch $i+1$
$\vdots$	$\vdots$
Patch $n-1$	Patch 0

- Agent moves forward one patch per time step:

$$f(x) = x + 1 + n \prod_{i=0, i \neq n-1}^{p-1} (i - x). \tag{5.2}$$

- Agent moves backward one patch per time step:

$$f(x) = x - 1 - n \prod_{i=1}^{p-1} (i - x). \tag{5.3}$$

- We can use these equations to construct a more general movement function $f(x, u)$, where $u \in \{-1, 0, 1\}$ specifies the direction of the agent by taking $u = -1$ when the agent moves back one step, $u = 0$ when the agent stays as is, and $u = 1$ when it moves one step forward. Note that $a^{p-1} = 1$ for all $a \neq 0 \in \mathbb{F}_p$.

$$f(x, u) = x + u + un(u+1)^{p-1} \prod_{i=0, i \neq n-1}^{p-1} (i-x) + un(u-1)^{p-1} \prod_{i=1}^{p-1} (i-x). \tag{5.4}$$

- Agent moves forward m patches per time step:

$$f(x, m) = x + m + n \sum_{j=1}^{m} \left(\prod_{i=0, i \neq n-j}^{p-1} (i - x) \right). \tag{5.5}$$

- Agent moves backward m patches per time step:

$$f(x, m) = x - m - n \sum_{j=0}^{m-1} \left(\prod_{i=0, i \neq j}^{p-1} (i - x) \right). \tag{5.6}$$

In order to show that Eq. (5.2) is the polynomial over $\mathbb{F}_p$ that describes moving one space forward on an n-by-n grid, we need to show that $f(x) = x + 1$ for all $x \neq n - 1$ and $f(x) = 0$ for $x = n - 1$. Note that $p = 0$ and $(p - 1)! = -1$ in $\mathbb{F}_p$.

- Case: $x \neq n - 1$: $f(x) = 1 + x + n(0) = 1 + x$.
- Case: $x = n - 1$:

$$
f(n - 1) = 1 + n - 1 + n \prod_{i=0, i \neq n-1}^{p-1} i - n + 1
$$

$$
= n + n(1 - n)(2 - n) \cdots (n - 2 - n + 1)
$$
$$
\times (n - n + 1)(n + 1 - n + 1) \cdots (p - 1 - n + 1)
$$
$$
= n + n(p + 1 - n)(p + 2 - n) \cdots (p - 1)(1)(2) \cdots (p - n)
$$
$$
= n + n(1)(2) \cdots (p - n)(p + 1 - n) \cdots (p - 1)
$$
$$
= n + n(p - 1)! = n + n(-1) = n - n
$$
$$
= 0.
$$

Exercise 5.30.   Show that Eqs. (5.3)–(5.6) indeed describe the movements as stated above.   ▽

5.9.2 Uphill and Downhill Movement

In many applications, agents can scan their close environment and move towards a desired resource. Netlogo has this behavior implemented as "uphill" and "downhill." "Uphill" moves an agent to the neighboring patch with the highest value of the desired variable. If no neighboring patch has a higher value than the current patch, the agent stays put. Since the variables are discrete, there may exist a tie between neighboring patches for highest (or lowest) concentration, in which case the agent moves toward the lowest arbitrarily numbered neighboring patch i. The polynomials describing the movement of agent x are the following.

- Uphill:

$$
f(x) = \sum_{i=1}^{8} \left((1 - (C - I_i)^{p-1}) l_i \prod_{j=0}^{i-1} (C - I_j)^{p-1} \right)
$$

$$
+ (1 - (C - I_0)^{p-1}) \cdot l_0 + \sum_{m=0}^{C-1} \left(\left(\prod_{k=m+1}^{C} \prod_{j=i}^{8} (k - I_i) \right)^{p-1} \right.
$$

$$
\cdot \left(\sum_{i=1}^{8} (1 - (m - I_i)^{p-1}) l_i \prod_{j=0}^{i-1} (m - I_j)^{p-1} \right.
$$

$$
\left. \left. + (1 - (m - I_0)^{p-1}) l_0 \right) \right), \tag{5.7}
$$

Table 5.2 Relative position of
neighbors to agent x.

l_8	l_1	l_2
l_7	$x = l_0$	l_3
l_6	l_5	l_4

- Downhill:

$$f(x) = \sum_{i=1}^{8} \left((1 - I_i^{p-1}) l_i \prod_{j=0}^{i-1} I_j^{p-1} \right)$$

$$+ (1 - I_0^{p-1}) \cdot l_0 + \sum_{m=1}^{C} \left(\left(\prod_{k=0}^{m-1} \prod_{j=i}^{8} (k - I_i) \right)^{p-1} \right.$$

$$\cdot \left(\sum_{i=1}^{8} (1 - (m - I_i)^{p-1}) l_i \prod_{j=0}^{i-1} (m - I_j)^{p-1} \right.$$

$$\left. \left. + (1 - (m - I_0)^{p-1}) l_0 \right) \right), \tag{5.8}$$

where the concentration ranges from 0 to C, C represents the highest possible concentration, I_i is the concentration level at neighboring patch i, and l_i is the relative location of patch i from the current patch, see Table 5.2. I_0 is the concentration of the current patch, $I_1, \ldots, I_8$ are the concentrations of its eight neighbors.

Next, we show that Eq. (5.7) describes the uphill movement.

$$f(x) = \left(1 - \prod_{i=0}^{8} (C - I_i)^{p-1} \right) \left(\sum_{i=1}^{8} \left((1 - (C - I_i)^{p-1}) l_i \prod_{j=0}^{i-1} (C - I_j)^{p-1} \right) \right.$$

$$+ \left(1 - (C - I_0)^{p-1} \right) l_0 \right) + \sum_{m=1}^{C-1} \left(\left(1 - \prod_{i=0}^{8} (m - I_i)^{p-1} \right) \right.$$

$$\times \prod_{k=m+1}^{C} \prod_{i=0}^{8} (k - I_i)^{p-1} \sum_{i=1}^{8} (1 - (m - I_i)^{p-1}) l_i$$

$$\left. \times \prod_{j=0}^{i-1} (m - I_j)^{p-1} + (1 - (m - I_0)^{p-1}) l_0 \right). \tag{5.9}$$

It is straightforward to see that Eq. (5.9) describes the movement to the neighboring patch with the highest concentration level: $1 - \prod_{i=0}^{8} (C - I_i)^{p-1}$ is 0 unless at least one of the $I_i = C$, i.e., one of the neighbors' concentrations level is the highest possible. In this case, $f(l) = l + i$. Indeed, the right hand of the first line evaluates to l_i for the i such that $I_i = C$ $(1 - (C - I_i)^{p-1})l_i$ is l_i if and only if $(I_i = C)$. If there are several neighbors with concentration level C, $f(l)$ should evaluate to the neighbor with the smallest index. This is assured by multiplying with $\prod_{j=0}^{i-1}(C - I_j)^{p-1}$, which evaluates to 0, if a neighbor with a smaller index has concentration C. The second and third line in Eq. (5.9) are equivalent to the first row, as they describe movement to patches with concentration levels lower than C. The term $\prod_{k=m+1}^{C}\prod_{i=0}^{8}(k - I_i)^{p-1}$ assures that the second summand evaluates to 0 if a patch has a higher concentration than m. The proof for the downhill movement (Eq. 5.8) is similar and left as an exercise.

Exercise 5.31. Based on the uphill and downhill polynomial, construct a polynomial $f(l)$ that evaluates to the maximum concentration of its eight neighbors and itself.▽

Exercise 5.32. Consider a 13 by 13 grid with torus topology. A rabbit moves two steps up and two to the right at every iteration. Construct a polynomial that describes a rabbit's movement. Use the polynomial to simulate the movement of a rabbit starting in the center of the grid.   ▽

Exercise 5.33. For the rabbit in Exercise 5.32, after how many iterations do you expect it to reach its starting position again? Confirm your answer by evaluating the polynomial.   ▽

Exercise 5.34. Construct the polynomial as in Exercise 5.32 for a grid of arbitrary size n by n.   ▽

Exercise 5.35. Consider a grid where each patch is covered with a *low, medium,* or *high* amount of grass. At each iteration, the rabbit moves to the neighboring patch with the most grass, i.e., to one of the eight neighboring patches or it stays on the same patch. Construct a polynomial describing the rabbit's movement based on the amount of grass on the neighboring patches.   ▽

Exercise 5.36. Consider the grid from Exercise 5.35. Rabbits eat grass, and the amount of grass on a patch decreases by one level for every iteration that a rabbit occupies it, i.e., a patch with *high* grass changes to *medium* grass, if occupied by a rabbit, *medium* to *low*, and *low* remains *low*. Construct a polynomial that describes the amount of grass on a patch.   ▽

We provide these general polynomials and simplification techniques to aid in the transformation of an ABM into a PDS. Whereas large agent-based models may be too complex for efficient analysis, we hope that the algebraic structure of a polynomial dynamical system can be used to expedite computation of optimal control.

5.10 SUMMARY

Agent-based models provide a very intuitive and convenient way to model a variety of phenomena in biology. The price we pay for these features is that the models are not explicitly mathematical, so that we lack mathematical tools for model analysis. For instance, many of these phenomena are connected to optimization and optimal control problems, as pointed out in this chapter, but no systematic methods are available for agent-based models to solve these. We have attempted here to do two things. Firstly, we described so-called heuristic local search methods, such as genetic algorithms, which can be applied directly to agent-based models. And we described a way in which one can translate an agent-based model into a mathematical object, in this case a polynomial dynamical system over a finite field. Many computational and theoretical tools are available for such systems. For instance, to compute the steady states of a polynomial system $F = (f_1, \ldots, f_n)$, one can solve the system of polynomial equations

$$f_1(x_1, \ldots, x_n) = x_1, f_2(x_1, \ldots, x_n) = x_2, \ldots, f_n(x_1, \ldots, x_n) = x_n.$$

There are several computer algebra systems available to solve such problems. To compute the steady states of an agent-based model, on the other hand, one is limited to extensive simulation, without any guarantee of having found all possible steady states.

The chapter provides a snapshot of ongoing research in the field. The approach via polynomial dynamical systems, for instance, is very promising, but still lacks appropriate algorithms that scale to larger models. In addition to searching for such algorithms, further research in model reduction is taking place, as outlined earlier in the chapter. At the same time, other mathematical frameworks, such as ordinary or partial differential equations and Markov models are being explored for this purpose. Much work remains to be done but, in the end, a combination of better algorithms, improvements in hardware, and dimension reduction methods is likely to provide for us a tool kit that will allow the solution of realistic large-scale optimization and optimal control problems in ecology, biomedicine, and other fields related to the life sciences.

5.11 SUPPLEMENTARY MATERIALS

All supplementary files and/or computer code associated with this article can be found from the volume's website http://booksite.elsevier.com/9780124157804

References

[1] Ledzewicz U, Naghnaeian M, Schattler H. Optimal response to chemotherapy for a mathematical model of tumor-immune dynamics. J Math Biol 2012;64:557–577.

[2] Sontag E.D. Mathematical control theory. New York, NY: Springer Verlag 2nd ed. 1998.

[3] Iglesias PA, Ingalls BP, editors.Control theory and systems biology. Cambridge, MA: The MIT Press 2009.

[4] Von Neumann J. The general and logical theory of automata. in: Jeffres LA. editor.Cerebral mechanisms of behavior -the Hixon symposium,vols.1–31 New York: John Wiley & Sons; 1951.

[5] Gardner M. The fantastic combinations of John Conway's new solitaire game life. Sci Am 1970;223:120–123.

[6] Wilensky U. NetLogo Center for connected learning and computer-based modeling, Northwestern University, Evanston, IL 2009. http://ccl.northwestern.edu/netlogo/urlhttp://ccl.northwestern.edu/netlogo/

[7] Laubenbacher R, Jarrah AS, Mortveit H, Ravi SS. Mathematical formalism for agent-based modeling. in: Meyers RA. editor. Encyclopedia of complexity and systems science,vols.160–176 Springer; 2009.

[8] Volker Grimm, Uta Berger, Finn Bastiansen, et al. A standard protocol for describing individual-based and agent-based models. Ecol Model 2006;198:115–126.

[9] Hinkelmann F., Murrugarra D, Jarrah AS. Laubenbacher R.A mathematical framework for agent based models of complex biological networks. Bull Math Biol 2010.

[10] Alan Veliz-Cuba. Jarrah Abdul Salam, Laubenbacher Reinhard. Polynomial algebra of discrete models in systems biology Bioinformatics. 2010;26:1637–1643.

[11] Kirk Donald E. Optimal control theory: an introduction. Englewood Cliffs, New Jersey: Prentice-Hall Inc; 1970.

[12] Goldberg DE. Genetic algorithms in search, optimization, and machine learning Reading. MA: Addison-Wesley; 1989.

[13] Holland John H. Adaptation in natural and artificial systems. Ann Arbor, Mich.: University of Michigan Press ;1975. [An introductory analysis with applications to biology, control, and, artificial intelligence 51].

[14] Koza John R. Evolution and co-evolution of computer programs to control independently-acting agents in: Proceedings of the First International Conference on Simulation of Adaptive Behavior,vols.366–375 MIT Press ;1991.

[15] Kirkpatrick S, Gelatt CD, Vecchi MP. Optimization by simulated annealing Science. 1983;220:671–680.

[16] Pennisi M, Catanuto R, Pappalardo F, Motta S. Optimal vaccination schedules using simulated annealing Bioinformatics. 2008;24:1740–1742.

[17] Glover Fred. Tabu search–part I. Informs J Comput 1989;1:190–206.

[18] Glover Fred. Tabu search–part II. Informs J Comput 1990;2:4–32.

[19] Fred Glover. Tabu search: a tutorial interfaces. 1990;20:74–94.

[20] Dorigo M, Maniezzo V, Colorni A. Ant system: optimization by a colony of cooperating agents Systems. Man, and Cybernetics, Part B: Cybernetics. IEEE Transactions on 1996;26:29–41.

[21] Marco Dorigo. Caro Gianni Di, Gambardella Luca M. Ant algorithms for discrete optimization. Artif Life 1999;5:137–172.

[22] Joslin David E, Clements David P. Squeaky wheel optimization. J Artificial Intelligence Res 1999;10:353–373 (electronic).

[23] Li Jingpeng, Parkes Andrew J, Burke Edmund K. Evolutionary squeaky wheel optimization: a new analysis framework evolutionary computation. 2011. published online.

[24] Stigler B, Jarrah A, Stillman M, Laubenbacher R. Reverse engineering of dynamic networks. Ann NY Acad Sci 2007;1115:168–177.

[25] Hinkelmann Franziska, Brandon Madison, Guang Bonny, et al. ADAM: analysis of discrete models of biological systems using computer algebra BMC Bioinformatics. 2011;12:295+.

[26] Reinhard Laubenbacher, Brandilyn Stigler. A computational algebra approach to the reverse engineering of gene regulatory networks. J Theor Biol 2004;229:523–537.

[27] Cox David A., Little John, O'Shea Donal. Ideals, varieties, and algorithms: an introduction to computational algebraic geometry and commutative algebra, 3/e (Undergraduate Texts in Mathematics). Secaucus, NJ, USA: New York:Springer-Verlag Inc; 2007.

[28] Jarrah Abdul, Laubenbacher Reinhard, Veliz-Cuba Alan. The dynamics of conjunctive and disjunctive boolean network models. Bull Math Biol 2010;72:1425–1447. 10.1007/s11538-010- 9501-z.

Neuronal Networks: A Discrete Model

Winfried Just*, Sungwoo Ahn† and David Terman‡

*Department of Mathematics, Ohio University, Athens, OH 45701, USA
†Department of Mathematical Sciences, Indiana University-Purdue University Indianapolis, IN 46202, USA
‡Department of Mathematics, Ohio State University, Columbus, OH 43210, USA

6.1 INTRODUCTION AND OVERVIEW

What do brains, ant colonies, and gene regulatory networks have in common? They are *networks* of individual *agents* (neurons, ants, genes) that interact according to certain rules. Neurons interact through electric and chemical signals at synapses, ants interact by means of olfactory and tactile signals, and products of certain genes regulate the expression of other genes. These interactions may cause a change of the state of an agent. It seems often possible to model such networks by considering only finitely many states for each agent. A neuron may fire or be at rest, an ant may be in the mood to forage, defend the nest, or tend to the needs of the queen, and a gene may or may not be expressed at any given time. If one also assumes that time progresses in discrete steps, then one can build a *discrete dynamical system* model for such networks.

Anyone who has ever closely looked at an ant colony will realize that, in general, there may be a significant amount of randomness in the agents' interactions, and for this reason one may want to conceptualize many biological networks as stochastic dynamical systems, as in the *agent-based models* of Chapter 4 in this volume. But if there is sufficient predictability in the interactions, a mathematically simpler deterministic model may be adequate. In this chapter we study in detail one such deterministic model for networks of neurons; more examples of models of this type are considered in several other chapters of this volume.

In a deterministic discrete dynamical system, the state of each agent at the next time step is uniquely determined by its current state and the current states of all agents it interacts with, according to the rules that determine the dynamics. The resulting sequence of network states is called a *trajectory*. If the system has only finitely many states, each trajectory must eventually enter a set of states that it will visit infinitely often. This set is called its *attractor*. The sequence of states visited prior to reaching the attractor is called the *transient* part of the trajectory, and the set of all initial states from which a given attractor will be reached is called its *basin of attraction*.

The *network connectivity* specifies who interacts with whom. In neuronal networks the connectivity corresponds to the wiring of the brain and can often be assumed to be fixed over time scales of interest. Our focus in this chapter is on how the network connectivity and two intrinsic parameters of individual neurons, the refractory period and firing threshold, influence the network dynamics, in particular the lengths and number of attractors, the sizes of their basins of attraction, and the lengths of transients.

The chapter is organized as follows. Section 6.2 reviews some basic facts about neuronal networks as well as some experimentally observed phenomena that motivated our work. In Section 6.3 we formally define our discrete models and review their basic properties. In Section 6.4 we study provable restrictions on the network dynamics for certain special connectivities, and Section 6.5 reviews some results on the dynamics of typical networks with given parameters. In Section 6.6 we consider an alternative interpretation of our models in terms of disease dynamics and discuss some general issues of choosing an appropriate mathematical model for a biological system. In Section 6.7 we discuss whether our models may be applicable to actual problems in neuroscience and review a result that guarantees an exact correspondence between our discrete models and more detailed ordinary differential equation (ODE) models of certain neuronal networks. In Section 6.8 we describe how the material presented here fits into the larger picture of some current research on the cutting edge of mathematical neuroscience.

We recommend that the reader attempts the exercises included in the main text right away while reading it. They are primarily intended to help the reader gain familiarity with the concepts that we introduce. The additional exercises in the online supplement [1] are less crucial for understanding the material and can be deferred until later. The online supplement also contains some additional material that is related to the main text and, most importantly, several projects for open-ended exploration. Most of these projects are structured in such a way that they start with relatively easy exercises and gradually lead into unsolved research problems.

6.2   NEUROSCIENCE IN A NUTSHELL

6.2.1   Neurons, Synapses, and Action Potentials

It is commonly believed that everything the brain does, and therefore everything we, as humans, do—from cognitive tasks such as thinking, planning, and learning to motor tasks such as walking, breathing, and eating—is the result of the collective electrical activity of neurons. There are roughly 10^{12} neurons in the human brain. Neurons communicate with other neurons at synapses and a brain has approximately 10^{15} synaptic connections; that is, on average, each neuron receives input from approximately 1000 other neurons. Whether or not a neuron fires an electrical signal, or *action potential*, depends on many factors. These include electrical and chemical processes within the neuron itself, properties of the synaptic connections and the underlying network architecture. A fundamental issue in neuroscience is to understand how these three factors interact to generate the complex activity patterns of populations of neurons that underlie all brain functions.

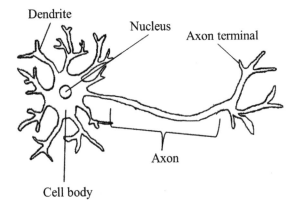

Dendrite

Nucleus

Axon terminal

Axon

Cell body

FIGURE 6.1

A schematic drawing of a neuron. More pictures of neurons can be found at [2].

A schema of a neuron is shown in Figure 6.1. Most neurons consist of dendrites, an axon, and a cell body (or soma). The dendrites spread out from the cell body in a tree-like manner and detect incoming signals from other neurons. In response to these incoming signals, the neuron may (or may not) generate an action potential (or nerve impulse) that propagates away from the soma along the axon. Many axons develop side branches that help bring information to several parts of the nervous system simultaneously.

All living cells are surrounded by a cell membrane that maintains a resting potential between the outside and the inside of the cell. In response to a signal, the membrane potential of a neuron may undergo a series of rapid changes, corresponding to the action potential. In order to generate an action potential, the initial stimulus must be above some threshold amount. Properties of the nerve impulse, including its shape and propagation velocity, are often independent of the initial (superthreshold) stimulus. Once a neuron fires an action potential, there is a so-called *refractory period*. During this time, it is impossible for the neuron to generate another action potential.

6.2.2 Firing Patterns

The firing pattern of even a single neuron may be quite complicated. An individual cell may, for example, fire action potentials in a periodic fashion, or it may exhibit bursting oscillations in which periods of rapid firing alternate with periods of quiescent behavior. More complicated patterns, including chaotic dynamics, are also possible. There has been a great deal of effort among mathematicians in trying to classify the types of firing patterns that can arise in single neuron models. For references, see [3].

In this chapter, we are primarily concerned with the collective behavior of a neuronal population. Examples of population rhythms include *synchrony,* in which every cell in the network fires at the same time and *clustering,* in which the entire population breaks up into subpopulations or clusters; every cell within a single cluster fires synchronously and different clusters are desynchronized from one another. By *dynamic clustering* we mean that there are distinct episodes in which some subpopulation of cells fire synchronously; however, membership within this subpopulation may change over time. That is, two neurons may fire together during one episode but not during a subsequent episode. Of course, more complicated population rhythms are also possible. Neurons within the network may be totally desynchronized from each other, or activity may propagate through the network in a wave-like manner.

6.2.3 Olfaction

These types of population rhythms have been implicated in many brain processes and it is critically important to understand how they depend on parameters corresponding to the intrinsic properties of cells, synaptic connections, and the underlying network architecture. Our example is concerned with olfaction; that is, the sense of smell. In the insect and mammalian olfactory systems, any odor will activate a subset of receptor cells, which then project this sensory information to a neural network in the antennal lobe (AL) of insects or olfactory bulb (OB) of mammals in the brain. Processing in this network transforms the sensory input to give rise to dynamic spatiotemporal firing patterns. These patterns typically exhibit dynamic clustering. Any given neuron might synchronize with different neurons at different stages, or episodes, and this sequential activation of different groups of neurons gives rise to a transient series of output patterns that through time converges to a stable activity pattern.

The role of these activity patterns in odor perception remains controversial. Distinct odors will stimulate different neurons, thereby generating different spatiotemporal firing patterns. It has been suggested that these firing patterns help the animal to discriminate between similar odors. That is, it may be difficult initially for an insect's brain to distinguish between similar odors, because similar odors stimulate similar subsets of neurons within the insect's AL. However, because of dynamic clustering, the subsets of neurons that fire during subsequent episodes change, so that the neural representations of the odors—that is, the subsets of neurons that fire during a specific episode—become more distinct. This makes it easier for the insect's brain to distinguish between the neural representations of the two odors.

Mechanisms underlying these firing patterns are poorly understood. One reason for this is that many biological properties of the AL and OB are still not known; these include intrinsic properties of neurons, as well as details of the underlying network architecture. Many of these details are extremely difficult to determine experimentally and computational modeling, along with mathematical analysis of these models, can potentially be very useful in generating and testing hypotheses concerning mechanisms underlying the observed patterns. In particular, a critical role for theoreticians is to classify those networks that exhibit population rhythms consistent with known biology and determine how activity patterns may change with network parameters.

This presents numerous very challenging problems for mathematicians. Any biological system is very complicated and it is not at all clear what details should be included in the model so that the model both accounts for the key biological processes, but is still amenable to mathematical or computational analysis. For a given network, one would like to classify all the attracting states and, since the olfactory system needs to be able to distinguish among many odors, it would be useful to understand under what conditions, on network architecture perhaps, a network exhibits a large number of attractors. Recall that in response to a given odor, the AL or OB responds with a transient series of dynamic clusters that converges to a stable firing pattern. It is possible that the animal recognizes and makes a decision on how to respond to the odor while this trajectory is still in its transient state, having not yet reached the stable firing pattern. This raises the mathematical question: How do properties of transients and attractors, including the length or time duration of each of these, depend on network properties such as the intrinsic properties of cells within the network and the network architecture? This question motivates most of the work presented in this chapter.

6.2.4 Mathematical Models

The seminal model in neuroscience is due to Hodgkin and Huxley and was published in 1952 [4]. This model describes the propagation of action potentials in the giant axon of a squid. The principles underlying the derivation of this model form the basis for the modeling of other cells throughout the nervous system. The Hodgkin-Huxley model consists of four differential equations. One of these is a partial differential equation that describes the evolution of the membrane potential. The other three equations are ordinary differential equations that are related to the opening of ionic channels in the cell membrane. It is the opening and closing of these channels that are responsible for changes in the membrane potential and, therefore, the generation of the action potential.

The Hodgkin-Huxley model is quite complicated and very difficult to analyze mathematically, especially if one considers networks of interacting neurons. For this reason, simpler models have been proposed. In the simplest models, each neuron has two states and time is discrete. A neuron is either "on" (firing an action potential) or "off" (at rest). There are then rules that describe when a neuron switches between these states and how the firing of one neuron impacts the state of other neurons in the network. Artificial neuronal network models of this type are appealing because they often allow for rigorous mathematical analysis and are easy to implement numerically as this chapter will outline in detail. But because of the models' simplicity and lack of biological detail, it is often not clear how to interpret results from such models biologically. A primary goal of the research program described in this chapter is to develop more realistic discrete-time models that are more closely related to the underlying biology. Section 6.8 describes some alternative models of this kind.

6.3 THE DISCRETE MODEL

Our notation will be mostly standard. The number of elements of a finite set A will be denoted by $|A|$, the symbol $\lfloor x \rfloor$ denotes the largest integer that does not exceed x, the least common multiple will be abbreviated by *lcm*, the greatest common divisor

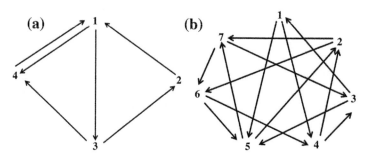

FIGURE 6.2

(a) A digraph with $V_D = [4]$ and (b) a digraph with $V_D = [7]$. Arcs $\langle i, j \rangle \in A_D$ are represented by arrows $i \to j$.

by *gcd*. For a positive integer n the symbol $[n]$ will be used as shorthand for the set $\{1, \ldots, n\}$.

Definition 6.1. A *directed graph* or *digraph* is a pair $D = \langle V_D, A_D \rangle$, where $A_D \subseteq V_D^2$. The set V_D is called the set of *vertices* or *nodes* of D, and the set A_D is called the set of *arcs* of D.

Digraphs are convenient models for all sorts of networks of interacting agents. The agents are elements of V_D, while an arc $\langle v, w \rangle \in A_D$ signifies that agent v may influence the state of agent w. Note that the interactions may not be symmetric; therefore directed graphs are usually more accurate models of networks than undirected graphs. Figure 6.2 gives a visual representation of two digraphs.

An arc of the form $\langle v, v \rangle$ is called a *loop*; a digraph that does not contain such arcs is called *loop-free*. In our intended interpretation, the set V_D stands for the set of neurons in the network, while an arc $\langle v, w \rangle \in A_D$ signifies that neuron v may send synaptic input to neuron w. Since we are not considering neuronal networks in which a neuron may send synaptic input to itself, the corresponding digraphs will be loop-free.

Moreover, it will be convenient for our purposes to consider only digraphs such that $V_D = [n]$ for some positive integer n.

Each of our network models will have an associated digraph D that specifies the *connectivity* of the network. While the connectivity is assumed here to remain fixed at all times, we are interested in the *dynamics* of our networks, that is, how the *state* of the network changes over time. Time t is assumed to progress in discrete steps or episodes, so that we will have a state $\vec{s}(t)$ of the network for each $t = 0, 1, 2, \ldots$ The *state $\vec{s}(t)$ of the network at time t* will simply be the vector $\vec{s}(t) = (s_1(t), s_2(t), \ldots, s_n(t))$, where $s_i(t)$ is the state of node $i \in V_D$ at time t. According to what was said in Section 6.2, at any given time step, a neuron i may either fire, or be at rest. We will assume that firing lasts exactly one episode, and neuron i needs a *refractory period* of p_i time steps before it can fire again, where p_i is a positive integer. We will allow the refractory periods p_i to differ from neuron to neuron. Each of our models will have a parameter

$\vec{p} = (p_1, \ldots, p_n)$ which specifies the refractory periods of all neurons. The state $s_i(t)$ of each neuron needs to specify whether or not the neuron fires at time t and whether it has rested long enough to fire again. We code this information as follows: $s_i(t) = 0$ means that the neuron fires at time t; more generally, a state $s_i(t) < p_i$ signifies that neuron i fired $s_i(t)$ time steps earlier, that is, in episode number $t - s_i(t)$ and is not yet ready to fire in episode $t + 1$; and $s_i(t) = p_i$ signifies that neuron i has reached the end of its refractory period and is ready to fire in episode $t + 1$.

When will a neuron i that is ready to fire ($s_i(t) = p_i$) actually fire in episode $t+1$? At time t, neuron i receives firing inputs from all neurons j with $\langle j, i \rangle \in A_D$ and $s_j(t) = 0$, and it will fire at time $t + 1$ if and only if there are sufficiently many such neurons. The exact meaning of "sufficiently many" may again differ from neuron to neuron and is conceptualized by a positive integer th_i that we call the ith *firing threshold*. The vector $th = (th_1, \ldots, th_n)$ will be another parameter of our models.

Now we are ready for a formal definition.

Definition 6.2.   A *neuronal network* is a triple $N = \langle D, \vec{p}, \vec{th} \rangle$, where $D = \langle [n], A_D \rangle$ is a loop-free digraph and $\vec{p} = (p_1, \ldots, p_n)$ and $\vec{th} = (th_1, \ldots, th_n)$ are vectors of positive integers. The state $\vec{s}$ of the system at time t is a vector $\vec{s}(t) = (s_1(t), \ldots, s_n(t))$, where $s_i(t) \in \{0, 1, \ldots, p_i\}$ for all $i \in [n]$. The state space of N will be denoted by St_N. The dynamics of N is defined as follows:

- If $s_i(t) < p_i$, then $s_i(t + 1) = s_i(t) + 1$.
- If $s_i(t) = p_i$ and there exist at least th_i different $j \in [n]$ with $s_j(t) = 0$ and $\langle j, i \rangle \in A_D$, then $s_i(t + 1) = 0$.
- If $s_i(t) = p_i$ and there are fewer than th_i different $j \in [n]$ with $s_j(t) = 0$ and $\langle j, i \rangle \in A_D$, then $s_i(t + 1) = p_i$.

A state $\vec{s}(0)$ at time 0 will be called an *initial state*. Each initial state uniquely determines a *trajectory* $\vec{s}(0), \vec{s}(1), \vec{s}(2), \ldots, \vec{s}(t), \ldots$ of successive states for all times t.

Example 6.3.   Consider the neuronal network $N = \langle D, \vec{p}, \vec{th} \rangle$, where D is the digraph of Figure 6.2a, $\vec{p} = (1, 1, 1, 2)$, $\vec{th} = (2, 1, 1, 1)$. Find $\vec{s}(1), \vec{s}(2), \vec{s}(3)$ for the following initial states: (a) $\vec{s}(0) = (0, 1, 1, 1)$, (b) $\vec{s}(0) = (1, 0, 0, 0)$, (c) $\vec{s}(0) = (0, 1, 1, 2)$.

Solution: a.   If $\vec{s}(0) = (0, 1, 1, 1)$, node 1 fires and thus cannot fire in the next time step, node 2 will fire in the next step since it is at the end of its refractory period and receives firing input from $th_2 = 1$ node, node 3 does not receive any firing input and will stay in state $p_3 = 1$, and node 4 will reach the end of its refractory period $p_4 = 2$. Thus we get $\vec{s}(1) = (1, 0, 1, 2)$. Now nodes 3 and 4 receive firing input from the required number of nodes, while nodes 1 and 2 do not receive any such input. Thus $\vec{s}(2) = (1, 1, 0, 0)$. Now node 1 receives firing input from $th_1 = 2$ nodes, while node 2 does not receive such input and nodes 3 and 4 must enter their refractory periods. It follows that $\vec{s}(3) = (0, 1, 1, 1)$.

b. If $\vec{s}(0) = (1, 0, 0, 0)$, then in the next step node 1 receives firing input from $2 \geqslant th_1$ nodes and will fire in step 1. Thus $\vec{s}(1) = (0, 1, 1, 1)$, and it follows from our calculations for part (a) that $\vec{s}(2) = (1, 0, 1, 2)$ and $\vec{s}(3) = (1, 1, 0, 0)$.

c. If $\vec{s}(0) = (0, 1, 1, 2)$, then $\vec{s}(1) = (1, 0, 1, 0)$. Now node 1 receives firing input from only $1 < th_1$ node and will not fire in the next step, but node 3 receives firing input from $1 \geqslant th_3$ node and will fire. Thus $\vec{s}(2) = (1, 1, 0, 1)$. Similarly $\vec{s}(3) = (1, 1, 1, 2)$.   □

Example 6.3 illustrates a number of important phenomena in the dynamics of our networks. First of all, notice that in part (a) we got $\vec{s}(3) = \vec{s}(0)$. Thus $\vec{s}(4) = \vec{s}(1), \vec{s}(5) = \vec{s}(2)$, and so on. The trajectory will cycle indefinitely through the set of states $AT = \{\vec{s}(0), \vec{s}(1), \vec{s}(2)\}$. The states in AT are called *persistent states* because every trajectory that visits one of these states will return to it infinitely often, and the set AT itself is called the *attractor* of $\vec{s}(0)$.

The meaning of the word "attractor" will become clear when we consider the solution to part (b). Notice that $\vec{s}(1)$ is in the same set AT as in the previous paragraph, so the trajectory will again cycle indefinitely through this set. In particular, the initial state $\vec{s}(0) = (1, 0, 0, 0)$ will never be visited again along the trajectory; it is a *transient state*. By Dirichlet's Pigeonhole Principle, this kind of thing must always happen in a discrete dynamical system with a finite-state space: Every trajectory will eventually return to a state it has already visited and will from then on cycle indefinitely through its attractor. The states that a trajectory visits before reaching its attractor occur only once. They form the *transient* (part) of the trajectory. Note that in part (b) of Example 6.3 the *length of the transient*, i.e., the number of its transient states, is 1; whereas in part (a) the length of the transient is 0.

The trajectories of the initial states in parts (a) and (b) will be out of sync by one time step; nevertheless, they both end up in the same attractor AT that has a length of 3.

In part (c) of Example 6.3, we get $\vec{s}(3) = (1, 1, 1, 2) = \vec{p}$. In this state, every neuron has reached the end of its refractory period and no neuron fires. Thus no neuron fires at any subsequent time, the length of the transient is 3, and we get $\vec{s}(3) = \vec{s}(4) = \ldots$ It follows that the set $\{\vec{p}\}$ is an attractor of length 1. Attractors of length 1 are called *steady state attractors* and their unique elements are called *steady states*. In contrast, attractors of length > 1 are *periodic attractors*.

Exercise 6.1.   Prove that $\vec{p}$ is the only steady state in any network $N = \langle D, \vec{p}, \vec{th} \rangle$.   ▽

The set of all initial states whose trajectories will eventually reach a given attractor AT is called the *basin of attraction of AT*. The initial states in parts (a) and (b) of Example 6.3 belong to the same basin of attraction, while the initial state in part (c) belongs to a different basin of attraction, namely the basin of the steady state.

Exercise 6.2.   Let $N = \langle D, \vec{p}, \vec{th} \rangle$ be a neuronal network and let $\vec{s}(0) \in St_N$ be a state such that $s_i > 0$ for all $i \in [n]$. Prove that $\vec{s}(0)$ is in the basin of attraction of the steady state attractor and determine the length of the transient.   ▽

Example 6.3(c) shows that the condition on $\vec{s}(0)$ in Exercise 6.2 is sufficient, but not necessary, for $\vec{s}(0)$ to be in the basin of attraction of the steady state attractor.

Definition 6.4.   Let $D = \langle V_D, A_D \rangle$ be a digraph. A *directed path* in D from v_1 to v_k is a sequence $(v_1, v_2, \ldots, v_k)$ of vertices $v_\ell \in V_D$ such that $\langle v_\ell, v_{\ell+1} \rangle \in A_D$ for all $\ell \in [k-1]$, and all vertices in the sequence, except possibly v_1 and v_k, are pairwise distinct. A *directed cycle* in D is a directed path $(v_1, v_2, \ldots, v_k)$ with $v_1 = v_k$. The *length* of a directed path (or cycle) is the number of arcs that connect its successive vertices.

For example, in the digraph of Figure 6.2b, the sequence $(1, 5, 2, 7, 3)$ is a directed path of length 4 and the sequence $(2, 7, 6, 4, 2)$ is a directed cycle of length 4. Two directed cycles will be considered disjoint if their set of vertices are disjoint. In particular, the directed cycles $(7, 3, 5, 7)$ and $(2, 7, 6, 4, 2)$ in this digraph are not disjoint despite the fact that they do not use any common arcs. We will usually consider directed cycles that differ by a cyclic shift as identical; for example, $(7, 3, 5, 7)$, $(3, 5, 7, 3)$, and $(5, 7, 3, 5)$ would be considered three representations of the same directed cycle.

Exercise 6.3.   Find all directed cycles in the digraph of Figure 6.2a. What is the maximum length of a directed path in this digraph?   ▽

For any neuronal network $N = \langle D, \vec{p}, \vec{th} \rangle$ we can define another digraph D_N whose vertices are the states and whose arcs indicate the *successor state* for each state, that is, $\langle \vec{s}, \vec{s}^* \rangle$ is an arc in this digraph if and only if $\vec{s}(t) = \vec{s}$ implies $\vec{s}(t+1) = \vec{s}^*(t)$. This digraph will be called the *state transition digraph* of the network. Note that if $\vec{p} = \vec{1} = (1, 1, \ldots, 1)$, then we can represent each state of a network N very conveniently by the set of all nodes that fire in this state. If the number of nodes is not too large, this allows us to actually draw D_N. Figure 6.3 gives an example.

Notice that the state transition digraph D_N of a network $N = \langle D, \vec{p}, \vec{th} \rangle$ is different from the *network connectivity* D. For example, D_N has many more nodes than D ($2^7 = 128$ vs. 7 in the example of Figures 6.3 and 6.2b). Moreover, while we are always assuming that the network connectivity D is loop-free, the state transition digraph contains the loop $\langle \vec{p}, \vec{p} \rangle$.

If $\vec{p} \neq \vec{1}$ the state transition digraph D_N must be drawn in such a way that each state is actually represented by its state vector. This may still be possible for very small n, as in Exercise 6.5.

State transition digraphs, as long as we can draw them, give a complete visual picture of the network dynamics. Let us take a closer look at Figure 6.3. The steady state is represented by the empty set of its firing nodes and forms a loop $\langle \emptyset, \emptyset \rangle$, that is, a directed cycle of length one. There are six directed cycles of length 2 each, and there is one directed cycle of length 5. These directed cycles of lengths larger than 1 correspond to periodic attractors of the network. All nodes outside of the union of these directed paths are transient states. For each transient state $\vec{s}$ there exists a unique directed path from $\vec{s}$ to a state in some attractor that visits no other persistent states. The basin of attraction of each attractor consists of all states from which the attractor can be reached via a directed path. The attractors $\{(124), (3567)\}$, $\{(147), (2356)\}$, $\{(157), (2346)\}$ and the steady state attractor $\{\emptyset\}$ form their own basins of attraction; the basins of

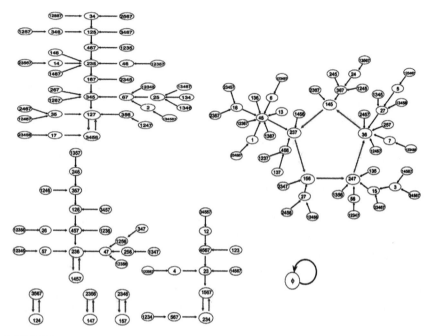

FIGURE 6.3

State transition digraph for the network $N = \langle D, \vec{1}, \vec{1} \rangle$ with connectivity D as in Figure 6.2(b). Adapted from [5].

attraction of the remaining four attractors are much larger and contain several transient states each. Possible transients show up as directed paths outside of the union of the attractors that end in a state that is succeeded by a state in the attractor. For example, $\{(34567),(12),(4567),(23)\}$ and $\{(12),(4567),(23)\}$ are transients for the trajectories with initial states coded by (34567) and (12), respectively, but $\{(34567),(12),(4567)\}$ cannot be the transient of any trajectory.

Exercise 6.4.   What is the maximum length of a transient for the system of Figure 6.3? List all initial states for which transients have this maximum length.   ▽

Exercise 6.5.

a. Draw the state transition digraph for the neuronal network N of Example 6.3. *Hint:* The network has 24 distinct states.

b. Find all attractors, their lengths, and the sizes of the basins of attraction for the neuronal network N of Example 6.3.

c. What is the maximum length of a transient in this network?   ▽

All our networks have exactly one steady state attractor. When does a network also have other periodic attractors? We have seen that such attractors correspond to directed cycles of length bigger than one in the state transition digraph D_N. Moreover, for the two examples with periodic attractors that we have explored so far, we also found directed cycles in the network connectivity D. Is the existence of directed cycles in D necessary for the existence of periodic attractors? Is it sufficient? Note that in essence these are questions about how the network connectivity D is related to the network dynamics. We will study this type of question in the next two sections. In Section 6.4 we will try to derive provable bounds on certain features of the dynamics such as maximum possible lengths of transients and attractors when D belongs to some special classes of digraphs. In Section 6.5 we will derive bounds that are true "on average" for connectivities D that are randomly drawn from certain probability distributions.

Numerical simulations are another powerful tool for exploring network dynamics. For a given network one could in principle use software to track the trajectory of every possible initial state until it visits the same state twice and record the length of all transients, all attractors, their lengths, and sizes of their basins of attraction. This is possible in practice when the number of nodes n is relatively small, but quickly becomes computationally infeasible since the number of states grows at least exponentially in n.

Exercise 6.6. Suppose $N = \langle D, \vec{p}, \vec{th} \rangle$ is a network with n nodes. Show that the size of the state space is

$$|St_N| = \prod_{i=1}^{n} (p_i + 1). \tag{6.1}$$

Conclude that if $\vec{p} = (p, \ldots, p)$ is constant, then $|St_N| = (p+1)^n$, and for arbitrary $\vec{p}$ we always have $|St_N| \geqslant 2^n$. $\triangledown$

Thus even networks with 30 nodes and $\vec{p} = \vec{1}$ contain about a billion different states. For larger networks one can still explore the trajectories of a few randomly chosen initial states, but it may be computationally infeasible to run the simulation until the same state will be reached twice. Exercise 6.17 of the online supplement [1] gives an illustration that even networks with very few nodes can have very long transients and attractors. Thus rigorous results about the dynamics of all but the smallest networks will require proving theorems. However, computer simulations can be very useful for generating conjectures for such results and giving us insight into the mechanisms that are responsible for interesting features of the dynamics. The online supplement for this chapter contains a number of projects that involve such numerical explorations.

6.4   EXPLORING THE MODEL FOR SOME SIMPLE CONNECTIVITIES

Recall that in Exercise 6.5 we explored the following numerical characteristics of network dynamics: lengths of the attractors, number of different attractors, sizes of their basins of attraction, and maximum lengths of transients. In this section we will derive bounds on these numbers for networks whose connectivities D belong to one of several important classes of digraphs. These bounds will be expressed as functions of the number n of neurons. We will usually need to make some assumptions on $\vec{p}$ and $\vec{th}$. Throughout this chapter p^* denotes max $\vec{p}$ and th^* denotes max $\vec{th}$. We are primarily interested in bounds that hold for all n and fixed values of p^* and th^*.

6.4.1 Acyclic Digraphs

A digraph D is *acyclic* if D does not contain any directed cycle. While none of the digraphs in Figures 6.2 and 6.3 is acyclic, Figure 6.4 gives an example of an acyclic digraph.

Lemma 6.5. *Let $N = \langle D, \vec{p}, \vec{th} \rangle$ be a neuronal network whose connectivity D is acyclic. Then $\{\vec{p}\}$ is the only attractor in N.*

Proof. Let N, D be as in the assumptions and assume $D = \langle [n], A_D \rangle$. Suppose there exists a periodic attractor AT in N, and let $\vec{s}(0)$ be a state in this attractor. Move the system forward to time $t = n$. By Exercise 6.2, some node fires in $\vec{s}(t)$, that is, there must exist $i \in [n]$ with $s_i(t) = 0$. Fix such $i = i(t)$. By the rules of the network dynamics, there must exist another node $i(t-1)$ with $\langle i(t-1), i(t) \rangle \in A_D$ such that $s_{i(t-1)}(t-1) = 0$. By recursion we can now construct a sequence of nodes $i(t), i(t-1), \ldots, i(0) \in [n]$ such that $s_{i(t-k)}(t-k) = 0$ for all $k \in \{0, \ldots, n\}$ and $\langle i(t-k), i(t-k+1) \rangle \in A_D$ for all $k \in [n]$. Since there are only n nodes total, there must exist $\ell < k$ such that $i(\ell) = i(k)$ and $(i(\ell), i(\ell+1), \ldots, i(k))$ is a directed cycle in D, which contradicts the assumption on D. □

Lemma 6.5 makes networks with acyclic connectivities rather uninteresting from our point of view, but investigating the class of acyclic digraphs will allow us to lay some groundwork for subsequent explorations. First of all, notice that Lemma 6.5 completely characterizes all four features of the dynamics that we are investigating here in that it implies points (a)–(d) of the following result.

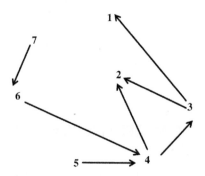

FIGURE 6.4

An acyclic digraph on [7]. Arcs $\langle i, j \rangle \in A_D$ are represented by arrows $i \to j$.

Theorem 6.6. *Let* $N = \langle D, \vec{p}, \vec{th} \rangle$ *be a network whose connectivity D is acyclic. Then*

a. *the maximum length of each attractor in N is 1,*
b. *the maximum number of different attractors in N is 1,*
c. *the size of the basin of attraction of the steady state attractor is* $|St_N| = \prod_{i=1}^{n}(p_i + 1),$
d. *the transients have length at most* $n + p^* - 1.$

Exercise 6.7. Prove part (d). *Hint:* Use arguments similar to the proof of Lemma 6.5 and the solution of Exercise 6.2.   ▽

The wording of Theorem 6.6 certainly looks overly formal for such a simple result, but it provides a nice template for the kinds of results we might hope to be able to prove for certain classes of digraphs. The bound in point (d) is sharp in the class of all networks with acyclic connectivities, which can be easily seen by considering the trivial case of a digraph with just one node and no arcs. However, this bound can be improved under some additional assumptions on the connectivity. Project 1 of the online supplement [1] gives some guidelines for readers who wish to prove stricter bounds under additional assumptions.

Let us take another look at Lemma 6.5. It tells us that the existence of a directed cycle in the connectivity D is a *necessary* condition for the existence of a periodic attractor in a network $N = \langle D, \vec{p}, \vec{th} \rangle$, but it does not answer the question whether the existence of a directed cycle in D is a *sufficient* condition. This type of question is best explored by looking first at the simplest possible connectivities that contain a directed cycle. Our next subsection will investigate the possible dynamics for these types of connectivities.

6.4.2 Directed Cycle Graphs

Let n be a positive integer. For convenience, we will write $i \% n$ for $((i-1) \bmod n) + 1$. Let $D_c(n)$ be the *directed cycle graph* on $[n]$ with arc set $A_c(n) = \{\langle i, (i+1)\%n \rangle : i \in [n]\}$. Throughout this subsection we will tacitly assume that we are given a neuronal network $N = \langle D, \vec{p}, \vec{th} \rangle$ with $D = D_c(n)$ for some positive integer n.

In such networks, node i gets input *only* from node $(i-1)\%n$. Thus, if $i \in [n]$ is such that $th_i > 1$, then $s_i(t) \neq 0$ for all $t > 0$ and each trajectory will eventually reach the steady state $\vec{p}$. Therefore, we will focus here exclusively on the case $\vec{th} = \vec{1}$ which allows for more interesting dynamics.

Let $\vec{s}(0)$ be a given initial state. For all $t \geqslant 0$ let

$$Act_t = \{i \in [n] : s_i(t) = 0\} \tag{6.2}$$

be the set of neurons that fire at time t. We call Act_t the *active set at time t*.

Note that for cyclic connectivities, a firing of node i at time t can induce at most one firing at time $t + 1$, and it follows that $|Act_t| \geqslant |Act_{t+1}|$ along any trajectory, which in turn implies that $|Act_t|$ is constant on any attractor. Thus throughout any attractor a firing of node i at time t *always will* induce a firing of node $(i + 1)\%n$ at time $t + 1$. If t_0 is such that $|Act_{t+t_0}| = |Act_{t_0}|$ for all $t \geqslant 0$, then for all $t \geqslant 0$ we have

$$Act_{t+t_0} = \{(i + t)\%n : i \in Act_{t_0}\}. \tag{6.3}$$

Exercise 6.8.   Prove that there exists a divisor m of n such that $Act_{t_0+m} = Act_{t_0}$. $\triangledown$

For node i to fire at time t, the rules of the dynamics require that it must be in state p_i at time $t - 1$. After node i has fired at time t (formally: $s_i(t) = 0$), its state will increase until it reaches the end of its refractory period: $s_i(t + 1) = 1, \ldots, s_i(t + p_i) = p_i$. This implies that $m \geqslant p_i + 1$ for m as in Exercise 6.8, that is, the distance between two firing cells is at least $p_i + 1$. In particular, this will be true for i of maximum refractory period $p_i = p^*$. These observations prove the following.

Proposition 6.7.   *Let $N = \langle D, \vec{p}, \vec{1} \rangle$ be a neuronal network with $D = D_c(n)$. Then the length $|AT|$ of any attractor is a divisor of n and $|Act_t| \leqslant \lfloor \frac{n}{p^*+1} \rfloor$ for all $t > 0$. Moreover, if $|AT| > 1$, then $|AT| \geqslant p^* + 1$.*

How many *different* attractors are there in N?

It follows from our proof of Proposition 6.7 that along any attractor the state $s_i(t)$ of any node will be uniquely determined by its refractory period and the set Act_t. We may consider each attractor with fixed $|Act_t| = k > 0$ as a necklace with k red beads and $n - k(p^* + 1)$ black beads. Pick a state $\vec{s}$ in the attractor. Each red bead corresponds to a block of $p^* + 1$ positions of $\vec{s}$ of the form $(*, \ldots, *, 0)$, where the wildcards $*$ correspond to the stages in the refractory period of the corresponding neuron, but not to firing. This type of block represents one firing together with the mandatory waiting period that must precede the next firing. Black beads correspond to optional extensions of the minimal waiting period between two consecutive elements of Act_t. There is a one-to-one correspondence between periodic attractors in N and this type of necklace. Now the well-known formula for the number of different such necklaces (e.g., Theorem 1 in [6]) plus the fact that there exists exactly one steady state attractor implies the following result .

Theorem 6.8.   *Let $N = \langle D, \vec{p}, \vec{1} \rangle$ be a neuronal network with $D = D_c(n)$. Then the number of different attractors is*

$$1 + \sum_{k=1}^{\lfloor \frac{n}{p^*+1} \rfloor} \frac{1}{n - kp^*} \sum_{a \in \{divisors\ of\ gcd(k, n-k(p^*+1))\}} \phi(a) \binom{\frac{n-kp^*}{a}}{\frac{k}{a}}, \tag{6.4}$$

where ϕ is Euler's phi function.

In particular, the number of attractors in neuronal networks $N = \langle D_c(n), \vec{p}, \vec{1} \rangle$ grows exponentially in n. (See Corollary 2 in [7].) In order to get a feel for the dependence of this number on n, the reader may want to do some numerical explorations as suggested in Exercise 6.18 [1].

How about the lengths of the transients? Recall that $|Act_t|$ is a nonincreasing function of $t \geqslant 0$. Thus, for any trajectory there must be a smallest time t_0 such that $|Act_t| = |Act_{t_0}|$ for all $t \geqslant t_0$. We can get a bound for the lengths of transients by carefully considering how t_0 is related to the refractory periods.

Exercise 6.9. Let $N = \langle D, \vec{p}, \vec{1} \rangle$ be a neuronal network with $D = D_c(n)$. Consider a trajectory and let t_0 be defined as above. Prove that:

a. The length of the transient is at most $t_0 + p^* - 1$.
b. If $t_0 > 0$, then there exists a node i with $s_i(t_0 - 1) = 0$ and $s_{(i+1)\%n}(t_0) > 0$.
c. If $t_0 > p^*$, then for all nodes i as in (b) we must have $p_i < p_{(i+1)\%n}$.   ▽

If all nodes have the same refractory period, then the scenario of point (c) of Exercise 6.9 cannot occur, and point (b) implies that $t_0 \leqslant p^*$. In view of point (a), this in turn implies the following result of [7].

Theorem 6.9. *Let $N = \langle D, \vec{p}, \vec{1} \rangle$ be a neuronal network with $D = D_c(n)$, and $\vec{p} = (p, \dots, p)$ is constant. Then the length of any transient is at most $2p - 1$.*

If $\vec{p}$ is not a constant vector, then the scenario of point (c) of Exercise 6.9 can occur, which adds significant complications. Exercise 6.19 [1] invites the reader to explore one such example and Theorem 3 of [7] gives a precise upper bound of the length of transients for all networks $N = \langle D_c(n), \vec{p}, \vec{1} \rangle$.

How about the sizes of the basins of attraction? At the time of this writing no complete characterization for all cases was known. We invite our readers to explore this problem on their own. Project 2 [1] gives some guidance.

6.4.3 Complete Loop-Free Digraphs

Both directed cycles and acyclic digraphs are characterized by "small" arc sets in the sense that directed cycle graphs contain just enough arcs to allow for a directed cycle that visits all nodes, and acyclic digraphs are missing the arcs that would close a directed cycle. However, the word "small" here does not mean the same thing as "few in number."

Exercise 6.10. Show that for every $n > 1$ there exists an acyclic digraph $D = \langle [n], A_D \rangle$ with arc set of size $\binom{n}{2}$, which is exactly half of the maximum number of arcs in any loop-free digraph with n vertices.   ▽

On the other extreme are *complete loop-free digraphs* $D_{\max}(n) = \langle [n], A_D \rangle$ whose arc set A_D contains every pair $\langle i, j \rangle$ with $i, j \in [n]$ such that $i \neq j$. In this subsection we will explore the possible dynamics for networks of the form $N = \langle D_{\max}(n), \vec{p}, \vec{th} \rangle$. As before, let us first consider the simplest case when $\vec{p} = \vec{th} = \vec{1}$.

Exercise 6.11. Let $N = \langle D_{\max}(n), \vec{1}, \vec{1} \rangle$ for some n.

a. Show that along any trajectory we always have $Act_{t+1} = [n] \setminus Act_t$, unless $Act_t = \emptyset$, in which case the system has reached the steady state.

b. Use the result of (a) to find the maximum length of any transient and any attractor, the number of different attractors, and the size of each basin of attraction.   ▽

When we allow for $\vec{p} \neq \vec{1}$, things become more interesting. The result of Exercise 6.11(a) generalizes as follows:

Proposition 6.10. *Let* $N = \langle D_{max}(n), \vec{p}, \vec{1} \rangle$ *for some n. Then along any trajectory we always have* $Act_{t+1} = \{i \in [n] : s_i(t) = p_i\}$ *if* $Act_t \neq \emptyset$, *and* $Act_{t+1} = \emptyset$ *whenever* $Act_t = \emptyset$.

Proposition 6.10 allows for an easy characterization of all features of the dynamics when $\vec{p}$ is constant and $\vec{th} = \vec{1}$.

Exercise 6.12.   Let $N = \langle D_{max}(n), \vec{p}, \vec{1} \rangle$ for some n, where $\vec{p} = (p, \dots, p)$ is constant.

a. Show that the basin of attraction of the steady state consists of all vectors $\vec{s}$ such that for at least one $q \in \{0, \dots p\}$ we have $\{i \in [n] : s_i = q\} = \emptyset$.
b. Show that every periodic attractor has length $p + 1$ and forms its own basin of attraction.
c. Find a formula that relates the number of different attractors to the number of functions that map $[n]$ onto $\{0, 1, \dots, p\}$.
d. Find a formula for the size of the basin of attraction of the steady state attractor in the spirit of (c) and also find a formula for the maximum length of any transient.   ▽

Let us introduce here some terminology that we will use later in this and the next section.

Definition 6.11.   Let $N = \langle D, \vec{p}, \vec{th} \rangle$ be a network, let $i \in V_D$ be a node, and let $\vec{s}(0) \in St_N$ be an initial state. We say that i is *minimally cycling (in the trajectory of* $\vec{s}(0)$) if for every $t \geqslant 0$ the implication $s_i(t) = p_i \Rightarrow s_i(t + 1) = 0$ holds.

Let AT be an attractor in N. We say that node i is *minimally cycling in AT* if it is minimally cycling in every trajectory that starts in AT. A node i is *active in AT* if it fires in at least one state of AT. We say that the attractor AT is *minimal* if every active node in AT is minimally cycling, and we call AT *fully active* if every node of the network is active in AT.

The following exercise gives a nice and useful condition on initial states whose trajectories contain at least one minimally cycling node.

Exercise 6.13.   Let $N = \langle D, \vec{1}, \vec{1} \rangle$ and let $\vec{s}(0) \in St_N$.

a. Prove that if there exists a directed cycle $C = (i_1, i_2, \dots, i_{2k}, i_1)$ in D such that $s_{i_j}(0) = j \bmod 2$ for all $j \in [2k]$, then there exists a minimally cycling node in the trajectory of $\vec{s}(0)$.
b. Is the condition in point (a) also necessary for the existence of a minimally cycling node?   ▽

It is easy to see that Proposition 6.10 implies the following:

Corollary 6.12. *Let* $N = \langle D_{max}(n), \vec{p}, \vec{1} \rangle$ *for some n. Then every periodic attractor is fully active and minimal.*

As long as $\vec{p} = (p, \dots, p)$ is constant, Exercise 6.12(b) implies that every periodic attractor has length $p + 1$, which is the minimum possible length of any periodic attractor in networks with constant refractory periods and arbitrary connectivity. The latter observation explains why we call such attractors minimal. If $\vec{p}$ is not constant though, as the next exercise shows, minimal attractors are longer than that.

Exercise 6.14. Let $N = \langle D_{max}(n), \vec{p}, \vec{1} \rangle$ for some n. Show that the length of any periodic attractor AT is equal to $lcm\{p_i + 1 : i \in [n]\}$. $\triangledown$

One can deduce from the above that in contrast to Proposition 6.7 for directed cycle graphs, there does not exist a universal upper bound for the attractor length of networks with complete loop-free connectivities. Exercise 6.20 [1] gives more details and also shows that the same remark applies to the number of different attractors.

If $\vec{th} \neq \vec{1}$, then it may no longer be true that every periodic attractor is minimal or fully active, see Exercise 6.21 [1]. The example in this exercise is somewhat pathological though: For networks $N = \langle D_{max}(n), \vec{p}, \vec{th} \rangle$ with n sufficiently large relative to p^* and th^*, a typical initial state will still belong to a fully active minimal attractor. Moreover, the same is true for typical network $N = \langle D, \vec{p}, \vec{th} \rangle$ whenever the connectivity digraph has sufficiently many arcs. We will prove these results and spell out a precise meaning of "typical" in Section 6.5.

6.4.4 Other Connectivities

There are a number of other interesting classes of digraphs for which one might explore the possible dynamics of corresponding networks. This leads to more challenging problems than for the classes of acyclic, directed cycle, and complete loop-free digraphs that we have discussed so far. This territory remains largely unexplored. The Wikipedia page [8] contains a gallery of named graphs, and the reader may want to explore these questions for selected connectivities that are digraph counterparts of some of these graphs.

In this subsection we attempt to give a flavor of this area of research and suggest projects for open-ended exploration. As a sample from the large menu of possible research directions we will focus on two problems.

When does the existence of a directed cycle in the network connectivity imply the existence of a periodic attractor in the network? By Lemma 6.5 the existence of a directed cycle in D is necessary. But it is not all by itself sufficient; the directed cycle must also be long enough.

Exercise 6.15. Let $N = \langle D_c(n), \vec{p}, \vec{1} \rangle$. Deduce from the proof of Proposition 6.7 that N has a periodic attractor only if the length n of the directed cycle is at least $p^* + 1$. Show that the latter condition is also sufficient for the existence of a periodic attractor in this type of networks. $\triangledown$

There is a nice generalization of the second part of Exercise 6.15 to arbitrary networks $N = \langle D, \vec{p}, \vec{1} \rangle$ with $\vec{p}$ constant: These networks have a periodic attractor if and only if the connectivity D has a directed cycle of length at least $p^* + 1$ (see Theorem 14 of [9]). In particular, if $\vec{p} = \vec{th} = \vec{1}$, then the existence of any directed cycle in D implies the existence of a periodic attractor in N. But the proof is rather more difficult than Exercise 6.15.

Exercise 6.35 [1] shows that the straightforward generalization of this result to networks with unequal refractory periods fails. When we allow for firing thresholds $th_i > 1$, then for many interesting classes of digraphs the existence of a directed cycle of any length no longer will be sufficient for the existence of a periodic attractor in N, even if $\vec{p} = 1$. The problem of characterizing all networks that have at least one periodic attractor under the most general assumptions on $\vec{p}$ and $\vec{th}$ remains open. Project 4 [1] invites the reader to embark on an open-ended exploration of this problem, starting with some interesting special cases.

The second problem we want to discuss here is finding optimal bounds for the maximal length of attractors. For the classes of connectivities we have studied so far, all attractors were relatively short (the maximal length is 1 for acyclic connectivities, n for directed cycle graphs; for complete digraphs when $\vec{th} = \vec{1}$ it is $lcm \{p_i + 1 : i \in [n]\}$, which is typically much less than n when there are only few distinct types of neurons and n is large). But in general, attractors can be much longer than n; some examples are given in Exercise 6.17 and Project 5 [1]. What is the upper bound on attractor length in the class of all networks $N = \langle D, \vec{p}, \vec{th} \rangle$ with n nodes, $p^* \leqslant P$ and $th^* \leqslant Th$, where P, Th are fixed? This is an open question even for $P = Th = 1$. Project 5 [1] reviews some partial results and gives guidance for readers who wish to explore this exciting problem.

6.4.5 Discussion: Advantages and Limitations of the Approach in this Section

The results of this section and the corresponding parts of [1] show that the study of relations between network connectivity and network dynamics is a fertile ground for exciting mathematical problems on the intersection of several areas of discrete mathematics, including graph theory, enumerative combinatorics, and number theory. The problems in our exercises and projects range from very easy ones to unsolved research problems. One can see that the problem of completely characterizing all possible dynamics becomes quickly very hard if we allow connectivities beyond the simplest ones. This may be good news for mathematicians, because solving these questions may well require development of new mathematical tools. It has even been claimed that this type of investigation will lead to an entirely new kind of science [10]. While the latter claim seems exaggerated, the hyperbole with which it has been promoted should not deter mathematicians' efforts to develop new tools for studying the dynamics of discrete-time finite-state networks.

The question arises to what extent the approach taken in this section is likely to yield results that are relevant to neuroscience. After all, with very few exceptions the connectivity of neuronal networks in actual organisms is not known. The one organism for which we have a complete wiring diagram of its neuronal network is the little roundworm *Caenorhabditis elegans*. Its hermaphrodite form has 302 neurons. The website [11] has a lot of fascinating information about this organism and its nervous system. For other organisms we have wiring diagrams of important parts of the nervous system, such as the stomatogastric ganglia of some crustaceans [12] which contain between 20 and 30 neurons.

The connectivities of networks that have been mapped are much more complicated than acyclic, directed cycle, or complete digraphs, and our results do not apply to them directly. However, let us assume that we have a trustworthy discrete-time finite-state model for the dynamics of such a model (we will discuss this assumption in some detail in Section 6.7). If the network is *very* small (like in the ganglia of crustaceans), one could use computer simulations as in the present section to completely characterize the dynamics. This will no longer be possible in a network like the one for *C. elegans* with its 302 neurons, since there would be at least 2^{302} states to consider. But simulations still could be used to explore the dynamics for a large sample of initial states. Theoretical investigations like the ones in this section are still useful for this type of exploration: Let, for example, $\vec{s}(0)$ be an initial state in a network $N = \langle D, \vec{1}, \vec{1} \rangle$, where D is strongly connected and has a few hundred nodes. Suppose that in simulations of the trajectory for, say, 1000 steps, we do not find (yet) an attractor, but we discover a node for which the result of Exercise 6.13 guarantees that it will be minimally cycling. While extending the simulation until the trajectory visits the same state twice may not be computationally feasible, by Proposition 12 of [7] we would know already that all nodes in the attractor will be minimally cycling and the attractor will have length 2.

What if the neuronal network in the organism of interest is too large to map all the connections? In that case, we cannot directly investigate the dynamics of the true network, but we can explore the "typical" dynamics of a network that is randomly drawn from the class of all networks that share certain properties for which we do have empirical confirmation. Two such properties that are often known are an estimate of the total number n of neurons and an estimate of the average number of connections that a given neuron makes to other neurons. In the next section and Projects 6 and 7 [1], we will introduce some methods for exploring "typical" dynamics. As it will become clear, such methods rely on theoretical results of the kind that we proved in the present section.

6.5 EXPLORING THE MODEL FOR SOME RANDOM CONNECTIVITIES

6.5.1 Erdős-Rényi Random Digraphs

The notion of a "typical" connectivity can be made mathematically rigorous by assuming that D is a digraph with vertex set $[n]$ that is randomly drawn from a given probability distribution $\mathcal{D}_n$. If Ch is a characteristic property of digraphs, then we can say that *a typical digraph drawn from these distributions has property Ch* if the probability that a digraph drawn from $\mathcal{D}_n$ has property Ch approaches 1 as $n \to \infty$. We can talk about *typical dynamics* of networks with connectivities drawn from the distributions $\mathcal{D}_n$ in a similar fashion.

The most straightforward implementation of this idea is obtained by assuming that $\mathcal{D}_n$ is the uniform distribution on the set of all loop-free digraphs $D = \langle [n], A_D \rangle$. For this distribution, consider a potential arc $\langle i, j \rangle$ with $i, j \in [n]$ and $i \neq j$. Notice that $\langle i, j \rangle$ is contained in the arc set of exactly half of all digraphs in $\mathcal{D}_n$. Thus $P(\langle i, j \rangle \in A_D) = 0.5$. Similarly, if $\langle i_1, j_1 \rangle, \dots, \langle i_\ell, j_\ell \rangle, \dots, \langle i_m, j_m \rangle$ are pairwise distinct potential arcs, then

$$P(\forall k \in [\ell] \langle i_k, j_k \rangle \in A_D \text{ and } \forall k \in [m] \setminus [\ell] \langle i_k, j_k \rangle \notin A_D) = 2^{-m}. \tag{6.5}$$

Thus the events $\langle i, j \rangle \in A_D$ are *independent*. This formula is very useful. For example, it immediately implies the following:

Proposition 6.13. *In a typical loop-free digraph drawn from the uniform distribution every two nodes will be connected by a path of length 2; in particular, the digraph will be strongly connected.*

Proof. If $i \neq j$ are different nodes, then

$$P(\forall k \in [n] \setminus \{i, j\} \ \langle i, k \rangle \notin A_D \text{ or } \langle k, j \rangle \notin A_D) = \frac{3^{n-2}}{4^{n-2}}. \tag{6.6}$$

Note that Eq. (6.6) gives the probability of the event that there does *not* exist a directed path of length 2 from i to j. Now consider the event F that for *some i, j* with $i \neq j$ there does not exist a path of length 2 from i to j. There are $n(n-1)$ ordered pairs $\langle i, j \rangle$ with $i, j \in [n], i \neq j$, and the probability of a union of events is always bounded from above by the sum of their probabilities. It follows from Eq. (6.6) that $P(F) \leqslant n(n-1)\frac{3^{n-2}}{4^{n-2}}$.

Note that $\lim_{n \to \infty} n(n-1)\frac{3^{n-2}}{4^{n-2}} = 0$. Thus $\lim_{n \to \infty} P(F^c) = 1$, where F^c denotes the complement of event F. $\qquad \square$

It follows from Eq. (6.5) that $\mathcal{D}_n$ is exactly the probability distribution of random digraphs on $[n]$ that can be constructed in the following way: For each potential arc $\langle i, j \rangle$, flip a fair coin. Include $\langle i, j \rangle$ in A_D if and only if the coin comes up heads. One can generalize this construction to the case where the coin is not fair and comes up heads with probability p, which can be an arbitrary number with $0 < p < 1$. The corresponding probability distributions will be denoted by $\mathcal{D}_n(p)$. These distributions give the classes of so-called *Erdős-Rényi* random loop-free digraphs.

Definition 6.14. Let $D = \langle V_D, A_D \rangle$ be a digraph and let $v \in V_D$. The *indegree* of v, denoted by $ind(v)$, is the number of $w \in V_D$ with $\langle w, v \rangle \in A_D$. The *outdegree* of v, denoted by $outd(v)$, is the number of $w \in V_D$ with $\langle v, w \rangle \in A_D$.

For example, in the digraph of Figure 6.2a, we have $ind(1) = outd(1) = 2 = outd(3)$, while $ind(3) = 1$. In a directed cycle graph $D_c(n)$ each node has indegree and outdegree 1; in a complete loop-free digraph $D_{max}(n)$ each node has both indegree and outdegree $n - 1$.

For nodes v in random digraphs, both $ind(v)$ and $outd(v)$ are random variables. It is easy to calculate the expected values of indegrees and outdegrees in Erdős-Rényi digraphs.

Proposition 6.15. *Let D be randomly drawn from $\mathcal{D}_n(p)$, and let $i \in [n]$. Then*

$$E(ind(i)) = E(outd(i)) = p(n - 1). \tag{6.7}$$

Proof.   This proof exemplifies a technique that is extremely useful in studying random structures. For every potential arc $\langle i, j \rangle$, define a random variable $\xi_{i,j}$ that takes the value 1 if $\langle i, j \rangle \in A_D$ and takes the value 0 if $\langle i, j \rangle \notin A_D$. Then $E(\xi_{i,j}) = p \cdot 1 + (1 - p) \cdot 0 = p$. Moreover,

$$ind(i) = \sum_{j \in [n] \setminus \{i\}} \xi_{j,i}, \quad outd(i) = \sum_{j \in [n] \setminus \{i\}} \xi_{i,j}. \tag{6.8}$$

Since the expected value of a sum of random variables is the sum of their expected values, we get

$$E(ind(i)) = \sum_{j \in [n] \setminus \{i\}} E(\xi_{j,i}) = p(n - 1) = \sum_{j \in [n] \setminus \{i\}} E(\xi_{i,j}) = E(outd(i)), \tag{6.9}$$

as required.   □

The average indegrees and outdegrees correspond to the average number of connections per neuron that can be estimated from empirical results even for those networks where we do not know the actual connectivity graph. Notice that since the uniform distribution $\mathcal{D}_n$ is equal to $\mathcal{D}_n(0.5)$, nodes in a random loop-free digraph that is drawn from this distribution will have average indegrees and outdegrees equal to $0.5(n - 1) \approx 0.5n$. However, empirical results indicate that in neuronal networks of actual organisms each neuron usually forms a lot fewer connections than that. For example, as we mentioned in Section 6.2, the human brain contains roughly 10^{12} neurons, while individual neurons receive on average input from approximately 10^3 other neurons. This makes distributions $\mathcal{D}_n(p)$ with relatively small p more promising candidates for predicting the dynamics of such networks than the uniform distribution. The probability p may in general depend on n, and we will often write $\mathcal{D}_n(p(n))$ instead of $\mathcal{D}_n(p)$ to emphasize this dependence.

There is a substantial literature on the study of Erdős-Rényi random *graphs,* which are objects similar to digraphs, but with undirected *edges* $\{i, j\}$ instead of directed arcs $\langle i, j \rangle$. These results tend to have straightforward translations into results on Erdős-Rényi random digraphs. Project 6 [1] gives some references to the literature on the subject. Many of these results go as follows: Suppose $p(n)$ is a function of the number of nodes n such that $\lim_{n\to\infty} p(n) = 0$, and Ch is a property of (di) graphs. Then there exists a threshold function $T(n)$ such that whenever $p(n)$ approaches zero faster than $T(n)$ we have $\lim_{n\to\infty} P(Ch(D)) = 0$ for D drawn from $\mathcal{D}_n(p(n))$ (that is, property Ch fails almost certainly for such random D), and whenever $p(n)$ approaches zero more slowly than $T(n)$ we have $\lim_{n\to\infty} P(Ch(D)) = 1$ for D drawn from $\mathcal{D}_n(p(n))$ (that is, almost all such random D have property Ch). This phenomenon is referred to as a *phase transition* with threshold function $T(n)$. The precise meaning of "faster" and "more slowly" may differ somewhat from property to property, and for any given property there are actually infinitely many functions that could serve as a threshold function in the above sense. One of them will be the function $T_{Ch}(n)$ such that $P(Ch(D)) = 0.5$ whenever D is randomly drawn from the distribution $\mathcal{D}_n(T_{Ch}(n))$, and we will refer to it as "the threshold function for Ch." It is usually impossible to find a precise formula for this function; one has to restrict oneself to the more modest goal of estimating how quickly it approaches zero as $n \to \infty$.

For example, if Ch is the property of being strongly connected, then a threshold function is $T(n) = \frac{\ln n}{2n}$ (see the discussion at the end of Project 6 [1]). Since $\frac{\ln n}{2n} < 0.5$ for all n, Proposition 6.13 is a special case of this more general result.

One can try to derive similar results for properties of the dynamics of networks $N = \langle D, \vec{p}, \vec{th} \rangle$, where D is randomly drawn from $\mathcal{D}_n(p(n))$. Results of this type were obtained in [13]; here we will briefly review the most important of these and then explore two other properties that were not studied in [13]. For ease of exposition we will restrict ourselves to the case where $\vec{p} = \vec{th} = \vec{1}$.

The dynamical properties we are interested in concern the trajectories of a "typical" initial state in such networks. Thus we assume that we draw a random digraph from $\mathcal{D}_n(p(n))$, and then we draw an initial state $\vec{s}(0)$ (independently of the choice of D) from the set $\{0, 1\}^n$ of all possible initial states with the uniform distribution.

Let $FAMA$ be the property that $\vec{s}(0)$ belongs to a fully active minimal attractor. In [13], we were able to show that if $\lim_{n\to\infty} \frac{np(n)}{\ln n} = 0$, then almost no initial states will belong to a fully active minimal attractor (not even a minimal one, actually), and if $\lim_{n\to\infty} \frac{np(n)}{\ln n} = \infty$, then almost all initial states will belong to a fully active minimal attractor. This indicates that $T_{FAMA}(n) \approx c_{FAMA} \frac{\ln n}{n}$ for some constant c_{FAMA}. Numerical simulations confirm that the phase transition is very steep, see Figure 6.5a. They also suggest that $c_{FAMA} \approx 2.95$ for $\vec{p} = \vec{th} = \vec{1}$; see Figure 6.5b.

This makes networks with highly connected Erdős-Rényi connectivities in which the average indegrees and outdegrees grow faster than $\ln n$ (see Proposition 6.15) very similar to networks with connectivities that are complete loop-free digraphs (compare with Corollary 6.12) and rather uninteresting, since in almost all trajectories every node will fire as soon as it reaches the end of its refractory period.

For slightly sparser connectivities, a typical initial state will not be in a minimal attractor, but will have a property that we label $AUTO$. It means there exists an *autonomous set* (see [13] for a precise definition) of minimally cycling nodes. Results in [13] indicate that there is a constant c_{AUTO} such that $T_{AUTO}(n) \approx \frac{c_{AUTO}}{n}$. Again, numerical simulations confirmed that the phase transition is very steep, as in

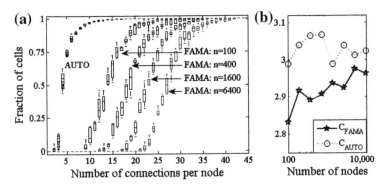

FIGURE 6.5

(a) Location of the phase transitions for networks of different sizes n. The connection probability is scaled so that the average number of connections per node is displayed on the horizontal axis. (b) Empirical values of c_{FAMA} and c_{AUTO} for different n. Adapted from [13].

Figure 6.5a. They also suggest that $c_{AUTO} \approx 3$ for $\vec{p} = \vec{th} = \vec{1}$; see Figure 6.5b. Thus for networks with connectivities of average indegrees and outdegrees $> c_{AUTO} \approx 3$, in the trajectory of almost every initial state the (vast) majority of nodes will be minimally cycling.

One should expect more interesting dynamics for connectivities with average indegrees and outdegrees <3. Consider the property $MNODE$ that the trajectory of $\vec{s}(0)$ contains at least one minimally cycling node, and the property NS that $\vec{s}(0)$ does not belong to the basin of attraction of the steady state attractor. Then $FAMA \Rightarrow AUTO \Rightarrow MNODE \Rightarrow NS$, and none of these implications can be reversed. It follows that for sufficiently large n the corresponding threshold functions should satisfy the inequalities

$$T_{NS}(n) \leqslant T_{MNODE}(n) \leqslant T_{AUTO}(n) \leqslant T_{FAMA}(n).$$

In particular, one would expect that $T_{NS}(n) \leqslant T_{MNODE}(n) < \frac{3}{n}$. Project 6 [1] invites the reader to explore the properties NS and $MNODE$. The project is structured in such a way that the reader will learn along the way some basic proof techniques for exploring random structures, in particular, the so-called *second-moment method*.

6.5.2 Discussion: Other Models of Random Digraphs

How relevant are results on random networks for neuroscience and real-world applications in general? As we already mentioned, usually we will not know the exact connectivity of a real-world network, but we may know some of its properties, such as the average number of connections per node. Studying random networks with the empirically observed properties at least gives us some idea what to expect. Moreover, such structures as neuronal networks of actual organisms have been designed by evolution (a stochastic process) over millions of years, so we should expect that the connectivity is at least somewhat random. But there is no absolute notion of randomness; this mathematical concept is always based on some distributions. However, the distributions $\mathcal{D}_n$ that define Erdős-Rényi random digraphs may not be the most realistic ones.

Recall our derivation of the formula (6.9) for the average indegrees and outdegrees in Erdős-Rényi digraphs. These were based on sums of independent random variables. Thus it follows from the Central Limit Theorem that for large n the ratio between the smallest indegree and the largest indegree of any node will be very close to 1 with probability arbitrarily close to 1; similarly for outdegrees. This is not what we observe in many real-world networks. In particular, some empirical studies have found evidence that the degree distributions in some actual neuronal networks are so-called *power law* or *scale-free* distributions, with a small number of highly connected neurons [14,15]. Erdős-Rényi digraphs are the wrong kind of random digraphs for modeling such networks.

Project 6 [1] gives a brief introduction to power law distributions and methods for randomly generating such digraphs with certain parameters. It invites the reader to explore the dependence of $P(NS)$, $P(MNODE)$, $P(AUTO)$, $P(FAMA)$ for networks with the resulting connectivities on the parameters of the degree distribution.

6.6 ANOTHER INTERPRETATION OF THE MODEL: DISEASE DYNAMICS

While our models are motivated by a problem in neuroscience and while we refer to our models N as "neuronal networks," there is nothing inherently "neuronal" about these structures. They are just mathematical objects. In general, a given mathematical object, such as a dynamical system, may serve as a model for any number of very different natural systems.

Imagine an infectious disease that spreads in a population of n individuals. At any given time, an individual can be in one of three categories: infected and prone to infect others (in the set I), healthy but susceptible to the infection (in the set S), or recovered and (temporarily or permanently) immune to the disease (in the set R). Infection may result from a contact between an infectious and a susceptible individual. These assumptions lead to the classical *SIR models of disease dynamics* (see [16] for a review). Notice that infectiousness is similar to the firing of a neuron: It can induce the same state at a subsequent time in another individual. In this section we will explore whether our networks might be suitable models for the dynamics of certain diseases.

The correspondence between a mathematical model and the underlying natural system is never perfect; mathematical modeling is always based on simplifying assumptions. The so-called *art of mathematical modeling* is in essence a knack for making simplifying assumptions that lead to models which are simple enough to allow exploration either by computer simulations or mathematical methods and yet incorporate enough detail to make realistic predictions about a natural system. The first question a modeler needs to address is which aspects of the natural system the model is supposed to predict. Here we will be interested only in the question of how the proportion of infected individuals in the population changes over time.

SIR models come in a variety of flavors; in particular, there are a lot of details to consider that differ from disease to disease. In actual modeling, these details are inferred from the available data and the model is constructed by deriving suitable assumptions from the data. Here we do not have the space to consider actual data for a disease; instead let us consider some modeling assumptions that one might have deduced:

1. An infected individual will be infectious for a time period T_0 with mean value $E(T_0)$ and very small standard deviation.
2. The time lag between infection and becoming infectious is very small relative to $E(T_0)$ so that it may be negligible.
3. After ceasing to be infectious, an individual will remain immune to the disease for a time period T_1 with $E(T_1) \approx pE(T_0)$ where p is a positive integer given by the data and the standard deviation σ of T_1 is small.
4. Contact between an infected and a susceptible individual during the time interval of infectiousness will result in a new infection with probability q.

These are all the assumptions we will be using here. But of course, these assumptions already gloss over a lot of details that might (or might not) significantly influence the disease dynamics. Before reading about how we build a model from these assumptions, you may want to take a few minutes to do the following:

Exercise 6.16.   List some details that are being ignored by the assumptions above but may significantly influence how the disease will spread in the population.   ▽

Since the length of time an individual remains infectious has "very small standard deviation" and the time lag between the moment of transmission and becoming infections may be negligible, we might try to work with a discrete-time model where the unit of time is chosen as T_0. Now consider a network $N = \langle D, \vec{p}, \vec{1} \rangle$ with $V_D = [n]$ and constant $\vec{p} = (p, \ldots, p)$. Let us interpret each $i \in [n]$ as an individual of a fixed population, interpret state $s_i(t) = 0$ as individual i being infectious at time t, state $s_i(t) = p$ as individual i being susceptible to infection and state $0 < s_i(t) < p$ as individual i being immune to infection. An arc $\langle i, j \rangle \in A_D$ indicates that individual i interacts with individual j. It is natural to assume here that A_D is symmetric, that is, either both arcs $\langle i, j \rangle, \langle j, i \rangle$ are in A_D or neither of them is, but as we will see in Project 8 [1] this assumption is not actually needed.

Will the dynamics of N correspond to the dynamics of the disease? Possibly. Notice that $s_i(t) = 0$ implies $s_i(t + \tau) \neq 0$ for $\tau \in [p]$, which nicely corresponds to being immune from the disease for a time period of pT_0.

However, in the above interpretation of N an individual j that is susceptible at time t *will* become infectious at time $t + 1$ whenever there is an arc $\langle i, j \rangle \in A_D$ with individual i being infectious at time t. This would imply that the transmission probability q is 1 and either i, j always interact during a time interval of length T_0 (when $\langle i, j \rangle, \langle j, i \rangle \in A_D$), or i, j never interact (when $\langle i, j \rangle, \langle j, i \rangle \notin A_D$). These assumptions are too extreme to be realistic. In practice, there will always be an element of stochasticity both in the structure of contacts and in the actual transmission of the disease that results from a given contact. We can build these effects into our model by altering it as follows. Assume that during each time period of length T_0 any two given individuals interact with probability q_0, and these interactions are independent. If a given interaction involves a susceptible and an infectious individual, a transmission will result with probability q. Thus we can model the dynamics of a random discrete dynamical system N_1 whose updating rules are identical to the model N above, except that the digraph D_t will not be fixed but will be randomly drawn anew at each time step from the distribution $\mathcal{D}_n(q_0 q)$ of Erdős-Rényi random digraphs.

Notice that N_1 treats the ratio $E(T_1)/E(T_0)$ exactly as an integer and treats the "small standard deviation" of T_1 that we get from the data as practically zero. This may be permissible, but one needs to be careful about such simplifications. Project 8 [1] invites the reader to explore whether or not these simplifications will lead to false predictions. More precisely, the project proposes a *suite* N_1, N_2, N_3, N_4 of progressively more elaborate and apparently more realistic models of the same natural system. In particular, the constructions used for models N_2 through N_4 incorporate the standard deviation of T_1 that is ignored by model N_1, and allow us to deal with situations where the empirically observed ratio $E(T_1)/E(T_0)$ is only approximately, but not exactly an integer, as will almost certainly be the case for real data sets. The more elaborate models are more difficult to study though. We will explore which of these models, if any, incorporates just the right level of detail.

Without giving away the correct answer for Project 8 [1], let us assume for the sake of argument that our simulations for model N_3 confirm the predictions of well-established models from the literature. Would this imply that model N_3 is "good enough" to make accurate predictions about the course of an actual disease? This kind of question cannot be answered in the affirmative with certainty; Nature always may hold some surprises in the form of hidden features of the real system that influence the dynamics but that a modeler may have overlooked. But it might be possible to prove mathematical theorems to the effect that a given simpler model such as our discrete model N_3 will give us the exact same answer to a given question than a more detailed one, such as an ODE version of the corresponding SIR model. We will not address this question for disease models, but in the next section we will describe such a theorem for models of certain neuronal networks. Theorems of this type would be extremely valuable, since discrete models are often easier to study, at least by simulations. For example, in disease dynamics the underlying network of contacts plays an important role, with some individuals having frequent contact with many others, whereas others may be more socially isolated. This phenomenon can be easily modeled in our framework by drawing D from a given distribution with unequal probabilities for different potential arcs $\langle i, j \rangle$, but it is more difficult to incorporate into a differential equations framework.

We want to point out though that such theorems could only address the suitability of a given simplification for a specified set of questions about model dynamics. Notice that there is at least one aspect of disease dynamics in which all our models are blatantly wrong: Assume, for example, that $E(T_0)$ is one week, which corresponds to one time step in our discrete models. Now, if one treats them too literally, our models would appear to say that each individual either gets sick on Monday and recovers on Sunday, or stays healthy the whole week. This is nonsense. Our models become useless for disease dynamics at time scales below one week.

This last observation brings us round circle back to neuroscience. Our models were inspired by the phenomenon of *dynamic clustering* where time appears to be partitioned into discrete *episodes* of roughly equal length. Some neurons fire while other neurons remain at rest during any given episode, and membership in the clusters of firing neurons changes from episode to episode. Notice that this phenomenon looks exactly like the dynamics of a hypothetical disease where groups of people will get sick on Mondays and recover on Sundays, with different groups being sick during different weeks and group membership changing over time. This sort of thing does not happen with diseases, but as mentioned in Section 6.2, it has been empirically observed in some recordings from actual neurons. It is a puzzling phenomenon, and one naturally wants to know what might account for it.

A literal reading of our discrete model would predict this phenomenon, but this does not mean that our discrete model *explains* dynamic clustering, because this pattern is already built into the assumptions of discrete time steps for our models. If one wants to *understand* why dynamic clustering will occur in some, but not in other, neuronal networks, one needs to consider models based on coupled differential equations which do not have a built-in assumption of discrete episodes and study conditions on these models under which the phenomenon will occur. The next section describes some results of this type and how they relate to the broader context of neuroscience.

6.7 MORE NEUROSCIENCE: CONNECTION WITH ODE MODELS

A natural question to ask at this point is: What does anything we have presented so far have to do with neuroscience? Put another way: How can the discrete-time dynamical system, and results concerning its dynamics, help us understand the workings of the brain? The discrete-time system is, after all, a rather simple model in which each "neuron" is represented by a finite number of states and there are rather simple rules for how the "firing" of one neuron impacts on the firing of other neurons. Real neurons, on the other hand, are very complicated cells with complex spatial geometries, the firing of an action potential involves both electrical and chemical processes and communication between neurons at synapses is surely more complicated than our discrete-time model would suggest.

There are many challenges facing mathematicians who wish to make an impact on our understanding of the brain. It is not clear, for example, how to model a neuronal system at an appropriate level of detail. If the model takes into account everything that is currently known about neurons, including their complex dendritic geometries, then the model would be too complicated to simulate numerically, let alone analyze mathematically. For this reason, simpler models have been proposed. These models usually take the form of systems of ordinary or partial differential equations; they take into account only those biological details that are believed to be important in generating the phenomena of interest. Even these models can be extremely difficult to analyze. They are highly nonlinear and involve numerous parameters (many of which are unknown experimentally). Moreover, if one considers any reasonably sized network, then the model will involve hundreds (if not thousands) of differential equations. Questions such as how the emergent population rhythm depends on network architecture present very difficult, yet exciting, challenges for mathematicians and other theoreticians.

A main focus of this chapter has been dynamic clustering, in which subpopulations of neurons fire in synchrony during distinct episodes and the membership of each subpopulation may change from episode to episode. Recall that this phenomenon has been observed in the antennal lobe (AL) of insects and olfactory bulb (OB) of mammals. Detailed differential equations models have been proposed for this phenomenon and these models have been quite useful in understanding biophysical mechanisms underlying this behavior. We note, however, that many (if not most) of the biophysical properties of these systems are not known; these include the intrinsic properties of neurons in the AL and OB, as well as details of the network connectivity. Computational studies have been useful in identifying what cellular and network properties may account for the observed behavior, leading to predictions that can, at times, be tested experimentally. However, as described above, a systematic study of these ODE models—one that can identify and classify those neuronal properties and network architectures that lead to some network behavior—has been lacking. There are simply too many nonlinear equations with too many unknown parameters.

Here we have described a discrete-time model, motivated by the phenomenon of dynamic clustering. This model is considerably simpler than the ODE models that arise in neuroscience and allows for a rigorous mathematical study of its dynamics. The question still remains, however: How can one translate results obtained from the discrete model to the biological system? This question is considerably easier to answer for the ODE models, because there is a more direct link between the biology and the ODE models: The ODE models are derived from physical principles and experimental measurements, so there is a direct link between, for example, changing a parameter in the model and changing some physical quantity in the system. In order for the discrete-time model to be useful, we would need, at a minimum, a direct correspondence between the discrete model and an ODE neuronal model, so that parameters in the discrete model, such as length of refractory periods, correspond directly to parameters in the ODE model and, therefore, to properties of the physical system. Ideally, we would like to prove a theorem that states that given an ODE model, say M, then over a wide range of parameter values and initial conditions, there is a discrete-time model, say N, such that M and N exhibit the same dynamics; that is, the corresponding discrete-time model N will make the same predictions as the ODE model M.

The previous sentence needs some clarification. Of course the discrete model N cannot literally make *the same* predictions as M because it can, at best, be a much simpler approximation of M and does not allow us to even consider such variables as actual membrane potentials. As we explained in the previous section, N can at best give the same answers to a narrowly specified set of questions. The question we are primarily interested in here is:

Question 1.   Which neurons in the network will fire during which episode along a given trajectory?

In [5], we considered a general class of ODE neuronal models and proved rigorous results for when there is a direct correspondence between the ODE model, M, and a discrete model, N. More precisely, we demonstrated that dynamic clustering occurs for all trajectories of M that start in a certain region U of the state space of M and there is a discrete-time model N that correctly predicts which neurons will fire during each episode, throughout any ODE trajectory that starts in U. We refer to such a situation by saying that N is *consistent with M on U*. Moreover, we may expect the set U, on which the two models are consistent, to be open and large enough to contain, for each possible initial state $\vec{s}(0)$ of N, a state of M that maps to $\vec{s}(0)$. In this case we will say that *M realizes N on U*. In [5] we proved the following:

Theorem 1.   *There exists a class of ODE models M for neuronal networks such that every model in this class realizes a corresponding discrete model on an open subset U of its state space and, conversely, every discrete network model $N = \langle D, \vec{p}, \vec{1} \rangle$ with constant vectors $\vec{p}$ is realized by some model M in this class.*

Figure 6.6 illustrates an example of an ODE model that realizes the network $N = \langle D, \vec{1}, \vec{1} \rangle$, where D is the digraph of Figure 6.2b. This network contains seven cells and the top left panel shows the membrane potentials of each of these cells for one particular solution. The middle left panel is a grey scale representation of this solution. Two other solutions are shown in the right panel. In order to generate the three solutions shown in Figure 6.6, we chose different initial conditions. Note that each cell's membrane potential typically lies in one of two "states"; the elevated or active state corresponds to the firing of an action potential and is represented by the dark rectangles in the grey scale representations. Moreover, there are discrete episodes in which some subpopulation of cells lie in the elevated state. These subpopulations change from episode to episode and the model exhibits dynamic clustering. The discrete model completely predicts which cells fire during which episode for a large class of initial conditions. The corresponding trajectories in the discrete model are represented by the active cells in each episode.

In the online supplement, we give a computer code for the ODE model that generates the solutions shown Figure 6.6. This code uses the software XPPAUT, which can be freely downloaded from the webpage [18]. The code is flexible enough so that one can explore ODE models that realize different networks. A discussion of why the model is designed as it is—that is, what the various dependent variables and parameters correspond to—is well beyond the scope of this chapter. The interested reader should consult either [3] or [5], where this model is described in detail.

As a consequence of Theorem 1, the results presented in this chapter are relevant for a general class of neuronal models. An important conclusion of our study of discrete-time models is that properties of the population rhythm, including

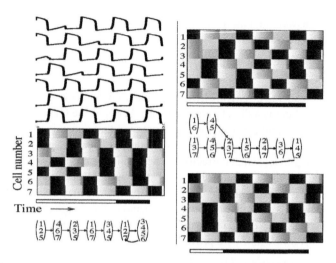

FIGURE 6.6

Simulations of an ODE model M that realizes the network $N = \langle D, \vec{1}, \vec{1} \rangle$, where D is the digraph of Figure 6.2b. The upper left panel shows the time courses of the membrane potentials of the seven cells in the network. The grey scale representations illustrate the emergence of discrete episodes with rather sharp boundaries. The corresponding trajectories in the discrete model N are represented by the active cells in each episode. Adapted from [17].

the number and size of attractors and length of transients, depend on *both* the intrinsic properties of the individual neurons, such as their refractory periods or firing thresholds, and the network architecture. A misguided assumption in much of the neuroscience literature is that if one understands how neurons are connected to each other in a given neuronal system, then one can determine the network's population dynamics. Here we have shown that this is typically not the case; the intrinsic properties of cells in the network may be equally important. Because of Theorem 1, this result translates immediately to biologically based ODE models for neuronal activity.

6.8 DIRECTIONS OF FURTHER RESEARCH

Much of the theory described in this chapter has been motivated by experimentally observed firing patterns in an insect's antennal lobe. However, most of the dynamic behaviors described in this chapter, such as synchrony and dynamic clustering, arise in many other neuronal systems. For example, these types of firing patterns have been observed in a brain region known as the basal ganglia in humans suffering from Parkinson's disease. See, for example [19,20]. In this case, too much synchrony among neurons within the basal ganglia represents a pathological state. In healthy patients, there is very little correlation or synchrony among neurons within the basal ganglia; however, in Parkinson patients, there is a significant increase in synchrony among these neurons.

Another brain region where these types of firing patterns arise is the thalamus, which plays an important role in sensory processing. Certain neurons within the thalamus are thought to be involved in the transition between sleeping and waking states [21,22]. Experiments have demonstrated that as we fall deeper into sleep, certain neurons within the thalamus become more synchronized with each other. Moreover, computational models for sleep rhythms suggest that neurons within the thalamus exhibit clustering during certain stages of sleep.

Finally, synchronous activity has been observed in cortical networks involved in working memory [23]. It is tempting to view the brain (or some brain region) as a large-scale dynamical system and a memory as an attractor of this dynamical system. If this is the case, then it would be important to understand how the number of attractors, and their basins of attraction, depend on network properties, including the intrinsic properties of agents (or neurons) in the system and network architecture. In particular, which networks are capable of exhibiting a very large number of attractors (since one would like to store a very large number of memories) and how do properties of attractors depend on changes in parameters that may, for example, occur during learning?

We note that the neuronal systems described above share many properties, including certain features of the neurons, synaptic connections, and network architecture. Of course, there must be some important differences between these systems, because they play such different functional roles in brain processing. However, from a mathematical viewpoint, it may be possible to consider a general class of neuronal networks that encompass each of these systems and develop an analytic framework that allows us to characterize the firing patterns that emerge.

Our hope is that the analysis presented in this chapter is a first step toward developing such a framework. However, many challenges remain. Here, we characterized a neuron by just two parameters: the firing threshold and the length of refractory period. Obviously neurons are much more complicated biological objects with a myriad of complex cellular processes. Whether or not a neuron fires an action potential may depend on many factors. The firing threshold, for example, may depend on an internal "state" of the neuron, such as the concentration of some ionic species which may change over time depending on how often the neuron fires. The strengths of synaptic connections may also change over time, again depending on the firing rates of neurons. It is not clear which of these details to include in a mathematical model, how to incorporate these details into the theoretical framework described here, and how difficult it would be to mathematically analyze discrete dynamical systems that do include these details. Some progress in this direction is presented in [17].

It is important to emphasize that significant progress can be made only if there is close dialog between mathematicians and experimental neuroscientists. It hardly seems likely that a single mathematical theory can be developed that takes into account every possible detail that may go into a neuronal network model, including complex properties of cells (such as their complex geometry) and all possible network architectures. Collaborations with biologists help guide mathematicians to better understand what details are important and what dynamic properties of the system may be of functional significance. Without some appreciation of the underlying biology, it is often difficult, if not impossible, to ask the most interesting questions and provide the most meaningful answers.

Students interested in pursuing work in mathematical neuroscience may want to start with reading [3], which gives a detailed description of how methods from non-linear dynamics have been used to address problems in neuroscience. The online supplement to this chapter also gives some additional information on other discrete models of neuronal networks and how they relate to ours.

6.9 SUPPLEMENTARY MATERIALS

The online appendix, additional files, and computer code associated with this article can be found, in the online version, at http://dx.doi.org/10.1016/B978-0-12-415780-4. 00024-7 and from the volume's website http://booksite.elsevier.com/9780124157804

References

[1] Just W, Ahn S, Terman D. Neuronal Networks: A Discrete Model - Online Supplement (Appendix 6). In: Mathematical Concepts and Methods in Modern Biology: Using Modern Discrete Models (Robeva R, Hodge T, Eds). Eslevier; 2013. http://booksite.elsevier.com/9780124157804

[2] Neuron, Wikipedia. http://en.wikipedia.org/wiki/Neuron.

[3] Ermentrout G, Terman D. Mathematical foundations of neuroscience. Springer; 2010.

[4] Hodgkin A, Huxley A. A quantitative description of membrane current and its application to conduction and excitation in nerve. J Physiol 1952;117:500–544.

[5] Terman D, Ahn S, Wang X, Just W. Reducing neuronal networks to discrete dynamics. Physica D 2008;237:324–338.

[6] Elashvili A, Jibladze M, Pataraia D. Combinatorics of necklaces and hermite reciprocity. J Algebraic Comb 1999;10:173–188.

[7] Ahn S, Just W. Digraphs vs. dynamics in discrete models of neuronal networks. DCDS-B 2012;17:1365–1381.

[8] Gallery of named graphs, Wikipedia. http://en.wikipedia.org/wiki/Gallery_of_named_graphs.

[9] Just, W., Ahn, S., and Terman, D. (2007). Minimal attractors in digraph system models of neuronal networks. MBI Technical Report 68, http://mbi.osu.edu/publications/reports2007.html.

[10] Wolfram S. A new kind of science. Wolfram Media; 2002.

[11] Wormatlas, http://www.wormatlas.org/.

[12] Stomatogastric ganglion, http://www.scholarpedia.org/article/Stomatogastric_ganglion.

[13] Just W, Ahn S, Terman D. Minimal attractors in digraph system models of neuronal networks. Physica D 2008;237:3186–3196.

[14] Cecchi GA., Rao AR, Centeno MV, Baliki M, Apkarian AV, Chialvo DR. Identifying directed links in large scale functional networks: application to brain fMRI. BMC Cell Biology 2007 8 (Suppl 1:S5).

[15] Varshney LR, Chen BL, Paniagua E, Hall DH, Chklovskii DB. Structural properties of the *Caenorhabditis elegans* neuronal network. PLoS Comput Biol 2011;7(2):e1001066. http://dx.doi.org/10.1371/journal.pcbi.1001066.

[16] Hethcote HW. The mathematics of infectious diseases. SIAM Rev 2000;42: 599–653.

[17] Ahn S, Smith B, Borisyuk A, Terman D. Analyzing neuronal networks using discrete-time dynamics. Physica D 2010;239:515–528.

[18] XPPAUT. http://www.math.pitt.edu/~bard/xpp/xpp.html.

[19] Bevan M, Magill P, Terman D, Bolam J, Wilson C. Move to the rhythm: oscillations in the subthalamic nucleus-external globus pallidus network. Trends Neurosci 2002;25:525–531.

[20] Terman D, Rubin J, Yew A, Wilson C. Activity patterns in a model for the subthalamopallidal network of the basal ganglia. J Neurosci 2002;22:2963–2976.

[21] Saper C, Chou T, Scammell T. The sleep switch: hypothalamic control of sleep and wakefulness. Trends Neurosci 2001;24:726–731.

[22] Steriade M, McCormick D, Sejnowski T. Thalamocortical oscillations in the sleeping and aroused brain. Science 1993;262:679–685.

[23] Durstewitz D, Seamans J, Sejnowski T. Neurocomputational models of working memory. Nature Neurosci 2000;3:1184–1191.

Predicting Population Growth: Modeling with Projection Matrices

Janet Steven* and James Kirkwood†

**Department of Biology, Sweet Briar College, Sweet Briar, VA 24595, USA*
†*Department of Mathematical Sciences, Sweet Briar College, Sweet Briar, VA 24595, USA*

7.1 INTRODUCTION

Biologists studying an organism in its natural environment often start with a description of a population's size and demographic structure. How many individuals are in the population? What major life stages does each individual pass through? How many offspring are generated? Once this initial description is complete, biologists frequently move on to examine the causes and consequences of the observed size and structure. Is the population shrinking or growing? Which demographic stage experiences the greatest mortality? How does survival at a particular stage affect overall growth? Matrix population models that utilize *projection matrices* function to bridge the gap between population structure and population dynamics by providing a mathematical framework that both summarizes and models a population [1].

A matrix-based model can incorporate the different stages an organism goes through during its life, and then predict both the overall growth of the population and the distribution of individuals across these life stages. A projection matrix generated from data collected in a natural population models transitions between stages for a given time interval and allows us to predict how many individuals will be in each stage at any point in the future, assuming that transition probabilities and reproduction rates do not change.

Projection matrices are widely used in ecology and conservation biology to address both theoretical and practical questions. They can reveal the life stages that have the greatest influence on population growth and are therefore potentially shaped most strongly by natural selection [2]. They have been used to examine the impact of herbivory on plant populations [3,4] and the relative importance of alternative mating systems [5]. But perhaps their greatest impact is in providing quantitative predictions for conservation management decisions; because projection matrices can use the current structure and size of a population to predict future growth, they are often used to assess extinction risk of the population [6,7]. Moreover, because the models can inform the relative importance of each life stage to growth, they are used in guiding decisions about hunting limits and choosing key life stages to target during conservation [8]. Projection matrices in combination with some knowledge of available habitat

Mathematical Concepts and Methods in Modern Biology. http://dx.doi.org/10.1016/B978-0-12-415780-4.00007-7

area can also help conservation biologists determine the size and number of reserves needed to protect a species from extinction [9].

In this chapter, we review the basics of stage-structured population growth models and explore the underlying matrix algebra that makes them possible. We provide code for MATLAB and R that assumes only a basic understanding of the software interface and work through an example from a population of wild ginseng threatened by over-harvesting. We close with a summary of additional analytical techniques involving projection matrices.

7.2 LIFE CYCLES AND POPULATION GROWTH

When studying wild populations, and especially species of conservation interest, biologists frequently want to predict whether the number of individuals in a population is growing, stable, or shrinking. The simplest models of population growth involve data on the number of births and deaths in a population within a given time frame. Frequently, biologists also collect additional data on the *life cycle* of an organism, such as how often it reproduces, how many offspring it makes, and how long it lives. Because each of these factors is often dependent on the age or size of an organism, biologists also include data that permits the separation of the life cycle into ages or stages. For example, the chance that an individual plant will survive and reproduce varies considerably with the age and size of the plant; the chance that a seed will survive and germinate, and then that the seedling will establish and grow, is typically very small. Many plants make thousands of seeds with only a few surviving to become new plants. But once a plant becomes established, often its chance of surviving to the next year is relatively high. Expanding a population growth model to include stages in the life cycle allows us to estimate the influence of each stage on population growth, and provides additional information for conservation management decisions.

Therefore, we need a mathematical model that describes a population in the following ways: (1) divides the life cycle of an individual into multiple stages, (2) keeps track of *transitions*, or how individuals move through those stages, (3) estimates the probability that an individual will die during a stage, and (4) keeps track of reproduction, both in terms of how many new individuals are made for each individual in a given stage and in terms of the stage in which new individuals appear.

First, we will construct a mathematical framework that does these four things. No model reflects the natural world with 100% accuracy, but describing population structure with a mathematical model will allow us to make observations and predictions that can guide our understanding of the dynamics in wild populations. Because the model is essentially a mathematical description of the population, we can use it to explore how the population might grow under certain conditions or how a single stage is affecting overall population growth. We will explain the mathematical theory that allows us to conduct these analyses, and then we will work through examples that use the models to predict the future of the population and identify the life stages that are critical for population growth.

7.3 DETERMINING STAGES IN THE LIFE CYCLE

In many models of population growth, life stages are defined based on morphological changes during growth, or changes in size. In some organisms, development leads to natural categories; seeds, seedlings, and reproductive plants, for example, or egg, larva, pupa, and adult in butterflies. In other organisms, sometimes it makes more sense to categorize individuals on the basis of age. Here we will only discuss models that use discrete categories based on life stage, but see [8] for an introduction to age-structured models (also called Leslie matrices). In the mathematical model we are building, the number of life stages must be finite (fixed). To simplify the mathematics, we impose a discrete time structure, where the time t can only take the values $t = 0, 1, 2, \ldots$ If t_0 is any non-negative integer value, any changes that occur in the interval between time t_0 and time $t_0 + 1$ are treated as occurring simultaneously at time $t_0 + 1$.

In between time t_0 and time $t_0 + 1$, an individual could die, reproduce, move to another life stage, or stay unchanged. In the most general version of the model, all four of these events are possible for every life stage. In addition, the new individuals that appear could start life in any life stage, and individuals could switch between any two life stages in one time interval. Biologically, this model seems unlikely; an adult chicken can't go back to being a chick, and a mature tree can't develop in a year. But for the purposes of building a general model, we are going to allow for all possible transitions and the possibility that new individuals appear in every life stage. When we develop models specific to a particular organism, this general framework can be restricted by setting biologically impossible transitions to zero.

7.4 DETERMINING THE NUMBER OF INDIVIDUALS IN A STAGE AT TIME $t_0 + 1$

If we know how many members there are in each stage at time t_0 and we know something about birth rates, death rates, and transitions to other stages, we can predict the number of members in each stage at time $t_0 + 1$. This is the first step in understanding whether the population is growing or shrinking and how each life stage contributes to that change. Mathematically, we divide the life cycle of the organism into k stages, which we designate $s_1, s_2, \ldots, s_k$. These stages could be based on changes in development or size; models often begin with a juvenile stage. The number of stages k is dependent on the biology of the system and the nature of the data. We designate the number of individuals in stage s_j at time t_0 as $n_j(t_0)$. Once we have defined the stages and number of individuals in each, we want to find $n_i(t_0 + 1)$, the expected number of individuals in stage s_i at time $t_0 + 1$. Note that we are considering $n_j(t_0)$ to be a known number and $n_i(t_0 + 1)$ to be a prediction. Finding this predicted number of individuals in a stage at time $t_0 + 1$ depends on knowing the number of individuals in *all* stages at time t_0 because each stage could potentially contribute to $n_i(t_0 + 1)$. Knowing $n_j(t_0)$ for all stages s_j, $j = 1, 2, \ldots, k$, we want to find the expected value of $n_i(t_0 + 1)$, $i = 1, 2, \ldots, k$. To do this, we must consider the ways an individual of stage s_j at time t_0 can produce an individual in stage s_i at time $t_0 + 1$. They are:

1. **Movement between stages.** An individual moves from stage s_j to stage s_i. If the probability for this transition in one unit of time is p_{ji}, then the expected number of members produced in stage s_i by a single member in stage s_j in one unit of time is p_{ji}. Note that i and j may be the same and p_{jj} is the probability that an individual in stage s_j stays in stage s_j after one time interval. In many models death is not considered to be a stage, and if this is the case, then the probabilities across all stages s_i will not necessarily sum to 1. The probabilities p_{ji} are called *transition probabilities*.
2. **Reproduction.** The individual in stage s_j produces new offspring, and the average number of offspring generated in one unit of time by one individual of stage s_j that begin life in stage s_i is denoted f_{ji}.

Therefore, if we know the following, we can determine the total number of members expected in stage s_i at time $t_0 + 1$:

1. the number of individuals s_j in each stage $j = 1, 2, \ldots, k$ at time t_0,
2. the probability p_{ji} that an individual in each of those stages will become an individual in stage s_i, $i = 1, 2, \ldots k$, after one time interval,
3. and the average number of offspring in stage s_i generated by a single individual in stage s_j, denoted f_{ji}.

Using the terms we defined above, we can express our ideas more precisely. Each member that is in stage s_j at time t_0 will produce an average of $p_{ji} + f_{ji}$ members in stage s_i at time $t_0 + 1$. Since there are $n_j(t_0)$ members in stage s_j at time t_0, the stage s_j will be expected to result in $n_j(t_0)[p_{ji} + f_{ji}]$ members in stage s_i at time $t_0 + 1$.

To get $n_i(t_0 + 1)$, that is the total number of members expected in stage s_i at time $t_0 + 1$, we sum over all stages s_j to get

$$n_i(t_0 + 1) = \sum_j n_j(t_0)[p_{ji} + f_{ji}]. \tag{7.1}$$

This formula will tell us how many individuals we expect in a single life stage after one time interval has passed. To keep track of the number of individuals in all life stages simultaneously, we turn to matrix algebra.

7.5 CONSTRUCTING A PROJECTION MATRIX

The framework provided by linear algebra allows us to build a population growth model that includes all the life stages we have defined and then use that model to gather information about the possible fate of the population. We can build a matrix that encompasses transitions among life stages, reproduction, and mortality for a specified time interval. We can also use the tools of linear algebra to describe properties of the projection matrix that are analogous to the biological characteristics of the population.

To generate a *projection matrix* for a population, we construct a square $k \times k$ matrix **A** that has an entry for each possible transition between the k life stages. The entry in the ith row and jth column of **A** is given by

$$(A)_{i,j} = \text{the average number of members in stage } s_i \text{ at time } t_0 + 1 \text{ that are}$$
$$\text{produced by a single member in stage } s_j \text{ at time } t_0.$$

Estimates for the probabilities p_{ji} and the average number of offspring f_{ji} are based on the relative number of individuals and offspring production by those individuals in each stage in natural populations, typically over multiple years (see [8] for more details on how this is done).

To illustrate the creation of a matrix from transition probabilities, we will work through an example from the biological literature on American ginseng (*Panax quinquefolius*). Ginseng is a plant native to the eastern United States and grows in forest understories. It is a perennial, and new leaves come up from an underground stem every year. This stem is enlarged and stores nutrients and starch. It also produces a variety of compounds that give ginseng a distinctive odor and taste, and give it value as an herbal medicine. The plants are often harvested from wild populations and the underground stems sold, sometimes for a considerable sum. Ginseng is becoming increasingly rare in many places, and its decline is likely a result of habitat loss and overharvesting. Therefore, it has been the subject of studies that use projection matrices to model the future viability of populations. The data we use for the examples below are from Charron and Gagnon [10].

The life cycle of ginseng lends itself easily to classification into life stages. Seeds spend a year and a half in the soil before germinating, and when they do germinate, they produce a seedling that becomes established in its first year and stores resources in its underground stem for the next year. At the end of the summer, it dies back above ground but the stem stays alive below ground. The next spring, the underground stem sends up a shoot with one compound leaf on it (compound leaves typically have five leaflets). The following year, the plant may make one leaf again, or it may make two leaves. In subsequent years, plants may make three or four leaves; rarely, plants produce five or six leaves. Plants sometimes flower when they have two leaves but more commonly wait until they have three or four leaves. The flowers will develop into fruits with seeds in them, and then these seeds will disperse away from the parent plant and become a new plant. The interval between time t_0 and time $t_0 + 1$ is a year, and because the data we are using were collected from the wild populations in the fall when the seeds were ripe, we can imagine the year starting at this time in the fall. In our model, we have six life stages: seed, seedling, one-leaved plant, two-leaved plant, three-leaved plant, and four-leaved plant.

An individual in each life stage has a certain probability of dying, staying in its stage, or moving into another stage. For example, we expect that some seedlings will become one-leaved plants in the next year, and some will die (the probability that a seedling will stay a seedling is 0). All nonzero transitions among life stages are illustrated in Figure 7.1.

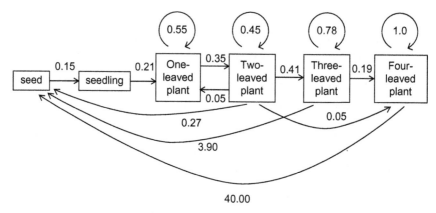

FIGURE 7.1

Life stages, transition probabilities, and reproduction (in italics) for a population of American ginseng. Values represent the average number of plants for every one plant that transition to another stage (or remain in the same stage) from one year to the next. In other words, for every 1 seed, 0.15 seedlings will be present in the next year, and for every 1 three-leaved plant 3.90 seeds will be present in the next year. Data taken from Population 2, 1986–1987 in [10].

With six life stages all possibly transitioning into all other stages, we have 36 different elements to keep track of, and we use a 6×6 matrix:

$$\begin{pmatrix} A_{1,1} & A_{1,2} & A_{1,3} & A_{1,4} & A_{1,5} & A_{1,6} \\ A_{2,1} & A_{2,2} & A_{2,3} & A_{2,4} & A_{2,5} & A_{2,6} \\ A_{3,1} & A_{3,2} & A_{3,3} & A_{3,4} & A_{3,5} & A_{3,6} \\ A_{4,1} & A_{4,2} & A_{4,3} & A_{4,4} & A_{4,5} & A_{4,6} \\ A_{5,1} & A_{5,2} & A_{5,3} & A_{5,4} & A_{5,5} & A_{5,6} \\ A_{6,1} & A_{6,2} & A_{6,3} & A_{6,4} & A_{6,5} & A_{6,6} \end{pmatrix}.$$

Each row and column in the matrix represents a life stage; the columns are for the life stage in a given year (time t_0), and the rows represent the life stages in the next year (time $t_0 + 1$). Each element in the matrix represents the average number of new plants in the next year that are produced by a single plant in the current year. In other words, if the value in the second column (representing seedlings) and third row (representing one-leaved plants) is $A_{3,2} = 0.21$, we expect for every one seedling at time t_0 there will be 0.21 one-leaved plants in the population at time $t_0 + 1$. To put it another way, 21% of the seedlings will survive to be one-leaved plants in the next year.

The first column in the matrix represents the transition from the stage of being a seed to all other possible stages. Seeds can only become seedlings in the first year; therefore all but one of the values in the first column will be zero. More formally, and drawing from the data presented in Figure 7.1:

The probability that a seed at time t_0:

remains a seed is 0, therefore $A_{1,1} = 0$,
becomes a seedling at time $t_0 + 1$ is 0.15, therefore $A_{2,1} = 0.15$,
becomes a one-leaved plant at time $t_0 + 1$ is 0, therefore $A_{3,1} = 0$,
becomes a two-leaved plant at time $t_0 + 1$ is 0, therefore $A_{4,1} = 0$,
becomes a three-leaved plant at time $t_0 + 1$ is 0, therefore $A_{5,1} = 0$,
becomes a four-leaved plant at time $t_0 + 1$ is 0, therefore $A_{6,1} = 0$.

Note the probabilities do not sum to 1. This is because not all seeds (in fact only 15%) survive to become seedlings in the following year.

From these probabilities, we know that the first column of the projection matrix is as shown in Table 7.1.

To generate the second column of the matrix, we must consider what happens to the seedlings between the first year (time t_0) and the second year (time $t_0 + 1$). Seedlings can either die or become plants with one leaf; they do not flower, so they do not make seeds, and they cannot become bigger than a one-leaved plant in just a year. Therefore:

The probability that a seedling at time t_0:

makes seeds at time $t_0 + 1$ is 0, therefore $A_{1,2} = 0$,
stays a seedling at time $t_0 + 1$ is 0, therefore $A_{2,2} = 0$,
becomes a one-leaved plant at time $t_0 + 1$ is 0.21, therefore $A_{3,2} = 0.21$,
becomes a two-leaved plant at time $t_0 + 1$ is 0, therefore $A_{4,2} = 0$,
becomes a three-leaved plant at time $t_0 + 1$ is 0, therefore $A_{5,2} = 0$,
becomes a four-leaved plant at time $t_0 + 1$ is 0, therefore $A_{6,2} = 0$.

From these probabilities, we know that the second column of the projection matrix is as shown in Table 7.2.

To generate the third column of the matrix, we consider what happens to plants with one leaf between time t_0 and time $t_0 + 1$. We know from the diagram in Figure 7.1 that one-leaved plants have a probability of staying a one-leaved plant in the next

Table 7.1 The first column of the projection matrix for a population of American ginseng. The values in this column are the probability of transitioning from the seed stage to another stage (data from [10]).

| | | Stage at time t_0 | | | | | |
		Seed	Seedling	One-leaved plant	Two-leaved plant	Three-leaved plant	Four-leaved plant
Stage	Seed	0	*	*	*	*	*
at	Seedling	0.15	*	*	*	*	*
time	One-leaved plant	0	*	*	*	*	*
t_0+1	Two-leaved plant	0	*	*	*	*	*
	Three-leaved plant	0	*	*	*	*	*
	Four-leaved plant	0	*	*	*	*	*

Table 7.2 The second column of the projection matrix for a population of American ginseng. The values in this column are the probability of transitioning from the seedling stage to another stage (data from [10]).

		Stage at time t_0					
		Seed	Seedling	One-leaved plant	Two-leaved plant	Three-leaved plant	Four-leaved plant
Stage	Seed	*	0	*	*	*	*
at	Seedling	*	0	*	*	*	*
time	One-leaved plant	*	0.21	*	*	*	*
t_0+1	Two-leaved plant	*	0	*	*	*	*
	Three-leaved plant	*	0	*	*	*	*
	Four-leaved plant	*	0	*	*	*	*

year of 0.55, and a 0.35 probability of becoming a two-leaved plant in the next year. They don't flower, so they can't make seeds, and they can't revert back to being seedlings. Therefore:

The probability that a one-leaved plant at time t_0:

> makes seeds at time $t_0 + 1$ is 0, therefore $A_{1,3} = 0$,
> becomes a seedling at time $t_0 + 1$ is 0, therefore $A_{2,3} = 0$,
> stays a one-leaved plant at time $t_0 + 1$ is 0.55, therefore $A_{3,3} = 0.55$,
> makes a two-leaved plant at time $t_0 + 1$ is 0.35, therefore $A_{4,3} = 0.35$,
> makes a three-leaved plant at time $t_0 + 1$ is 0, therefore $A_{5,3} = 0$,
> makes a four-leaved plant at time $t_0 + 1$ is 0, therefore $A_{6,3} = 0$.

The fourth column of the matrix represents transitions from the two-leaved stage to other stages. There are several possible fates for a two-leaved plant; they occasionally transition back into one-leaved plants between one year and the next, they often stay two-leaved plants in the next year, they sometimes become three-leaved plants in the next year, and they rarely become four-leaved plants in the next year. Occasionally, a two-leaved plant flowers, which means it is generating seeds that are present in the following year. Therefore, based on the data from Figure 7.1:

> The average number of seeds per plant generated by two-leaved plants between time t_0 and time $t_0 + 1$ is 0.27, therefore $A_{1,4} = 0.27$.
> The probability that a two-leaved plant is a seedling at time $t_0 + 1$ is 0, therefore $A_{2,4} = 0$.

> The probability that a two-leaved plant is a one-leaved plant at time $t_0 + 1$ is 0.05, therefore $A_{3,4} = 0.05$.

> The probability that a two-leaved plant is a two-leaved plant at time $t_0 + 1$ is 0.45, therefore $A_{4,4} = 0.45$.

The probability that a two-leaved plant is a three-leaved plant at time $t_0 + 1$ is 0.41, therefore $A_{5,4} = 0.41$.

The probability that a two-leaved plant is a four-leaved plant at time $t_0 + 1$ is 0.05, therefore $A_{6,4} = 0.05$.

If we continue to build the matrix in the same fashion from Figure 7.1, we get the complete projection matrix in tabular form, presented in Table 7.3.

The same information can be described by the following projection matrix:

$$\begin{pmatrix} 0 & 0 & 0 & 0.27 & 3.90 & 40.00 \\ 0.15 & 0 & 0 & 0 & 0 & 0 \\ 0 & 0.21 & 0.55 & 0.05 & 0 & 0 \\ 0 & 0 & 0.35 & 0.45 & 0 & 0 \\ 0 & 0 & 0 & 0.41 & 0.78 & 0 \\ 0 & 0 & 0 & 0.05 & 0.19 & 1.0 \end{pmatrix}. \tag{7.2}$$

While the general model allows for individuals in all life stages to potentially transfer to any other life stage within one time interval and reproduce offspring that appear in any other life stage, the biological reality of our study species means several transitions do not actually occur. Therefore, we have a number of zeroes in our matrix, and only the first row reflects the creation of offspring. In addition, the probability that a four-leaved plant stays a four-leaved plant is an interesting case; because no four-leaved plants died or reverted to a smaller stage over the course of the study, the probability is one, and it appears that the four-leaved plants are immortal. This cannot happen in reality, but because these probabilities are our best estimates based on data and we have no additional information from which to estimate the "true" number, we will leave it at one. Gathering additional years of data from this population could generate a more biologically realistic estimate.

It is also useful to point out that all matrix elements except those involving the production of seeds are transition probabilities and are thus less than or equal to one. However, for the first row, the values represent the average number of seeds

Table 7.3 The projection matrix for a population of American ginseng (data from [10]).

		Stage at time t_0					
		Seed	Seedling	One-leaved plant	Two-leaved plant	Three-leaved plant	Four-leaved plant
Stage	Seed	0	0	0	0.27	3.90	40.00
at	Seedling	0.15	0	0	0	0	0
time	One-leaved plant	0	0.21	0.55	0.05	0	0
t_0+1	Two-leaved plant	0	0	0.35	0.45	0	0
	Three-leaved plant	0	0	0	0.41	0.78	0
	Four-leaved plant	0	0	0	0.05	0.19	1.0

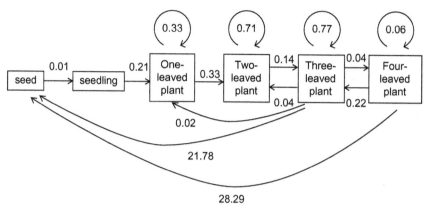

FIGURE 7.2

Life stages, transition probabilities, and reproductive values for Population 3 in [10].

made by plants in the other stages; therefore they may be greater than 1 and they are values that represent *reproduction*.[1] As previously mentioned, if we sum the transition probabilities for a particular (non-seed) stage at time $t_0 + 1$, they do not necessarily total to 1; that is because the probabilities only reflect the fraction of individuals that survive from one stage to the next.

Exercise 7.1. Construct the projection matrix for the ginseng population diagrammed in Figure 7.2. ▽

7.6 PREDICTING HOW A POPULATION CHANGES AFTER ONE YEAR

We have constructed a projection matrix to summarize how plants growing in a wild population are changing from year to year; the matrix can now serve as a model for the wild population, and we can manipulate the model to learn more about the population. First, we will use this matrix to predict the number of individuals we expect in each stage after one year, given a starting number of individuals in each stage chosen by us. The number of individuals in each stage at time can be expressed as a *vector*; for example, if we have a population of ginseng with 800 seeds, 90 seedlings, 56 one-leaved plants, 23 two-leaved plants, 31 three-leaved plants and 11 four-leaved

---

[1] It is somewhat easy to confuse the projection matrix with the transition matrix of Markov chains, because some entries in the projection matrix are transition probabilities. We caution against trying to make connections between the two.

plants, the vector would look like this:

$$\begin{pmatrix} 800 \\ 90 \\ 56 \\ 23 \\ 31 \\ 11 \end{pmatrix}. \qquad (7.3)$$

Here we go through the steps of conducting the calculations for predicting the number of two-leaved plants in the population at time $t_0 + 1$, and then we present the matrix algebra to calculate the number of individuals in all life stages simultaneously.

Using the projection matrix (7.2) that we constructed above, we can predict the number of two-leaved plants at time $t_0 + 1$ (the next year) by applying Eq. (7.1). Recall that $n_j(t_0)$ is the number of members in stage j at time t_0, the entry $(A)_{i,j}$ is the average number of members that a single member in stage j produces in stage i between t_0 and $t_0 + 1$, and that two-leaved plants are in stage 4. So

$$n_4(t_0 + 1) = n_1(t_0) \times A_{4,1} + n_2(t_0) \times A_{4,2} + n_3(t_0) \times A_{4,3} + n_4(t_0)$$
$$\times A_{4,4} + n_5(t_0) \times A_{4,5} + n_6(t_0) \times A_{4,6}.$$

Written out in words, this becomes:

(number of seeds at time t_0) × (probability that a seed becomes a two-leaved plant between t_0 and $t_0 + 1$) +

(number of seedlings at time t_0) × (probability that a seedling becomes a two-leaved plant between t_0 and $t_0 + 1$) +

(number of one-leaved plants at time t_0) × (the probability that a one-leaved plant becomes a two-leaved plant between t_0 and $t_0 + 1$) +

(number of two-leaved plants at time t_0) × (the probability that a two-leaved plant stays a two-leaved plant between t_0 and $t_0 + 1$) +

(number of three-leaved plants at time t_0) × (the probability that a three-leaved plant becomes a two-leaved plant between t_0 and $t_0 + 1$) +

(number of four-leaved plants at time t_0) × (the probability that a four-leaved plant becomes a two-leaved plant between t_0 and $t_0 + 1$) $= 800 \times 0 + 90 \times 0 + 56 \times 0.35 + 23 \times 0.45 + 31 \times 0 + 11 \times 0 = 29.95$.

Therefore, we predict that in the next year the population will have 29.95 two-leaved individuals.

To expand our predictions to include every life stage in the following year, we will make full use of matrix algebra. Note that what we have done in the previous paragraph is multiply the vector (7.3) by the fourth row of the projection matrix. To determine the number of individuals in every stage at time $t_0 + 1$, we can multiply the entire matrix by the vector. We can represent this group of equations as a matrix equation

$$
\begin{pmatrix} n_1(t_0+1) \\ n_2(t_0+1) \\ n_3(t_0+1) \\ n_4(t_0+1) \\ n_5(t_0+1) \\ n_6(t_0+1) \end{pmatrix} = \begin{pmatrix} A_{1,1} & A_{1,2} & A_{1,3} & A_{1,4} & A_{1,5} & A_{1,6} \\ A_{2,1} & A_{2,2} & A_{2,3} & A_{2,4} & A_{2,5} & A_{2,6} \\ A_{3,1} & A_{3,2} & A_{3,3} & A_{3,4} & A_{3,5} & A_{3,6} \\ A_{4,1} & A_{4,2} & A_{4,3} & A_{4,4} & A_{4,5} & A_{4,6} \\ A_{5,1} & A_{5,2} & A_{5,3} & A_{5,4} & A_{5,5} & A_{5,6} \\ A_{6,1} & A_{6,2} & A_{6,3} & A_{6,4} & A_{6,5} & A_{6,6} \end{pmatrix} \begin{pmatrix} n_1(t_0) \\ n_2(t_0) \\ n_3(t_0) \\ n_4(t_0) \\ n_5(t_0) \\ n_6(t_0) \end{pmatrix}. \quad (7.4)
$$

If we let

$$
\mathbf{n}(t_0) = \begin{pmatrix} n_1(t_0) \\ n_2(t_0) \\ n_3(t_0) \\ n_4(t_0) \\ n_5(t_0) \\ n_6(t_0) \end{pmatrix},
$$

and we indicate vectors and matrices in bold, then Eq. (7.4) can be written in matrix form as

$$
\mathbf{n}(t_0+1) = \mathbf{A}\mathbf{n}(t_0). \quad (7.5)
$$

To determine the number of individuals in every stage after a year for our example, we use the following equation, which multiplies the matrix (7.2) by the vector (7.3):

$$
\mathbf{n}(t_0+1) = \begin{pmatrix} 0 & 0 & 0 & 0.27 & 3.90 & 40.00 \\ 0.15 & 0 & 0 & 0 & 0 & 0 \\ 0 & 0.21 & 0.55 & 0.05 & 0 & 0 \\ 0 & 0 & 0.35 & 0.45 & 0 & 0 \\ 0 & 0 & 0 & 0.41 & 0.78 & 0 \\ 0 & 0 & 0 & 0.05 & 0.19 & 1.00 \end{pmatrix} \begin{pmatrix} 800 \\ 90 \\ 56 \\ 23 \\ 31 \\ 11 \end{pmatrix}.
$$

Matrix multiplication by hand is tedious and there are many software systems that can be used to carry out matrix computations. Here we give the commands for loading the matrix **A** and the vector **n** into MATLAB or R and performing the multiplication. In MATLAB, the projection matrix **A** can be defined with the following command:

```
A = [0 0 0 0.27 3.90 40.00; 0.15 0 0 0 0 0;
     0 0.21 0.55 0.05 0 0; 0 0 0.35 0.45 0 0;
     0 0 0 0.41 0.78 0; 0 0 0 0.05 0.19 1.0]
```

Similarly, we can define the vector **n** with the following command:

```
n = [800; 90; 56; 23; 31; 11]
```

To multiply the matrix **A** by the vector **n**, we use the command:

```
A*n
```

To conduct the same analysis in R, first define matrix (7.2) as **A** using:

```
A <- matrix(c(0,  0,  0,  0.27,  3.90,  40.00,  0.15,
              0,  0,  0,  0,  0,  0,  0.21,  0.55,  0.05,
              0,  0,  0,  0,  0.35,  0.45,  0,  0,  0,  0,
              0,  0.41,  0.78,  0,  0,  0,  0,  0.05,
              0.19,  1.0),  nrow = 6,  byrow = T)
```

Vector (7.3) can be entered with the command:

```
n <- matrix (c (800, 90, 56, 23, 31, 11), nrow = 6)
```

To multiply **A** by **n**, use:

```
A%*%n
```

The resulting vector is then

$$\mathbf{n}(t_0 + 1) = \begin{pmatrix} 567.11 \\ 120.00 \\ 50.85 \\ 29.95 \\ 33.61 \\ 18.04 \end{pmatrix}. \tag{7.6}$$

The interpretation of the vector components is presented in Table 7.4. Note that while we expect fewer seeds and seedlings in the next year, the number of three- and four-leaved plants has increased.

Exercise 7.2.

a. Using the projection matrix you generated in Exercise 7.1, find the predicted number of members in each stage at time $t_0 + 1$ if the initial distribution of individuals is the same as those given by vector (7.3) above.

b. Compare the distribution of individuals across stage for the Population 2 example above with the distribution you just calculated for Population 3. Although in our simulation both populations started with the same number of individuals in each stage, we do not expect the same distribution after a year. How are the two different? What are the implications of your findings for the future growth of each population?   ▽

Table 7.4 Predicted number of individuals at time $t_0 + 1$ for a population of American ginseng based on the computation from Eq. (7.5).

Stage at Time t_0+1	Number of Individuals
Seeds	567.11
Seedlings	120.00
One-leaved plants	50.85
Two-leaved plants	29.95
Three-leaved plants	33.61
Four-leaved plants	18.04

7.7 THE STABLE DISTRIBUTION OF INDIVIDUALS ACROSS STAGES

In some cases, it might be helpful to predict the number of individuals in each stage one time interval into the future, but most of the time we want to make predictions further into the future. From the projection matrix we can compute the predicted average number of members at any later time. For example, if we wanted to predict the number of individuals in each stage after two time intervals, we could calculate the number in each stage after one interval and then multiply the resulting vector by the transition matrix again. Mathematically, we can write

$$\mathbf{n}(t_0 + 2) = \mathbf{An}(t_0 + 1) = \mathbf{A}\big(\mathbf{An}(t_0)\big) = \mathbf{A}^2\mathbf{n}(t_0).$$

The vector of the number of individuals in each stage at time t_0 is multiplied by the projection matrix once for each time interval; in this case, since there are two time intervals, the projection matrix is raised to the second power.

To predict the distribution of individuals over stages at any time, we multiply the initial vector of the number of individuals in each stage class by the projection matrix to a power equal to the number of time intervals. Therefore, for any positive integer k

$$\mathbf{n}(t_0 + k) = \mathbf{A}^k\mathbf{n}(t_0). \tag{7.7}$$

After each generation, the population will grow or shrink in overall size, but the *proportion* of individuals in each stage at time $t_0 + k$ becomes very similar to the *proportion* at time $t_0 + k + 1$ for large values of k. For example, if we predict the number of individuals in each stage class for a population with projection matrix (7.2) and the initial distribution of vector (7.3) after 10 years, we get the following distribution:

$$\mathbf{n}(t_0 + 10) = \begin{pmatrix} 4609.7 \\ 591.5 \\ 204.2 \\ 94.6 \\ 91.1 \\ 125.6 \end{pmatrix}.$$

For the same population and initial distribution, we predict the following distribution after 11 years:

$$\mathbf{n}(t_0 + 11) = \begin{pmatrix} 5406.0 \\ 691.4 \\ 241.2 \\ 114.0 \\ 109.8 \\ 147.7 \end{pmatrix}.$$

To compare these two vectors in terms of the *relative* distribution of individuals across stages, we can standardize by dividing each value in the vector by the total number of individuals in all stages. If we do so, we get:

$$\frac{\mathbf{n}(t_0 + 10)}{5716.7} = \begin{pmatrix} 0.8064 \\ 0.1035 \\ 0.0357 \\ 0.0165 \\ 0.0159 \\ 0.0220 \end{pmatrix} \quad \text{and} \quad \frac{\mathbf{n}(t_0 + 11)}{6710.1} = \begin{pmatrix} 0.8057 \\ 0.1030 \\ 0.0359 \\ 0.0170 \\ 0.0164 \\ 0.0220 \end{pmatrix}.$$

While the total number of individuals in the population is changing, the relative proportion of individuals in each stage changes very little between years after 10 years have passed. After a long period of time (as $k \to \infty$), this distribution approaches a stable state. The *stable distribution* provides useful information about a population, and we next investigate the method by which it is calculated.

7.8 THEORY SUPPORTING THE CALCULATION OF STABLE DISTRIBUTIONS

We begin with reviewing some key concepts from linear algebra and explore the theoretical structure that makes the calculation of the stable distribution possible. Readers without a linear algebra background can find the details on linear algebra basics in any linear algebra text, including matrix algebra, determinants, bases of linear spaces, and linear combinations of vectors (see, e.g., [11]). We will return to the application of the model to biological data in Section 7.9.

7.8.1 Eigenvalues and Eigenvectors

If $\mathbf{A}$ is an $n \times n$ matrix, the nonzero vector $\mathbf{v}$ is an *eigenvector* of $\mathbf{A}$ with *eigenvalue* λ if

$$\mathbf{A}\mathbf{v} = \lambda\mathbf{v}. \tag{7.8}$$

In other words, multiplying an eigenvector by its associated eigenvalue gives the same answer as multiplying the eigenvector by the entire matrix. This characteristic of eigenvalues and eigenvectors makes them handy tools for exploring the properties of matrices.

To find the eigenvalues and eigenvectors of a matrix we use the following equation:

$$\det(\mathbf{A} - \lambda\mathbf{I}) = 0.$$

The expression $\det(\mathbf{A} - \lambda\mathbf{I})$ is a polynomial with λ being the variable. This polynomial is called the *characteristic polynomial* for $\mathbf{A}$. When we set the characteristic polynomial equal to 0, we obtain the eigenvalues of $\mathbf{A}$. An $n \times n$ matrix has at most n eigenvalues, each with its own eigenvectors. After each eigenvalue λ has been determined, then its corresponding eigenvector(s) is (are) found by solving the equation $\mathbf{A}\mathbf{v} = \lambda\mathbf{v}$.

For example, suppose that a certain matrix $\mathbf{A}$ is defined as:

$$\mathbf{A} = \begin{pmatrix} 1 & 0 & 0 \\ 0 & -4 & 0 \\ 0 & 0 & -4 \end{pmatrix}.$$

Then

$$\mathbf{A} - \lambda \mathbf{I} = \begin{pmatrix} \lambda - 1 & 0 & 0 \\ 0 & \lambda + 4 & 0 \\ 0 & 0 & \lambda + 4 \end{pmatrix},$$

and

$$\det(\mathbf{A} - \lambda \mathbf{I}) = (\lambda - 1)(\lambda + 4)^2.$$

Therefore, the eigenvalues of $\mathbf{A}$ are $\lambda = 1$ and $\lambda = -4$.

Each eigenvalue has at least one eigenvector. The total possible number of linearly independent (that is, different no matter how they are scaled) eigenvectors for an eigenvalue is the *algebraic dimension* of the eigenvalue. In our example above, the eigenvalue 1 has only one linearly independent eigenvector, and the eigenvalue -4 has a maximum of two linearly independent eigenvectors. However, there could be fewer such eigenvectors. The actual number of eigenvectors for each eigenvalue can be determined only by more calculations (see [11]). The *geometric dimension* of the eigenvalue is the number of linearly independent eigenvectors of an eigenvalue that actually exist. Thus, the geometric dimension is less than or equal to the algebraic dimension. The set of linear combinations of the eigenvectors for an eigenvalue is its *eigenspace*. Fortunately for us, if the matrix we are using to describe the population meets certain characteristics, we can isolate a single eigenvalue with one associated eigenvector (and therefore an eigenspace of 1) that describes the growth and stable distribution of our population. The Perron-Frobenius theorem makes this possible.

7.8.2 The Perron-Frobenius Theorem

In calculating the stable distribution that is generated by a particular projection matrix, we will take advantage of a theorem in matrix algebra; the *Perron-Frobenius theorem*, named after the German mathematicians who proved it in the early 20th century. The theorem describes the properties of a matrix that satisfies a certain set of conditions (see [12] for a proof). If the projection matrices generated from biological data meet these conditions, the theorem allows us to use eigenvalues and eigenvectors to predict population growth and the stable distribution of individuals across stages. First, we will examine the theorem and its assumptions, and then we will explain how to apply it to calculating the stable distribution.

The Perron-Frobenius theorem states:

If $\mathbf{A}$ is a positive $n \times n$ matrix, that is if $(A)_{i,j} > 0$ for all i and j, then

a. There is an eigenvalue r of $\mathbf{A}$ for which r is positive and $r > |\lambda|$ for any other eigenvalue λ of $\mathbf{A}$. (Note that $\mathbf{A}$ may have complex eigenvalues.)

b. The algebraic dimension, and thus the geometric dimension, of the eigenspace of r is 1.
c. There is an expression of the eigenvector of r in which all of the entries are positive.

In summary, the theorem states that positive, square matrices have a single positive eigenvalue that is larger than the others, and associated with that eigenvalue is a single eigenvector with only positive entries. We can use this eigenvalue and its associated eigenvector to calculate the stable distribution of individuals across stages for a projection matrix after an infinite number of time intervals. This distribution and its eigenvalue provide information about the possible future growth and structure of a population.

The assumptions of the theorem are not always met by models built from biological examples, but the theorem still applies if certain additional assumptions are met. Projection matrices are structured to have the same number of columns and rows, because we allow for the possibility that every stage could transition to every other stage. Therefore, the assumption of a square matrix is met. A matrix $\mathbf{A}$ is *positive* when $(\mathbf{A})_{i,j} > 0$ for all i and j. (In other words, a matrix is positive when every element in the matrix is greater than 0.) Our example matrix (7.2) is clearly not a positive matrix; it has a number of elements that are 0. Fortunately, the Perron-Frobenius theorem still holds if the matrix is nonnegative, irreducible, and primitive.

A *nonnegative* matrix may contain zeroes but has no negative elements; that is $(\mathbf{A})_{i,j} \geq 0$ for all i and j. Models constructed from biological data do not contain negative values because organisms can only contribute neutrally or positively to growth; therefore, we expect all the matrices we construct to be nonnegative.

A matrix is *irreducible* if there is a path through the life cycle that connects every stage to every other stage, or more generally, beginning at any stage i it is possible to get to the stage j in a finite number of steps. If a matrix contains a stage that acts only as a sink and does not contribute to reproduction or any other life stage, the matrix is *reducible*. Reducible matrices are uncommon but they can occur when a post-reproductive life stage is included in the model (Figure 7.3). (See [1] for more information on working with reducible models.) Our ginseng example is irreducible because every individual has the potential to contribute to reproduction, either directly or by passing through other life stages first.

A nonnegative, irreducible matrix must also be *primitive* for the Perron-Frobenius theorem to still hold true. A primitive matrix is a matrix that, when raised to a sufficiently high power, contains only positive elements. In other words, there is a positive integer m for which $(\mathbf{A}^m)_{i,j} > 0$ for every entry in the matrix $\mathbf{A}$. Again, it is uncommon for a projection matrix constructed from biological data to fail to be primitive, but it can occur if organisms only reproduce after they reach a certain stage and then die before reproducing again (Figure 7.4). If a matrix contains a self-loop (that is, individuals in a stage can stay in that stage in the following time step), it is primitive. We can also raise the matrix to a higher power and see whether the elements all become positive to determine whether it is primitive. Reducible matrices will also fail

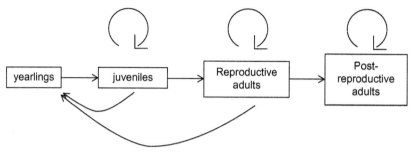

FIGURE 7.3

Life stages, transitions between stages, and reproduction for a pod of killer whales. Adults are divided into two stages; reproductive adults, which contribute to the "yearling" stage by having offspring, and "post-reproductive adults," which do not reproduce or revert to other stages, making the matrix reducible. Modified from [13].

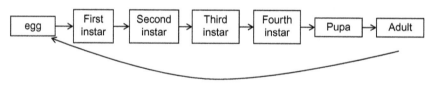

FIGURE 7.4

Life stages, transitions between stages, and reproduction for *Chironomus riparius*, a midge species. Adults have wings, and lay their eggs in water. The eggs hatch into larvae, which pass through four larval stages before pupating. Only adults reproduce, and adults reproduce only once, resulting in a transition matrix that is not primitive. Modified from [14].

this test; at least one element will always remain 0 because it is not connected back into to the rest of the matrix through any path.

If we raise matrix (7.2) to the fifth power, we get:

$$A^5 = \begin{pmatrix} 0.0427 & 0.7475 & 6.2298 & 15.873 & 23.214 & 40.119 \\ 0.0004 & 0.0427 & 0.5339 & 1.8309 & 3.0004 & 6.0000 \\ 0.0061 & 0.0258 & 0.1197 & 0.3387 & 0.7246 & 2.3562 \\ 0.0085 & 0.0397 & 0.1281 & 0.0831 & 0.1603 & 0.8820 \\ 0.0080 & 0.0650 & 0.3191 & 0.3597 & 0.3063 & 0.1808 \\ 0.0020 & 0.0261 & 0.1926 & 0.4391 & 0.6164 & 1.0220 \end{pmatrix}.$$

Thus, the matrix in our ginseng example is primitive. It is also nonnegative and irreducible; therefore we can continue using the Perron-Frobenius theorem to determine the stable distribution of individuals across stages.

7.8.3 Raising a Matrix to a Power in MATLAB and R

After you have entered a matrix and named it A, the MATLAB command for this calculation is:

```
A^5
```

In R, you can multiply the matrix A by itself five times using the following commands:

```
B <- A%*%A
C <- B%*%A
D <- C%*%A
D%*%A
```

Exercise 7.3.   The life stages, transitions, and reproduction for chan, a South American weed, are illustrated in Figure 7.5:

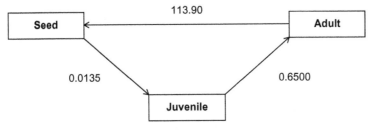

FIGURE 7.5

Life stages, transitions between stages, and reproduction (in italics) for chan (*Hyptis suaveolens*), an annual savanna weed. Seeds germinate in late spring to early summer and grow until the end of the wet season in December and January, when they produce seeds and die. Modified from [15].

  a. Build a matrix from the figure.
  b. Is the matrix positive? Why or why not? If it is not positive, is it nonnegative?
  c. Is the matrix irreducible? Why or why not?
  d. Raise the matrix to the fifth power. Can you claim it is primitive?   ▽

7.8.4 Finding the Stable Distribution

To determine the stable state of a population, we want to know the vector that describes the relative distribution of individuals across stages as the number of time intervals goes to infinity. While in reality no population will exist an infinite amount of time, this distribution can act as a descriptor of the population and gives us the ability to make comparisons among life stages within the distribution and with stable distributions from other populations.

Here, we will work through the steps to explain why the eigenvector associated with the largest eigenvalue allows us to determine the stable distribution of individuals across stage classes. Suppose that $\mathbf{A}$ is a projection matrix that meets the assumptions of the Perron-Frobenius theorem and that $\mathbf{v}_a$ is any vector. We can describe the vector $\mathbf{v}_a$ using the eigenvectors that form a *basis* for this vector space. A set of vectors is a basis for a vector space V provided any vector in V can be written as a linear combination of $\{\mathbf{v}_1, \mathbf{v}_2, \ldots, \mathbf{v}_n\}$ in exactly one way. Suppose that $\{\mathbf{v}_1, \mathbf{v}_2, \ldots, \mathbf{v}_n\}$ is a basis of eigenvectors and let λ_i denote the eigenvalue of $\mathbf{v}_i$. Thus,

$$\mathbf{v}_a = c_1\mathbf{v}_1 + c_2\mathbf{v}_2 + \cdots + c_n\mathbf{v}_n,$$

where $c_1, c_2, \ldots, c_n$ are real-valued constants.

Multiplying by the matrix $\mathbf{A}$ from the left, we get:

$$\mathbf{A}\mathbf{v}_a = \mathbf{A}(c_1\mathbf{v}_1 + c_2\mathbf{v}_2 + \cdots + c_n\mathbf{v}_n) = c_1\mathbf{A}\mathbf{v}_1 + c_2\mathbf{A}\mathbf{v}_2 + \cdots + c_n\mathbf{A}\mathbf{v}_n.$$

Because $\mathbf{v}_i$ is an eigenvector of $\mathbf{A}$ and $\mathbf{A}\mathbf{v}_i = \lambda_i\mathbf{v}_i$, we have:

$$c_1\mathbf{A}\mathbf{v}_1 + c_2\mathbf{A}\mathbf{v}_2 + \cdots + c_n\mathbf{A}\mathbf{v}_n = c_1\lambda_1\mathbf{v}_1 + c_2\lambda_2\mathbf{v}_2 + \cdots + c_n\lambda_n\mathbf{v}_n.$$

Recall from Eq. (7.7) that we can raise $\mathbf{A}$ to the kth power to determine the distribution of individuals across stages after units of time. Similarly,

$$\mathbf{A}^k\mathbf{v}_a = c_1\lambda_1^k\mathbf{v}_1 + c_2\lambda_2^k\mathbf{v}_2 + \cdots + c_n\lambda_n^k\mathbf{v}_n. \tag{7.9}$$

Now we need to determine what happens when k goes to infinity. Since the *dominant eigenvalue* is greater than the absolute value of each of the other eigenvalues, dividing another eigenvalue by the dominant eigenvalue will always generate a value less than one. If we raise that value to the kth power, we expect it to approach zero as k approaches infinity. Expressed mathematically, we mean:

Since $\lambda_1 > |\lambda_i|$ for $i > 1$, we have $\left| \left(\frac{\lambda_i}{\lambda_1}\right) \right| < 1$ for $i > 1$ and so $\left(\frac{\lambda_i}{\lambda_1}\right)^k \to 0$ as $k \to \infty$.

Therefore, if we divide both sides of Eq. (7.9) by λ_1^k, we get

$$\frac{1}{\lambda_1^k}\mathbf{A}^k\mathbf{v}_a = c_1\mathbf{v}_1 + c_2\left(\frac{\lambda_2}{\lambda_1}\right)^k\mathbf{v}_2 + \cdots + c_n\left(\frac{\lambda_n}{\lambda_1}\right)^k\mathbf{v}_n.$$

When we take the limit as $k \to \infty$, all terms except the one containing the dominant eigenvector will give a limit of zero. Thus,

$$\lim_{k\to\infty} \frac{1}{\lambda_1^k}\mathbf{A}^k\mathbf{v}_a = c_1\mathbf{v}_1. \tag{7.10}$$

Applied to the vector $\mathbf{n}(t_0)$ of the number of individuals at each of the stages at time t_0, Eq. (7.10) shows that the eigenvector associated with the dominant eigenvalue describes a long-range, and stable, distribution of individuals across stages. More specifically, if λ_1 is the dominant eigenvalue for $\mathbf{n}(t_0)$, combining Eqs. (7.7) and (7.10) yields

$$\lim_{k\to\infty} \frac{n(t_0 + k)}{\lambda_1^k} = \lim_{k\to\infty} \frac{1}{\lambda_1^k}\mathbf{A}^k\mathbf{n}(t_0) = c_1\mathbf{v}_1. \tag{7.11}$$

As before, it is important to note that this distribution does not give us the proportion (relative distribution) of individuals at each of the life stages. To obtain the relative distribution at the steady state, each of the elements of the vector $c_1 v_1$ will have to be divided by the sum of all the elements.

From a practical perspective, Eq. (7.11) states that to find the relative distribution at the steady state, one needs to compute the dominant eigenvalue of the projection matrix, calculate an eigenvector for the dominant eigenvalue, and then normalize it by dividing each of its components by the sum of its elements.

To illustrate the convergence from Eq. (7.11), we turn again to the ginseng example with projection matrix A from Eq. (7.2) and distribution vector $n(t_0)$ from Eq. (7.3). Based on Eq. (7.10), we expect that if we calculate the dominant eigenvalue λ_1 for the matrix and evaluate the expression $\frac{1}{\lambda_1^k} A^k n(t_0)$ for large values of k, we will obtain approximations for both the steady state distribution of individuals across stages and of the eigenvector associated with the dominant eigenvalue.

Using software for the computations (see next section for the appropriate MATLAB and R commands), we can find that the dominant eigenvalue for matrix A is $\lambda = 1.184$. We can then compute the proportions of individuals across stages k years into the future from:

$$\frac{A^k n(t_0)}{\lambda^k}.$$

If we let $k = 100$, we find

$$\frac{A^{100} n(t_0)}{\lambda^{100}} = \begin{pmatrix} 848.37 \\ 107.47 \\ 36.98 \\ 17.63 \\ 17.89 \\ 23.25 \end{pmatrix}.$$

In order to compare this vector with others, we must normalize it to a sum of 1. The sum of the entries of the vector above is 1051.59, so therefore

$$\frac{1}{1051.59} \begin{pmatrix} 848.37 \\ 107.47 \\ 36.98 \\ 17.63 \\ 17.89 \\ 23.25 \end{pmatrix} = \begin{pmatrix} 0.806 \\ 0.102 \\ 0.053 \\ 0.017 \\ 0.017 \\ 0.022 \end{pmatrix}.$$

Letting $k = 200$ gives

$$\frac{A^{200} n(t_0)}{\lambda^{200}} = \begin{pmatrix} 855.80 \\ 108.41 \\ 37.31 \\ 17.79 \\ 18.05 \\ 23.45 \end{pmatrix}.$$

The sum of the entries of the vector above is 1060.81, and

$$\frac{1}{1060.81} \begin{pmatrix} 855.80 \\ 108.41 \\ 37.31 \\ 17.79 \\ 18.05 \\ 23.45 \end{pmatrix} = \begin{pmatrix} 0.806 \\ 0.102 \\ 0.053 \\ 0.017 \\ 0.017 \\ 0.022 \end{pmatrix}.$$

The entries for the normalized vectors for $k = 100$ and $k = 200$ are actually equal to 6 decimal places, illustrating the convergence established by Eqs. (7.10) and (7.11). Furthermore, based on Eq. (7.8), if this distribution is the same as the eigenvector associated with the dominant eigenvalue, we should be able to multiply it by either **A** or λ and obtain the same vector. When we do this, we find

$$A \begin{pmatrix} .806 \\ .102 \\ .053 \\ .017 \\ .017 \\ .022 \end{pmatrix} = \begin{pmatrix} .955 \\ .121 \\ .042 \\ .199 \\ .020 \\ .026 \end{pmatrix}, \text{ and } 1.184 \begin{pmatrix} .806 \\ .102 \\ .053 \\ .017 \\ .017 \\ .022 \end{pmatrix} = \begin{pmatrix} .955 \\ .121 \\ .042 \\ .199 \\ .020 \\ .026 \end{pmatrix},$$

so the equilibrium state is the normalized eigenvector for the dominant eigenvalue.

7.9 DETERMINING POPULATION GROWTH RATE AND THE STABLE DISTRIBUTION

Now that we have explained the mathematics behind determining the stable distribution of individuals across stages, we can begin to interpret the biological meaning of eigenvalues and eigenvectors. As shown above, the normalized long-range stable distribution of individuals across states is the same as the normalized eigenvector associated with the dominant eigenvalue. This eigenvector is also sometimes called the *right eigenvector*. Therefore, if we can calculate this eigenvector from a projection matrix and normalize it so that the sum of the entries is 1, we get the proportion of individuals in each stage when the population is in a stable state.

The dominant eigenvalue itself also provides useful biological information; it is an estimate of *population growth*. If the eigenvalue is larger than 1, the population is predicted to grow, and if it is less than 1, the population will diminish. This estimate is useful when trying to understand the future of a population, but it assumes that the projection matrix stays constant over time, which is biologically unrealistic. As a population grows, competition and resource availability are likely to change survival and reproduction rates, resulting in changes in the projection matrix. In addition, a constant projection matrix ignores any changes in the environment from year to year. The estimate of growth provided by the dominant eigenvalue, therefore, is more a measure of what the population is capable of, given its current state, than a measure of what

Table 7.5 Steady state distribution of individuals across stages for a population of American ginseng (data from [10]).

Stage	Number of Individuals
Seeds	0.8067
Seedlings	0.1022
One-leaved plants	0.0352
Two-leaved plants	0.0168
Three-leaved plants	0.0170
Four-leaved plants	0.0221

will actually happen in the future. In addition, the quality of our estimates is directly related to the accuracy of the initial data that were used to generate the projection matrix. Estimating projection matrices from multiple years of data and incorporating environmental uncertainty into the model with a stochastic approach can lead to more biologically accurate estimates; see [8] for more details.

We can calculate the dominant eigenvalue and its associated eigenvector using MATLAB and R, as demonstrated below. When we do so for our ginseng example, we find that the estimated growth rate of the population is 1.1841. Because this value is larger than one, we expect that if the projection matrix did not change this population would grow. Through the computations below, we obtain again the steady state distribution of individuals across stages (see Table 7.5). As expected, they are the same as the approximated values we obtained for large k in the previous sections. We predict that if the population illustrated in Figure 7.1 were to reach a stable state, the majority of individuals in the population would be seeds; this is not surprising from a biological perspective, given the low rate at which seeds become seedlings. It is also interesting to note that we predict more four-leaved individuals than two- and three-leaved individuals.

7.9.1　Calculating Eigenvalues and Eigenvectors in MATLAB

We can define the projection matrix (7.2) as A in MATLAB with the following command:

```
A = [0 0 0 0.27 3.90 40.00; 0.15 0 0 0 0 0;
     0 0.21 0.55 0.05 0 0; 0 0 0.35 0.45 0 0;
     0 0 0 0.41 0.78 0; 0 0 0 0.05 0.19 1.0]
```

We can generate all eigenvalues of A using the command eig (A), and we want to give the vector of eigenvalues a name so we can use it in later commands, so we define eig (A) as E with the command

```
E = eig (A)
```

The largest eigenvalue is the dominant eigenvalue, and the growth rate of the population. To identify the largest eigenvalue, first we ask MATLAB to identify the index of the maximum eigenvalue, then we rename the maximum eigenvalue "lambda" with the following two commands:

```
imax = find (E==max(E));
lambda = E(imax)
```

To calculate the right eigenvector, which gives the stable distribution of individuals across stages, we use the following command to generate eigenvectors and their associated eigenvalues:

```
[W,d]=eig (A)
```

(We are using the same eig (A) command as before, but now we are requesting MATLAB to calculate right eigenvectors, which we define as W, and the associated eigenvalues, which MATLAB gives us as diagonals in the matrix that we have labeled d.)

Next, we must pull out the dominant eigenvector and find its associated eigenvalue using:

```
eigvals=diag(d);
imax = find(eigvals==max(eigvals));
w=W(:,imax)
```

Next, we need to rescale the dominant right eigenvector so that it sums to 1 and represents proportions of individuals across stage classes:

```
dominant=w/sum(w)
```

7.9.2 Calculating Eigenvalues and Eigenvectors in R

If you are using R to conduct your calculations, first download and install the popbio package.

We can define our projection matrix (7.2) as A in R with the following command:

```
A <- matrix (c(0, 0, 0, 0.27, 3.90, 40.00,
               0.15, 0, 0, 0, 0, 0, 0, 0.21, 0.55,
               0.05, 0, 0, 0, 0, 0.35, 0.45, 0, 0, 0,
               0, 0, 0.41, 0.78, 0, 0, 0, 0, 0.05,
               0.19, 1.0), nrow = 6, byrow = T)
```

To identify the dominant eigenvalue, which is the growth rate of the population, we use the command

```
lambda (A)
```

To get the stable distribution of individuals across stages, standardized to total to 1, we use the command

```
stable.stage (A)
```

Note that the R commands are tailored specifically for generating values directly related to projection matrices in biology, rather than presenting all of the eigenvalues and eigenvectors for the matrix. If you would like to see all the eigenvalues and eigenvectors for your matrix, use the command

```
eigen (A)
```

Exercise 7.4.   Calculate the dominant eigenvalue and the dominant right eigenvector for the transition matrix you generated in Exercise 7.1, and compare it with the values from the example above. How do the two populations differ in their growth rate? How do they differ in their stable distribution across stages?   ▽

7.10 FURTHER APPLICATIONS OF THE PROJECTION MATRIX

In this chapter, we have introduced you to the basic structure and theory behind using projection matrices in estimating population growth and structure. Biologists and mathematicians have gone much further in exploring methods to account for natural variability in populations. *Sensitivity* analysis uses partial derivatives of matrix elements to determine the relative impact of each element on λ and therefore examines the importance of each life stage on overall population growth [1]. Computer simulations are used to generate multiple estimates of growth rates under different environmental states, providing a better understanding of how robust a population is to environmental change [8]. Models can also incorporate spatial data on patches or multiple populations, providing information on the relative contribution of each patch or population to overall growth [8]. The flexibility and power of the projection matrix in population growth models has led to many important discoveries, contributions, and practical applications in ecology and conservation biology. We encourage you to explore the topic further through the literature cited here, or better yet, collect your own data from a wild population and apply the tools you learned here to predict its future growth.

References

[1] Caswell H. Matrix population models: construction, analysis, and interpretation. 2nd ed. Sunderland, MA: Sinauer Associates; 2001.
[2] Burns JH, Blomberg SP, Crone EE, Ehrlen J, Knight TM, Pichancourt J-B, et al. Empirical tests of life-history evolution theory using phylogenetic analysis of plant demography. J Ecol 2010;98: 334–44.
[3] Steets JA, Knight TM, Ashman T-L. The interactive effects of herbivory and mixed mating for the population dynamics of *Impatiens capensis*. Am Nat 2007;170:113–27.

[4] Knight TM, Caswell H, Kalisz S. Population growth rate of an understory herb decreases non-linearly across a gradient of deer herbivory. Forest Ecol Manag 2009;257:1095–103.

[5] Le Corff J, Horvitz CC. Population growth versus population spread of an ant-dispersed neotropical herb with a mixed reproductive strategy. Ecol Model 2005;188:41–51.

[6] Lande R. Demographic models of the northern spotted owl (*Strix occidentalis caurina*). Oecologia 1988;75:601–7.

[7] Crone EE, Menges ES, Ellis MM, et al. How do plant ecologists use matrix population models? Ecol Lett 2010;14:1–8.

[8] Morris WF, Doak DF. Quantitative conservation biology: theory and practice of population viability analysis. Sunderland, MA: Sinauer Associates; 2002.

[9] Menges ES. Population viability analyses in plants: challenges and opportunities. Trends Ecol Evol 2000;15:51–6.

[10] Charron D, Gagnon D. The demography of northern populations of *Panax quinquefolium* (American ginseng). J Ecol 1991;79:431–45.

[11] Lay D. Linear algebra and its applications. 3rd ed. Boston, MA: Addison Wesley; 2002.

[12] Horn RA, Johnson CR. Matrix analysis. Cambridge, England: Cambridge University Press; 1985.

[13] Brault S, Caswell H. Pod-specific demography of killer whales (*Orcinus orca*). Ecology 1993;74:1444–54.

[14] Charles S, Ferreol M, Chaumot A, Péry ARR. Food availability effect on population dynamics of the midge *Chironomous riparius*: a Leslie modeling approach. Ecol Model 2004;175:217–29.

[15] Schwarzkopf T, Trevisan MC, Silva JF. A matrix model for the population dynamics of *Hyptis suaveolens*, an annual weed. Ecotropicos 2009;22:23–6.

Metabolic Pathways Analysis: A Linear Algebraic Approach

Terrell L. Hodge

Department of Mathematics and College of Arts and Sciences Dean's Office,
Western Michigan University, Kalamazoo, MI 49008, USA

8.1 INTRODUCTION

To quote from a well-known biochemistry textbook [1], "Metabolism is the overall process through which living systems acquire and utilize free energy to carry out their various functions." Metabolism is enacted through metabolic pathways: chains of consecutive enzymatic reactions that produce specific products for use by an organism. As explored by biologists and biochemists, there are hundreds of such "chains of reactions" fitting together in many complex (and sometimes not well-understood) ways. For example, the single bacterium *Escherichia coli* is known to have 600–700 metabolic reactions. Standard (bio)chemical diagrams of such systems of reactions for multiple cellular reactions can easily take up wall-sized charts across multiple walls. To get a sample of this (with relatively uncomplicated diagrams) in the case of *E. coli*, go to the Kegg reference pathway site `http://www.genome.jp/kegg/pathway/map/map01100.html`. Similar diagrams exist for human and animal cell metabolism, with even more complexity; a portion is shown below in Figure 8.1. The *metabolites* in a metabolic pathway are usually taken to be the substrates, intermediates, and reactants in a chain of reactions.

So why the interest? Cellular metabolism is the complex set of chemical reactions that enable a cell to extract energy and other necessities for life from nutrients, and to build the new structures it needs to live and to reproduce. While it may not provide a metaphysical answer to the question "Why are we alive?" metabolism certainly provides a physical answer to the question of how our cells, and hence ourselves, are able to exist, to grow, and, ultimately, what fails and results in death. The study of cellular metabolism is at the heart of numerous questions and basic research about health, such as aging, and on the emerging sidelines as a consideration for others, such as autism. The degree of interconnectedness of our bodies and our environment, through metabolic interactions in the cells of our gut and numerous other cells[1] that we host there, has emerged as a hot research topic, via the study of the microbiome (resp., metabalome), like a biosphere of the gut (resp., a complete profile at a metabolic level). Such research suggests that a broad spectrum of modern

---

[1]Ten times the number of our own!

Mathematical Concepts and Methods in Modern Biology. http://dx.doi.org/10.1016/B978-0-12-415780-4.00008-9

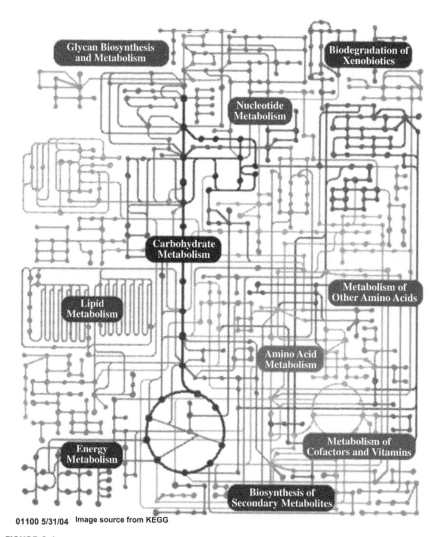

01100 5/31/04 Image source from KEGG

FIGURE 8.1

A representation of a portion of the metabolic pathways for human and animal cell metabolism, from http://www.genome.jp/kegg-bin/show_pathway?org_name=map&mapno=01100&mapscale=1.0&show_description=show.

diseases, such as diabetes, may be the result of having metabolic processes that are not functioning properly, perhaps due to a lack or imbalance of what were evolutionarily fine-tuned contributions of these non-human cells in our germ-killing, antibiotic present, antibacterial-soap world. See, e.g., [2] for a recent accessible introduction.

Even setting aside such new speculations, metabolic processes have been adjusted and tinkered with profitably for some time, and not only for treatment of disease. *Metabolic engineering* is defined as directed modification of cellular metabolism and properties through the introduction, deletion, and modification of metabolic pathways by using recombinant DNA and other molecular biological tools. Currently, green alternatives for many compounds produced chemically using oil are being sought, as well as more green methods. Biochemical production methods of forcing biological organisms or components to overproduce certain desired compounds are one such alternative, achievable through metabolic engineering. Some of the organisms used as production hosts include *E. coli*, *Mycobacterium tuberculosis*,[2] and *Saccharomyces cerevisiae* (yeast). Most of us are quite familiar with the benefits of yeast metabolism, but, as another example, biochemical engineering processes "feed" glucose and corn steep liquor to *E. coli* or other bacteria and generate, through metabolic processes, succinic acid, a precursor to production of pharmaceuticals, fine chemicals, biodegradable polymers, and more. A goal is to understand better the host of metabolic pathways in organisms, and use this knowledge to increase flux through helpful reactions (so produce more output) or even to discover previously unsuspected reaction chains that might produce the desired metabolites in some other fashion. Helpful models would also allow us to test alternative hypotheses, say by computer, more cheaply than running multiple experiments, and might give insight into which types of experiments would be the most useful. One would also hope to obtain a more global perspective, a *systems perspective* that, for example, allows one to see and predict the effects of multiple interconnected reactions at a less reductionist level than reaction by reaction.

How might one explore and understand these interconnected reactions, this *biochemical reaction network*, in a systematic and computational way? What does it mean to have such a system (beyond drawing cartoons)? In order for the cell, and hence the body as a whole, to be in a living, thriving state, it must generally be able to maintain some balance (homeostasis). Each reaction will transform a fixed set of inputs into a fixed set of outputs, but the "flow" or "flux" through a reaction describes how that transformation or flow through the reaction is occurring. If one focuses on a portion of these biochemical reactions that form a (sub)network of interest, then in a balanced state, the total concentration of all chemical compounds in that (sub)system is not changing. In such a state, what are the chains of reactions for which the "total flux", the combined measure of flux in all the reactions, is not changing, i.e., is 0? It is the exploration of this query and setup *mathematically*, through applying standard tools from linear algebra to the so-called *stoichiometry matrix* of the reaction system, that will be the focus of this chapter. In this way, we have an opportunity to see how a mathematical model for metabolic networks, and certain pathways within them, can be constructed. Mathematical models like this have the potential to help us understand, clarify, and make predictions about the very complex inner workings of

---

[2]Note: Tuberculosis infects about 2 billion people—one third of the Earth's population! It's satisfying to think one can turn this threat on its ear and use knowledge of biology to get *M. tuberculosis* to produce useful compounds.

cellular processes in common to all creatures, as studied by clinicians and doctors, biologists, chemists, engineers, and many others.

8.2   BIOCHEMICAL REACTION NETWORKS, METABOLIC PATHWAYS, AND THE STOICHIOMETRY MATRIX

8.2.1   Stoichiometric Matrix I: Nullspaces, Linear Dependence, and Spanning Sets

Suppose one wishes to consider a finite sequence of chemical reactions, involving m chemical compounds $C_1, C_2, \ldots, C_m$.

Recording a reaction in standard chemistry notation, *each reaction* would take the form

$$a_1 C_{i_1} + a_2 C_{i_2} + \cdots + a_k C_{i_k} \longrightarrow a_{k+1} C_{i_{k+1}} + \cdots + a_t C_{i_t},$$
$$1 \leqslant i_1, \ldots, i_t \leqslant m, a_{i_1}, \ldots, a_{i_t} \in \mathbb{R}. \tag{8.1}$$

That is, in a given equation in the list, some of the possible compounds will be used (the C_{i_j}s from among the possible choices $C_1, \ldots, C_m$) in some amounts (the numbers a_j, one for each compound C_{i_j} involved in the reaction). Let's consider this concretely in some steps of glycolysis, a cellular metabolic pathway that is part of the even more complex pathway of cellular respiration. For example, a possible first step in the metabolic pathway of glycolysis in humans is the glucokinase reaction:

$$\underset{\text{(glucose)}}{\text{GLC}} + \underset{\text{(adenosine triphosphate)}}{\text{ATP}} \longrightarrow \underset{\text{(glucose 6-phosphate)}}{\text{G6P}} + \underset{\text{(adenosine diphosphate)}}{\text{ADP}}. \tag{8.2}$$

In reaction (8.2), the biochemical molecules/compounds that are input, i.e., the *substrates* GLC and ATP, and those that are output, i.e., the *products* G6P and ADP, are called metabolites. The addition of ATP yielding ADP is a step repeated again in glycolysis and is very important—it is essentially a way for the cell to release and recapture energy in a controlled fashion. Alternatively, it's not uncommon for biologists to think of the reaction step (8.2) simply as

$$\text{GLC} \longrightarrow \text{G6P},$$

that is, glucose is transformed, in the presence of ATP, into glucose 6-phosphate, with ADP being tossed off in the process. In this view, the reaction captures glucose for the cell, transforming it into a product that cannot wander back off outside the cell wall, whence further reactions in glycolysis continue its transformation. Such shifting of perspective, e.g., from (8.2) to just $GLC \rightarrow G6P$, is a common step in modeling (bio)chemical reaction systems—what constitutes a "metabolite" depends upon what one's interest in the process is. For this reason, we will use the term *metabolite* to mean any compound of interest involved in a biochemical reaction, a use somewhat more loose than by biologists.

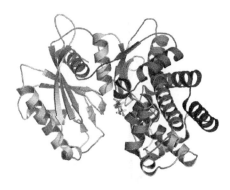

FIGURE 8.2

The enzyme glucokinase, from   http://commons.wikimedia.org/wiki/File:Glucokinase-1GLK.png?useFormat=mobile.

Generally, in each biochemical reaction step in a metabolic system, the reaction is *enzymatic*, in that enzymes, special proteins, catalyze these biochemical reactions. The hexokinases are enzymes that carry out this step from glucose to G-6-phosphate. Figure 8.2 [3] shows a particular hexokinase, glucokinase. Beyond its importance in glycolysis, functional mutations in glucokinase are responsible for Type II diabetes, so there is special interest in targeting this enzyme. (As for many proteins, incorrect folding can lead to disease.)

By many common ways of describing glycolysis, there are 9 further reactions in the glycolysis pathway, with the total of 10 reactions involving a total of 17 (different) compounds as metabolites (some as substrates, some as products, and most as both). For a movie that illustrates this process step-by-step, along with the reaction chain of gluconeogenesis that produces glucose, rather than breaking it down, do see `http://wps.prenhall.com/esm_horton_biochemistry_4/37/9594/2456197.cw/-/2456228/index.htm`. For a static picture, see Figure 8.3 [4]; the reader is encouraged to look on the Web or in texts like [1] for others to see a multiplicity of perspectives.

Coming back to the reactions themselves, if we were to label the compounds with the generic variables above, we would need 10 chemical equations in 17 variables, $C_1, \ldots, C_{17}$. Two additional examples of these chemical reactions involved in glycolysis are:

$$(1, 3\text{BPG}) + \text{ADP} \longrightarrow (3\text{PG}) + \text{ATP}, \tag{8.3}$$

and

$$(\text{PEP}) + \text{ADP} + H^+ \longrightarrow (\text{Pyr}) + \text{ATP}, \tag{8.4}$$

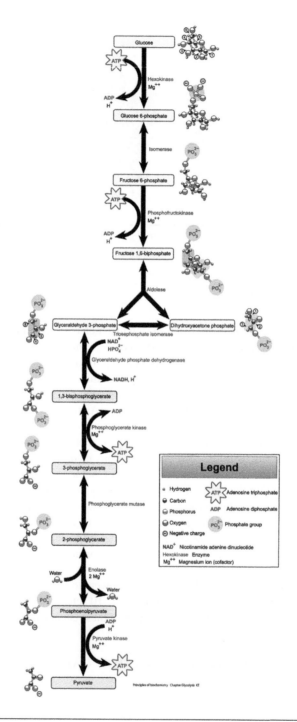

FIGURE 8.3

A representation of glycolysis, from http://upload.wikimedia.org/wikipedia/commons/archive/a/a0/20091114111506%21Glycolysis.svg.

where (if you care to know)

$$
\begin{array}{lll}
1,3\text{BPG} & \text{is} & 1,3-\text{biphosphoglycerate,} \\
3\text{PG} & \text{is} & 3-\text{phosphoglycerate,} \\
\text{PEP} & \text{is} & \text{phosphoenolpyruvate, and} \\
\text{Pyr} & \text{is} & \text{pyruvate.}
\end{array}
$$

Now, each (bio)chemical equation (8.1) can be rewritten mathematically as follows:

$$a_1 C_{i_1} + a_2 C_{i_2} + \cdots + a_k C_{i_k} = a_{k+1} C_{i_{k+1}} + \cdots + a_t C_{i_t}. \qquad (8.5)$$

Necessarily, each $a_j > 0$, since a_j is the number of molecules of the metabolite C_{i_j}. Equivalently, the equality (8.5) can be rewritten as

$$-a_1 C_{i_1} + (-a_2)C_{i_2} + \cdots + (-a_k)C_{i_k} + a_{k+1}C_{i_{k+1}} + \cdots + a_t C_{i_t} = 0, \quad (8.6)$$

with the convention that the number a_j of molecules of the metabolite C_{i_j} is altered by a minus sign when C_{i_j} is a substrate (an "input," or a metabolite "consumed by the reaction"), and taken to be positive when C_{i_j} is a product (an "output," or a metabolite "produced by the reaction"). For example, placing Eq. (8.2) in the form Eq. (8.6) gives

$$(-1)(\text{GLC}) + (-1)(\text{ATP}) + (1)(\text{G6P}) + (1)(\text{ADP}) = 0. \qquad (8.7)$$

It is not uncommon for the coefficients of these (mathematically written) reaction equations to be simply ± 1; in fact, equations arising from biochemical networks[3] can often be represented in this form.

Exercise 8.1.   Rewrite Eq. (8.3) in the form of Eq. (8.7) and call the result Eq. (8.8).

$$\qquad (8.8)$$

Now do the same for Eq. (8.4), and call the result Eq. (8.9). ∇

$$\qquad (8.9)$$

The data in the three Eqs. (8.7)–(8.9) can be encoded in a matrix, a stoichiometric matrix, which we now define.

Definition 8.1.   The *stoichiometric matrix*, for a system of n biochemical reactions (8.1), is an $m \times n$ matrix $S \in M_{m,n}(\mathbb{R})$, where m is the total number of metabolites $C_1, \ldots, C_m$ in the n equations. The stoichiometry matrix S has exactly one row for each metabolite appearing in the system. Each column vector of the matrix S corresponds to a reaction, by recording the coefficients of the metabolites taken from the reaction equation in the form (8.6). If a metabolite does not appear in a reaction, the corresponding column entry is taken to be 0.

---

[3] As opposed to "elemental equations," reaction equations associated to the decomposition of compounds into their basic chemical elements, like hydrogen, oxygen, etc.

Example 8.1.   In Eqs. (8.7)–(8.9), there are a total of 9 metabolites. Thus, S will have rows $[S]_{i*}$ with each corresponding to a C_i, and columns $[S]_{*j}$ with each corresponding to an equation in the form of Eq. (8.6). Letting

$$C_1 = \text{GLC},$$
$$C_2 = \text{ADP},$$
$$C_3 = \text{G6P},$$
$$C_4 = \text{ATP},$$
$$C_5 = 1, 3\text{BPG},$$
$$C_6 = 3\text{PG},$$
$$C_7 = \text{PEP},$$
$$C_8 = \text{Pyr, and}$$
$$C_9 = \text{H}^+,$$

then the first reaction Eq. (8.7) gives rise the first column

$$[S]_{*1} = \begin{bmatrix} -1 \\ 1 \\ 1 \\ -1 \\ 0 \\ 0 \\ 0 \\ 0 \\ 0 \end{bmatrix}.$$

Exercise 8.2.   Use Eqs. (8.8) and (8.9) to fill in the remaining two columns of the 9 by 3 stoichiometric matrix S for the system given by (8.7)–(8.9):

$$S = \begin{bmatrix} -1 \\ 1 \\ 1 \\ -1 \\ 0 \\ 0 \\ 0 \\ 0 \\ 0 \end{bmatrix}.$$

∇

Exercise 8.3.

a. The stoichiometric matrix S you found in Exercise 8.2 corresponds to a system of only *some* of the equations needed to represent the full glycolysis network. What would be the shape of the stoichiometric matrix for the full glycolysis network?

b. As analyzed in [6], the full metabolic network of the human red blood cell involves 51 reactions (and their "fluxes," in biochemical parlance) and 29 metabolites. What would be the shape of the stoichiometric matrix S for this metabolic system[4]?   ▽

As one mathematical model,[5] one can also attempt to capture metabolic systems, and biochemical reaction networks in general, pictorially, through graphs. In these graphs, there is one vertex (i.e., node) for every metabolite. Edges join two metabolites if they are linked in a (bio)chemical reaction, and a value associated to each edge is the "flux," a measure of the level of activity through the reaction (or rate at which the reaction is occurring). The direction of an arrow represents the direction of the reaction, and reversible reactions are pictured through oppositely oriented double edges (see Figure 8.4). A large picture of multiple metabolic networks was given earlier (Figure 8.1); a simpler example of such a graph as given in [6] is pictured as graph (A) in Figure 8.4.

Such a graph as graph (A) in Figure 8.4 might correspond to looking at only part of a larger system (just as glycolysis appears in the much larger network of cellular metabolism). In this case, one can draw a boundary around the relevant part of the system, and include arrows to represent flows into or out of the subsystem from the larger system, pictured as graph[6] (B) of Figure 8.4. There will be *internal fluxes*, corresponding to reactions in the particular system under consideration, and *external fluxes*, those involving inputs or outputs from parts of the system not under direct focus, but that, given the connectedness of systems, cannot be ignored. In [6], it is argued that a useful biochemical convention is to first represent the exchange fluxes as arrows "going out" of the boundary, even if the direction in a chemical reaction sense is the opposite. Hence, one sees the arrow conventions and labelings[7] of internal fluxes (by $v_1, \ldots, v_7$) and external fluxes ($b_1, \ldots, b_4$) in (B) of Figure 8.4.

For each vertex in the graph of the biochemical reaction network, one may write a "balance" or "node" equation in the internal and external fluxes. In this, the formal sum of fluxes "going in" to a fixed node must equal the formal sum of fluxes "going out." Thus, using the second of the two graphs above (graph (B)) at node A, one obtains

$$v_1 + b_1 = 0;$$

---

[4]No computations are necessary to answer this, but see the final section for an associated project wherein one can use free software to explore this example concretely, in more detail.

[5]See the final section for the associated project which explores shortcomings and identifies another model.

[6]By many formal definitions, the result is no longer formally a graph in the mathematical sense, since there are arrows which do not join two nodes, but we will continue to abuse terminology and call this a graph.

[7]This labeling of some fluxes by v_is and some by b_js is a notational convenience in what follows. However, the total flux vector will still be denoted by **v**, with first seven entries the internal fluxes, and last four entries the external fluxes.

(A)

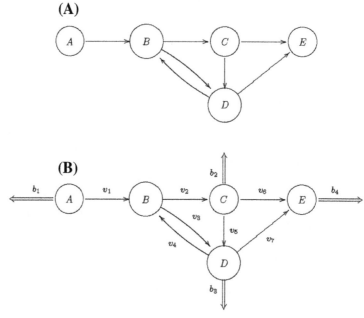

(B)

FIGURE 8.4

A graphical example from [6] of a possible metabolic network (graph A), also shown with (re)labelings for internal and external fluxes added as per conventions in [6] (graph B).

at node B, one obtains

$$v_2 + v_3 = v_1 + v_4.$$

Alternatively, one can assign the following conventions: label the flux for each incoming edge as positive, and label the flux for each outgoing edge negative; then the two node/balance equations above can be rewritten as:

$$\text{Node } A: \quad -v_1 - b_1 = 0, \tag{8.10}$$
$$\text{Node } B: \quad v_1 + v_4 - v_2 - v_3 = 0. \tag{8.11}$$

Exercise 8.4. Complete the list of node/balance equations begun above, i.e.,

$$\text{Node } A \quad : \quad -v_1 - b_1 = 0,$$
$$\text{Node } B \quad : \quad v_1 + v_4 - v_2 - v_3 = 0.$$
$$\text{Node } C \quad :$$
$$\cdots \quad : \quad \cdots \tag{8.12}$$

$\triangledown$

In the form of Eq. (8.12) above, the sequence of balance/node equations corresponds to a homogeneous linear system in the 11 variables $v_1, \ldots, v_7, b_1, \ldots, b_4$, that is, in the flux variables. There is one equation for each node.

Exercise 8.5. Write down the coefficient matrix for the homogeneous linear system described in Exercise 8.4.   ▽

Wonderfully, the coefficient matrix you just found is just the stoichiometric matrix S for the system! Writing the total flux vector $\mathbf{v} = (v_1, \ldots, v_{11})^T := (v_1, \ldots, v_7, b_1, \ldots, b_4)^T$ (so that $v_8 := b_1$, $v_9 := b_2$, $v_{10} := b_3$, $v_{11} := b_4$, and the other v_is are the internal fluxes as before), the homogeneous system described by the node/balance equations has the form

$$S\mathbf{v} = 0. \qquad\qquad (8.13)$$

This time, however, we have obtained S by focusing on the rows, instead of the columns.

Exercise 8.6. How many reaction equations occur in the system described by this graph/stoichiometric matrix S?   ▽

Recall that the nullspace $N(S)$ of S is the set of all solutions to the homogeneous system (8.13). Equation (8.13) is interpretable as a statement of conservation of mass [6]. Each vector $\mathbf{v}$ in the nullspace $N(S)$ describes the relative distribution of "fluxes," and the variable entries $v_1, \ldots, v_n$ of each flux vector $\mathbf{v}$ give values that represent the activity of the individual reactions, indicated by their flow rates. Thus, the product $S\mathbf{v}$ assigns the flux throughout the entire metabolic system represented by S [6, p. 4194]. Flux vectors $\mathbf{v}$ satisfying $S\mathbf{v} = 0$ correspond to steady-state solutions to a "dynamic mass-balance equation" $\frac{d\mathbf{X}}{dt} = S\mathbf{v}$, where $\mathbf{X} = (X_1, \ldots, X_m)^T$ for X_i the concentration of the ith metabolite (compound C_i), and $\frac{dX_i}{dt} = a_{i,1}v_1 + \cdots + a_{i,n}v_n$ is the change in concentration of the ith metabolite, and as before, v_j the rate ("flux") of reaction j.

Exercise 8.7.

a. Let $\mathbf{v} = (1, 1, 0, 0, 0, 1, 0, -1, 0, 0, 1)^T$. Show that $\mathbf{v} \in N(S)$.

b. Represent the total flux vector $\mathbf{v}$ from part (a) by shading in Figure 8.5 the corresponding edges of the graph which have nonzero fluxes. (Recall that a zero entry means there is no flux, hence no involvement, of a particular reaction.) What do you notice about the resulting walk in the graph?

c. Let $\mathbf{v}' = (1, 1, 0, 0, 0, 0, 0, 0, -1, 1, 0, 0)^T$. Repeat the instructions for parts (a) and (b) above, using Figure 8.5.

d. Check that $\mathbf{v}$ and $\mathbf{v}'$ are linearly independent vectors.

e. Using your observations from parts (a) to (d), find and sketch the graphical interpretation of another total flux vector, $\mathbf{v}''$, which is linearly independent from $\mathbf{v}$ and $\mathbf{v}'$ (that is, show $\{\mathbf{v}, \mathbf{v}', \mathbf{v}''\}$ is a linearly independent set). Use Figure 8.5.

f. Now, find a total flux vector $\mathbf{w}$ for which $\{\mathbf{v}, \mathbf{v}', \mathbf{w}\}$ form a linearly dependent set. Do you have a graphical interpretation for $\mathbf{w}$?   ▽

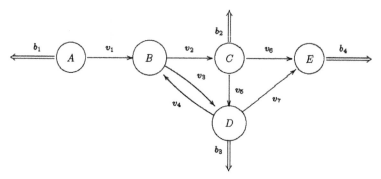

FIGURE 8.5

Graph for Exercise 8.7 (b).

Exercise 8.8.

a. Find the entire space $N(S)$ for the stoichiometric matrix S you found in Exercise 8.5. Before beginning, make a prediction as to the number of free variables you will find, and try to complete the rest of the accompanying statement below.

My prediction for the number of free variables is: _____, and the relation to the minimal number of vectors spanning $N(S)$ is: _____.

Express your solution as linear combinations of vectors, with coefficients the free variables, that is, express $N(S)$ as a span of a set of vectors. (Feel free to use your calculator here to find the rref E_S of S.)

b. Argue (without a lot of computation) that the spanning set of vectors for $N(S)$ that you found in part (a) forms a linearly independent set.

c. As in Exercise 8.7(b), give the graphical interpretation of each element of the spanning set for $N(S)$ by using as many copies of Figure 8.5 as needed. Explain whether these results are consistent with your expectations formed from Exercise 8.7. (We will explore this outcome further in some later problems.)   ▽

Exercise 8.9.

a. For the stoichiometric matrix S as in Exercise 8.2, find the nullspace $N(S)$ of S.

b. *Then*, explain the biochemical significance of the result of part (a).

c. Can you hazard a guess also as to the graphical significance of the result of part (a)? [Hint: Take a look at the glycolysis network represented in Figure 8.3.]

d. Would you expect the same result as in part (a) if you used instead the stoichiometric matrix S for the entire metabolic network given by glycolysis? Be as specific as you can about the results you would expect, without doing any actual computations.   ▽

8.2.2 More on the Nullspace of the Stoichiometric Matrix: Spanning with Biochemical Pathways and Base Changing

Exercise 8.10.   The vectors (total flux vectors **v**) you found in Exercise 8.8, spanning the nullspace $N(S)$ of the stoichiometry matrix S, include two which have at least one negative entry in a variable associated to an internal reaction. List these two vectors.   ▽

Since any reversible reaction was already broken down graphically into a pair of double edges (oppositely oriented), and otherwise the arrows of internal reactions correspond to directions of the associated chemical equations, a negative value on an internal flux gives a somewhat nonsensical interpretation. Flux vectors with this problem (one or more negative flux distributions for internal reactions) represent biochemically impossible outcomes, while flux vectors without this problem represent chemically feasible pathways through the metabolic system [6]. This corresponds to what you have seen pictorially in Exercise 8.8(c).

Consequently, one goal of [6] is to find a biologically legitimate spanning set of flux vectors **v** for the nullspace $N(S)$ of a stoichiometric matrix S, so that each **v** corresponds to a biochemically valid pathway through the metabolic system described by S. The mechanism [6] employ is base-changing, as we shall now discuss. First:

Exercise 8.11.

  a. Recall the definition of a basis of a vector space here.
  b. List a basis for the vector space $N(S)$, for the stoichiometric matrix S you found in Exercise 8.5.   ▽

In seeking a "good basis" for the nullspace $N(S)$ of a stoichiometric matrix S [6], look for a set of total flux vectors that simultaneously:

  1. all represent biochemically valid pathways,
  2. span $N(S)$, that is, each possible flux vector in $N(S)$ is a linear combination of the basis vectors, and
  3. form a linearly independent set: no one flux vector in the basis set (or its graphical interpretation) can be expressed nontrivially as a linear combination of the remaining ones.

Exercise 8.12.   A quick check on your understanding: Is the basis you found for Exercise 8.11 a "good basis", in the sense above? Is a "good basis" a basis?   ▽

To create a "good basis" as in [6], the authors further require the following conditions be satisfied by a set of total flux vectors for the nullspace $N(S)$ of any stoichiometric matrix S:

  4. All coordinates corresponding to internal fluxes will be positive, that is,

$$v_i \geqslant 0 \text{ (all internal fluxes).}$$

5. For the coordinates corresponding to exchange fluxes values,

$$b_j \geqslant 0 \text{ (if the metabolite is exiting the system only),}$$
$$b_j \leqslant 0 \text{ (if the metabolite is entering the system only).}$$

Let

$$B = \begin{bmatrix} 1 & 1 & 1 & 0 & 0 & 0 \\ 1 & 0 & 1 & -1 & 1 & 0 \\ 0 & 1 & 0 & 1 & -1 & 1 \\ 0 & 0 & 0 & 0 & 0 & 1 \\ 0 & 0 & 0 & 0 & 1 & 0 \\ 1 & 0 & 0 & -1 & 0 & 0 \\ 0 & 0 & 0 & 1 & 0 & 0 \\ \hline -1 & -1 & -1 & 0 & 0 & 0 \\ 0 & 0 & 1 & 0 & 0 & 0 \\ 0 & 1 & 0 & 0 & 0 & 0 \\ 1 & 0 & 0 & 0 & 0 & 0 \end{bmatrix}, \tag{8.14}$$

where the dashed line in B simply represents a partition of B into two blocks, but can otherwise be ignored. Set $z_j = [B]_{*j}$ and take $\mathcal{B} = \{z_1, \ldots, z_6\}$. We claim $\mathcal{B}$ is a basis for the nullspace $N(S)$ as in Exercise 8.11.[8] Writing down the matrix B, instead of the list $\mathcal{B}$, is just a compact way to represent this basis. (The basis $\mathcal{B}$ is still really the set of *columns* of B, though!) The entries in each part of the columns above the dashed line in B correspond to the internal flux variables $v_1, \ldots, v_7$, while the entries below the dashed line correspond to the external flux variables $b_1, \ldots, b_4$. Observe that the basis $\mathcal{B}$ given by B is not a "good basis".

Exercise 8.13.

a. Set $y_i = z_i$, $1 \leqslant i \leqslant 3$, $y_4 = z_4 + z_1$, $y_5 = z_5 + z_2$, and $y_6 = z_5 + z_6$. Write out a matrix P with columns $[P]_{*j} = y_j$, $1 \leqslant j \leqslant 6$. (Use a partitioning scheme like that for B above.) Let $\mathcal{P} = \{y_1, \ldots, y_6\}$.
b. Argue that $\mathcal{P}$ spans $N(S)$, for S as in Exercises 8.5 and 8.8.
c. Show that $\mathcal{P}$ is a set of linearly independent vectors.
d. Parts (b) and (c) show $\mathcal{P}$ form a basis of $N(S)$. Finally, check that $\mathcal{P}$ satisfies the additional properties required to make a "good," that is, biochemically feasible, "basis" for $N(S)$, as per the added conditions from [6] listed previously.
e. Illustrate the paths corresponding to the total flux vectors given by the basis vectors $\mathcal{P}$ using copies of Figure 8.5.
f. Find a nontrivial total flux vector $\mathbf{x}$ which does not correspond to a basis vector in $\mathcal{P}$, and represent it as a linear combination of paths (i.e., total flux vectors) from $\mathcal{P}$.

▽

---

[8] Hopefully, this looks familiar up to relabeling. Thus, we will not check that $\mathcal{B}$ is a basis for $N(S)$ here.

The basis P found above is an example of "base-changing" from a mathematically valid basis B of $N(S)$ to one that is mathematically *and* biochemically valid. Question to ask at this juncture include:

Is it always possible to find a mathematically valid basis for the nullspace $N(S)$ of a stoichiometric matrix S? If so, is it always possible to find a biochemically valid basis P, starting from a mathematically valid basis B of the nullspace of a stoichiometric matrix S? The first we can answer; the second is a research question which we will illustrate in another example.

Changing bases of a vector space can also be viewed in terms of "changing coordinates." A set $B = \{z_1, \ldots, z_t\}$ of basis vectors for an (arbitrary) vector space W can be viewed as a set of "coordinates" for W. This literally means that any vector $w \in W$ has a unique expression as a linear combination in terms of the elements of B. More precisely, there are real numbers $w_1, \ldots, w_t$ so that

$$w = w_1 z_1 + \cdots + w_t z_t,$$

and if $w' \in W$ is any other vector, then $w' = w$ if and only if, in the corresponding expression

$$w' = w_1' z_1 + \cdots + w_t' z_t,$$

one has $w_1 = w_1', \ldots, w_t = w_t'$.

One can represent $w \in W$ by listing the coefficients w_j of w in the linear combination to give

$$w = w_1 z_1 + \cdots + w_t z_t$$

as a vector $w = (w_1, \ldots, w_t)^T$. In this way, every vector w in W corresponds uniquely to a "point" $(w_1, \ldots, w_t)$.

Exercise 8.14.

a. Suppose $W = \mathbb{R}^2$, the usual Euclidean plane. Let e_1 be the standard unit vector (vector of length one) in the direction of the positive x-axis, and let e_2 be the standard unit vector in the direction of the positive y-axis. Use geometric properties of vector operations in Euclidean space to sketch the vector $(-4)e_1 + 5e_2$. To what ordered pair does this linear combination of vectors correspond?

b. However, now replace e_1 by $b_1 := 2e_1 + 3e_2$, and let $b_2 = e_2$. Let $w \in \mathbb{R}^2$ be the vector whose coordinates are $(-4, 5)$ in the new coordinates given by $\{b_1, b_2\}$. Sketch w using the old coordinate system $\{e_1, e_2\}$.   ▽

Going back to metabolic pathways and stoichiometric matrices, complete the following exercise.

Exercise 8.15.

a. Define $\mathcal{A}$ to consist of the elements of P and the vector x as in Exercise 8.13(f). Find a vector in the corresponding system $N(S)$ which has two distinct representations as linear combinations of elements of $\mathcal{A}$. (This reflects a general principle: One cannot add a new vector to a set that is already a basis and still have a basis. Be able to justify this!)

b. Express each vector of $\mathcal{B}$ (coming as before from (8.14)) as an element of $N(S)$, with coordinates coming from the basis $\mathcal{P}$.

$\triangledown$

Exercises 8.13 and 8.15 explore two bases, namely the sets of vectors $\mathcal{B}$ and $\mathcal{P}$, for the nullspace $N(S)$ of the stoichiometric matrix S as in Exercises 8.5 and 8.8. As per Exercises 8.13 and 8.15, dim $(N(S)) = 6$, so any vector in $N(S)$ can be expressed uniquely in six coordinates taken either with respect to $\mathcal{B}$, or with respect to $\mathcal{P}$. For example, viewing $N(S)$ as a subspace of Euclidean space $\mathbb{R}^{11}$, we previously checked that

$$\mathbf{v} := (1, 1, 0, 0, 0, 1, 0, -1, 0, 0, 1)^T \in N(S).$$

(As per usual conventions, the coordinates here are in terms of the standard basis $\{\mathbf{e}_1, \ldots, \mathbf{e}_{11}\}$ of $\mathbb{R}^{11}$.) However, in terms of coordinates with respect to the ordered basis $\mathcal{B} = \{\mathbf{z}_1, \ldots, \mathbf{z}_6\}$ of $N(S)$ as a six-dimensional vector space,

$$\mathbf{v} = \mathbf{z}_1 = (1, 0, 0, 0, 0, 0)^T_\mathcal{B}. \tag{8.15}$$

Likewise, since $\mathbf{y}_1 = \mathbf{z}_1$, in terms of coordinates with respect to the ordered set $\mathcal{P} = \{\mathbf{y}_1, \ldots, \mathbf{y}_6\}$, one has

$$\mathbf{v} = (1, 0, 0, 0, 0, 0)^T_\mathcal{P}. \tag{8.16}$$

On the other hand,

$$\mathbf{w} := (0, -1, 1, 0, 0, -1, 1, 0, 0, 0, 0)^T \in N(S) \subset \mathbb{R}^{11},$$

satisfies

$$\mathbf{w} = \mathbf{z}_4 = (0, 0, 0, 1, 0, 0)^T_\mathcal{B}, \tag{8.17}$$

but

$$\mathbf{w} = (-1, 0, 0, 1, 0, 0)^T_\mathcal{P} = \mathbf{y}_4 - \mathbf{y}_1. \tag{8.18}$$

A *change-of-basis matrix*[9] from the basis $\mathcal{B}$ to the basis $\mathcal{P}$ is a square 6×6 matrix $A_{\mathcal{B},\mathcal{P}}$ with the following property:
If

$$\mathbf{u} := (u_1, \ldots, u_6)^T_\mathcal{B},$$

then the matrix product $\mathbf{q} := A_{\mathcal{B},\mathcal{P}}\mathbf{u}$ gives the coordinate expression for $\mathbf{u}$ in the basis $\mathcal{P}$, i.e.,

$$\mathbf{q} = (q_1, \ldots, q_6)^T_\mathcal{P}.$$

The matrix $A_{\mathcal{B},\mathcal{P}}$ can be created by setting each of its columns $[A]_{*j}$, $1 \leq j \leq 6$, to be the basis vector $\mathbf{z}_j \in \mathcal{B}$ written in terms of coordinates of $\mathcal{P}$. Thus, in this case, by Eqs. (8.15) and (8.16),

$$[A]_{*1} = (1, 0, 0, 0, 0, 0)^T,$$

while by Eqs. (8.17) and (8.18),

$$[A]_{*4} = (-1, 0, 0, 1, 0, 0)^T.$$

---

[9] Also called a "transition matrix" in some texts, while in others, this term is reserved for Markov chain processes.

Exercise 8.16.

a. Complete the remaining four columns of the change-of-basis matrix $A_{\mathcal{B},\mathcal{P}}$, from basis $\mathcal{B}$ to $\mathcal{P}$, using the results of Exercise 8.13:

$$A_{\mathcal{B},\mathcal{P}} = \begin{bmatrix} 1 & -1 \\ 0 & 0 \\ 0 & 0 \\ 0 & 1 \\ 0 & 0 \\ 0 & 0 \end{bmatrix}.$$

b. Use $A_{\mathcal{B},\mathcal{P}}$ to find the coordinates of $\mathbf{u} := (1, 3, 0, -1, 2, 1)_{\mathcal{B}}^T$ in terms of $\mathcal{P}$. *Also*, what vector of $N(S)$ does $\mathbf{u}$ represent, expressed in terms of the standard coordinates in $\mathbb{R}^{11}$?

c. Use $A_{\mathcal{B},\mathcal{P}}$ to find the coordinates of an arbitrary vector $\mathbf{c} := (c_1, c_2, c_3, c_4, c_5, c_6)_{\mathcal{B}}^T$ in terms of $\mathcal{P}$. $\triangledown$

In the next exercise, you will repeat the previous exercise, but switch the roles of $\mathcal{B}$ and $\mathcal{P}$.

Exercise 8.17.

a. Using the results of Exercise 8.15, now find the change-of-basis matrix $A_{\mathcal{P},\mathcal{B}}$ from basis $\mathcal{P}$ to basis $\mathcal{B}$ of $N(S)$.

b. Use $A_{\mathcal{P},\mathcal{B}}$ to find the coordinates of $\mathbf{v} := (1, 3, 0, -1, 2, 1)_{\mathcal{P}}^T$ in terms of $\mathcal{B}$. *Also*, what vector of $N(S)$ does $\mathbf{v}$ represent, expressed in terms of the standard coordinates in $\mathbb{R}^{11}$?

c. Use $A_{\mathcal{P},\mathcal{B}}$ to find the coordinates of an arbitrary vector $\mathbf{d} := (d_1, d_2, d_3, d_4, d_5, d_6)_{\mathcal{P}}^T$ in terms of $\mathcal{B}$.

d. Compute the matrix products $A_{\mathcal{B},\mathcal{P}} A_{\mathcal{P},\mathcal{B}}$ and $A_{\mathcal{P},\mathcal{B}} A_{\mathcal{B},\mathcal{P}}$, and explain your answer. $\triangledown$

Exercise 8.18.   The "Rank + Nullity Theorem"[10] says that, for an arbitrary matrix $A \in M_{m,n}(\mathbb{R})$,

$$\dim (N(A)) + \text{rank}(A) = n,$$

where the dimension $\dim (N(A))$ of the nullspace of A is often called the *nullity* of A, and rank (A), the rank of A, is the dimension of the row space of A, equivalently, the number of nonzero rows in E_A.

Suppose for parts (a) - (d) and (f), $S \in M_{m,n}(\mathbb{R})$ is a generic stoichiometric matrix:

a. *Recall:* What is the biological meaning of the number n of columns of S?
b. *Recall:* What is the biological meaning of the number m of rows of S?

---

[10]This is sometimes just called the "Rank Theorem."

c. *Recall:* What is the biological meaning of a vector in $N(S)$, and of dim $(N(S))$, the number of (linearly independent) vectors spanning $N(S)$? What about the columns of S?

d. For the metabolic system captured by S, under what circumstances is the number of total flux vectors needed to describe the metabolic pathways (via linear combinations) in the system the same as the difference between the number of metabolites in the system and the number of chemical reactions in the system? [Hint: Use your previous answers and the Rank + Nullity Theorem.]

e. Were the circumstances you identified in part (d) of this exercise met in Exercise 8.8? What about in Exercise 8.9?

f. By definition of the rank of a matrix, rank(S) satisfies the inequality rank$(S) \leqslant m$. As per [7, p. 298], rank$(S) < m$ whenever conservation relationships hold in the system, for example, that ATP + ADP equals some constant value for the whole system. Using this, give a restatement (in English) of the biological meaning of the Rank + Nullity Theorem for a stoichiometric matrix, tying in your results from part (d).

g. Schilling et al. [7, pp. 298–299] presents an analysis of the reaction scheme of a metabolic system consisting of the glyoxylate cycle and related reactions, as pictured therein. The outcome of solving for the nullspace $N(S)$ of a stoichiometric matrix S for this metabolic system[11] results in a nullspace $N(S)$ with dim $(N(S)) = 3$. A basis for $N(S)$ is also pictured below, organized as the columns of the matrix B. For reasons of space, the actual matrix shown is B^T; the columns correspond to the following metabolite abbreviations: Eno, Acn, Sdh, Fum, Mdh, AspC, Gdh, Pyk, AceEF, GltA, Icd, Icl, Mas, AspCon, Ppc, GluCon:

$$B^T = \begin{bmatrix} 2 & 1 & 1 & 1 & 2 & 1 & 1 & 2 & 2 & 1 & 0 & 1 & 1 & 1 & 0 & 0 \\ 1 & 1 & 1 & 1 & 2 & 0 & 0 & 2 & 2 & 1 & 0 & 1 & 1 & 0 & -1 & 0 \\ 3 & 2 & 1 & 1 & 2 & 0 & 1 & 3 & 3 & 2 & 1 & 1 & 1 & 0 & 0 & 1 \end{bmatrix}.$$

Using the data from matrix B above, that is, from $N(S)$ (and without attempting to compute S itself[12]), answer the following questions:

i. How many free variables must have appeared in E_S?

ii. If there were no conservation relationships holding among metabolites in the system, how many reactions were there?

$\triangledown$

8.2.3 Conclusion

Biologically "good" bases for $N(S)$ lead to the notion of "extreme paths". For example, for S as in Exercise 8.18(g), it is possible to find a "good basis" for $N(S)$: check that by setting $\mathbf{b_i}$ to be the ith row of B^T, and taking $\mathbf{p_1} = \mathbf{b_1} - \mathbf{b_2}$, $\mathbf{p_2} = \mathbf{b_2}$, $\mathbf{p_3} = \mathbf{b_3} - \mathbf{b_2}$, and $\mathbf{p_4} = \mathbf{p_3}$, one obtains a "biologically good basis." In [7] one can see a graph of

---

[11] As explored in earlier exercises in this module.
[12] This could be made into an additional project.

the associated system with nodes indexed by the metabolites listed, and the problem that $\mathbf{b}_2$ poses biochemically. (But given your experience to date, you should guess what that will be!) Since any three linearly independent vectors in $\mathbb{R}^n$, $n \geq 3$ span a three-dimensional vector space, there is a geometric interpretation $\mathbf{p}_1$, $\mathbf{p}_2$, $\mathbf{p}_3$, and $\mathbf{p}_4$ as the edges of a convex cone, the "flux cone" of S in the space. (Do see [7] for representative pictures and more discussion.) The edges $\mathbf{p}_i$ determine all possible total flux vectors for the system S represents, in that any positive convex linear combination of them (i.e., point in the flux cone) is a total flux vector. This leads to the notion of the $\mathbf{p}_i$ as so-called *extreme pathways*. Although they are biologically a bit difficult to describe, their cone yields all total flux vectors for which the system has no changes in concentrations of the metabolites, so metabolites are conserved, and we arrive back to the ideas we discussed in Section 8.1.

The papers we have noted so far represent an early attempt to apply linear algebraic ideas to this area of metabolite conservation. In a series of papers over the years, their authors and their colleagues have employed other concepts in linear algebra [8] and more significant linear algebraic techniques (e.g., the SVD [9]). They have combined linear algebraic techniques with notions from convex analysis and other areas to try to address questions we have raised here, and to tackle biochemical reaction systems that are much more sophisticated than the tutorial examples in this chapter, taken from their early papers, address. The reader is encouraged to check http://gcrg.ucsd.edu/Researchers for a wealth of developments. In this context, there has been developed a program `expa` (free) for computing extreme pathways, and subsequently much more extensive software, the COBRA toolbox http://opencobra. sourceforge.net/openCOBRA/Welcome.html (requires Matlab or Python). A significant problem that arises early on, when working with more sophisticated reaction systems, is that our naive intuition linking paths in the graph/cartoon of the biochemical reaction system to extreme paths and the stoichiometry quickly breaks down or can become stretched in a biologically incorrect direction. A final section of this chapter, which could be broken apart as a project unto itself, includes a companion tutorial for the free download `expa` and gives the interested reader a chance to explore these issues and a potential solution to them by working with hypergraphs instead of graphs. These additional materials will also allow you to explore large (and hence, more realistically interesting) biochemical reaction systems, as was proposed would be useful in the introduction. Papers using related linear algebraic techniques combined with convex analysis and differential equations have enabled researchers to isolate new metabolic pathways for *E. coli* [10], to identify potential disease mechanisms in red blood cells [5], and a host of other biomechanical engineering and other applications, see, e.g., the survey [11]. Other papers explore the links between varying versions of biologically "special" sets of vectors (such as the "extreme flux modes" of [13], to name just one) e.g., [12] and modeling of biochemical reaction networks using many other types of modeling tools, such as graph theoretic and other algebraic approaches to system dynamics, e.g., [14,15], algebraic geometric methods, e.g., by [16], and more. It is even possible to use these ideas, in combination with aspects of phylogenetics discussed in Chapter 10 of this volume, to explore the evolution of metabolic pathways [17, 18].

8.3 EXTREME PATHS AND MODEL IMPROVEMENTS

This final section was coauthored by Robert J. Kipka and Terrell L. Hodge.

In this section (which could be a stand-alone project), we'll use software to find extreme paths. These are particular elements of the nullspace of a stoichiometric matrix for which all nullspace elements are nonnegative linear combinations. This section will allow us to explore some of our modeling assumptions in greater depth and to do some simple analysis of the biochemistry of a red blood cell. The objectives of this section are:

a. To reflect on the advantages and disadvantages of representing biochemical processes using directed graphs.
b. To consider *directed hypergraphs* as a possible alternative and to interpret extreme pathways using directed hypergraphs.
c. To develop some basic familiarity with the *ExPA* program for finding extreme paths of complex biochemical systems.

The first thing we'll do is download *ExPA* and make sure it can run on your computer.

8.3.1 Downloading and Installing expa.exe

The software we'll use for this section is available for free download at the following Website:

    http://gcrg.ucsd.edu/Downloads/ExtremePathwayAnalysis.

The program is meant to be run from a terminal window. If you're using a PC, you can download an additional file from http://booksite.elsevier.com/ 9780124157804 that allows you to run the program without having to deal with the terminal window. If you're using a Mac, things are more complicated. At the time of this writing, the program will not run at all on OSX 10.7 or later. If you're using OSX 10.6 or older, the program will probably run but you will have to use a terminal window. We encourage you to do so! Very little terminal use is required and the benefits far outweigh the small amount of difficulty the terminal window represents.

8.3.1.1 *Instructions for a PC*

To install *ExPA* on your PC, follow the following steps in the order they're given.

1. Open a Web browser and go to http://gcrg.ucsd.edu/Downloads/ ExtremePathwayAnalysis.

2. You should download the latest version of expa.exe. At the time of this writing, the name of your download is "A new ExPA program." Right-click on the colored text which says "A new ExPA program" and choose "Save target as..."
3. Choose a folder that can find again and save the file in this folder.
4. Find the file you just downloaded. If the file is still zipped (which it likely is), right-click on it and select "Extract all... " Choose the folder you'd like to place expa.exe in.
5. Once you've downloaded the file and unzipped it, it's ready to use from the command prompt.

Using ExPA on a PC from a Command Prompt

1. Open up "My Computer," open up the "C" drive, and create a folder called "expa."
2. Find the folder that was created during the unzipping process, above. It should contain "expa.exe." Copy the contents of this folder into the expa folder you created on the C drive.
3. Open up a command prompt. This can probably be found in your start menu under "Accessories."
4. In the command prompt, type in "cd c:\expa". This changes your working directory to the expa folder we created earlier.
5. To see if the program runs, type in "*expa rbc.expa*". A file called "paths.txt" should appear.
6. For our purposes, to run *ExPA* you'll only need to type in "*expa*" followed by the file name of your input file. Output is always written to "paths.txt." The program only needs to be *run* from the command prompt. The easiest way to edit the input and output files is with a text editor like Notepad.

Using ExPA on a PC without the Command Prompt

1. Link to `http://booksite.elsevier.com/9780124157804` and download the "run-expa.bat" file. *Place this file in the folder that expa created when you unzipped it.* This is crucial. If you're not certain about what folder that is, look for the files "test.expa" and "rbc.expa" and place the download in the same folder as these files.
2. Open up the expa folder and right click on "test.expa." Select "Rename." Rename the file "source.txt." It is important that you replace the ".expa" part of the name with ".txt" or things will not work properly. The operating system might give you a warning when you do this, but it's perfectly safe.
3. You're ready to use *ExPA* without the command prompt. To see if it works, double-click on the "`run-expa.bat`" file. A new file called "paths.txt" should appear. If it does not appear, try refreshing the folder (an option under right-click). If it still does not appear, something is wrong.

4. *To use expa without the command prompt, all input has to be placed in this "source.txt" file.* When you double-click on "`run-expa.bat`," the program reads what's written in "source.txt" and writes its output to "paths.txt."

8.3.1.2 *Instructions for a Mac*

The instructions for getting *ExPA* to run on a Mac are not as detailed as those for a PC. This is because the program may not even run on a Mac. At the time of this writing the program will not work on OSX10.7 and may not work on OSX10.6 without additional effort. To install *ExPA* on a Mac, take the following steps:

1. Open a Web browser and go to `http://gcrg.ucsd.edu/Downloads/ExtremePathwayAnalysis`.
2. Download "A new Mac ExPA program" to your desktop.
3. Open up your "Applications" folder and create a new folder called "expa."
4. Copy the contents of the folder created by the download to the "expa" folder you created in "Applications."
5. Open a terminal window and type in "`cd    /Applications/expa`". Press enter. This will change your working directory.
6. Type in "`ls`" and press enter. This command lists the contents of the directory you're working in. Check that "macexpa.out" is in the list of files. If it's not, check that you copied the *contents* of the downloaded folder (not the whole folder) to the "expa" folder in "Applications."
7. Once you've found "macexpa.out" using the terminal, type in "`chmod    u + x macexpa.out`". This only needs to be done once. The program is now ready to be used.
8. To test the program, type in "`./macexpa.out./rbc.expa`". The program should create a file called "`paths.txt`". This is how the program is run. Simply navigate to the `/Applications/expa` directory and type in "`./macexpa.out`" followed by the name of your input file. The program only needs to be *run* from the terminal window. The easiest way to edit input files or read output files is with a text editor.

8.3.2 Analyzing a Modeling Decision: Directed Graphs

In representing biochemical reactions as directed graphs, we've chosen to represent metabolites as vertices and place edges between vertices whenever the metabolites are biochemically linked. For example, $A \longrightarrow B + C$ appears in a directed graph as in Figure 8.6.

Of course, there's nothing that forces us to make this choice; this is a modeling decision and comes with both strengths and weaknesses. In this section we'll explore this decision in a little bit greater depth and consider an alternative representation. At the end, we'll use our new understanding to analyze red blood cell metabolism using *ExPA*.

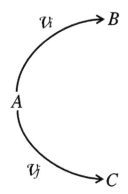

FIGURE 8.6

Graph for $A \longrightarrow B + C$.

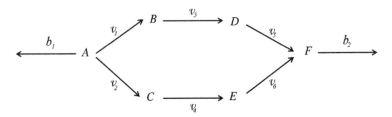

FIGURE 8.7

Graph for system (8.19), external fluxes added.

Let's start by looking at the following abstract biochemical process:

$$\begin{cases} A \longrightarrow B + C, \\ B \longrightarrow D, \\ C \longrightarrow E, \\ D + E \longrightarrow F. \end{cases} \qquad (8.19)$$

Suppose that A is an input to the system and F is an output. According to our directed graph conventions, this system has the following representation as in Figure 8.7.

To analyze this system with *ExPA*, we need to create an input file. An input file for *ExPA* provides the program with a list of internal fluxes (labeled in the chapter as v_i) and exchange fluxes (labeled in the chapter as b_i). A biochemical reaction can be *reversible* or *irreversible*. All of the reactions in our abstract system are irreversible. Based solely on our directed graph, our input file should look like Table 8.1.

Table 8.1

(Internal Fluxes)					
v1	I	−1	A	1	B
v2	I	−1	A	1	C
⋮					
(Exchange Fluxes)					
A	Input				
F	Output				

The I is for "irreversible."

Exercise 8.19. The vertical dots in the input pictured in Table 8.1 should be replaced with the data for v_3 through v_6. Create an input file[13] for *ExPA*, based on Figure 8.7 that will model our abstract system.   ▽

Exercise 8.20. Use *ExPA* to find the extreme paths in the representation from Exercise 8.19.   ▽

Exercise 8.21. Now consider the system

$$\begin{cases} A \longrightarrow B, \\ A \longrightarrow C, \\ B \longrightarrow D, \\ C \longrightarrow E, \\ D \longrightarrow F, \\ E \longrightarrow F. \end{cases} \qquad (8.20)$$

Represent (8.20) as a directed graph and compare your answer to the graph for (8.19).   ▽

Exercise 8.22. Write down at least one more biochemical system which will give you the directed graph from Exercise 8.21.

Your answer to the previous exercise highlights a possible weakness of our decision to model processes using directed graphs. In particular, a directed graph may not perfectly capture the stoichiometry of a biochemical process. For this reason, it can be interesting to use *hypergraphs*. Whereas in a directed graph, each edge must start at a unique vertex and end at a unique vertex, in a directed hypergraph an edge (really "hyperedge") is allowed to have both multiple starting points and multiple endpoints. For example, a hypergraph representation of (8.19) would look like Figure 8.8, with r_1 and r_4 hyperedges that are not just edges.

Exercise 8.23. When we represent a biochemical process using a hypergraph, what do the "hyperedges" represent? In what ways is this the same as or different from the edges in a directed graph?   ▽

---

[13] Note: If you're using a PC and plan on running the program without opening the command prompt, this input file has to be named "source.txt" and has to be placed in the same directory as "*expa.exe*". If you intend to use a command prompt or terminal window, the input file can be given any name.

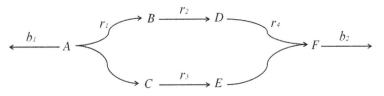

FIGURE 8.8
Hypergraph for system (8.19).

Some immediate questions regarding our software arise. In particular, can the software deal with hypergraphs? And if so, how will we represent hypergraphs as input for *ExPA*? It turns out that *ExPA* handles everything beautifully. In fact, because *ExPA* does its analysis on the stoichiometric matrix, all the program requires is the stoichiometry of the biochemical process. And so in order to have *ExPA* find the extreme paths of this process, all we need to do is enter internal fluxes in the obvious way. For example, r_1 will be given as in Table 8.2.

Exercise 8.24.   Create an input file for *ExPA* which represents Figure 8.8. List the extreme paths given by the program and explain in laymen's terms why your answer makes sense.   ▽

Exercise 8.25.   Write down the stoichiometric matrix for (8.19) and explain how your input file for *ExPA* relates to this matrix.   ▽

Now let's look at something a little more complex. Consider the following abstract biochemical process:

$$\begin{cases} A \longrightarrow B + C, \\ D \longrightarrow E + F, \\ B + E \longrightarrow G, \\ C + E \longrightarrow H, \\ F \longrightarrow I, \\ H + I \longrightarrow J + K, \\ G + J \longrightarrow L. \end{cases} \quad (8.21)$$

Suppose that A and D are system inputs, while I, J, K, and L are system outputs.

Exercise 8.26.   Draw a hypergraph representation for (8.21).   ▽

Exercise 8.27.   Using only the hypergraph or representation (8.21), see if you can find the proper inputs to create one molecule of J as an output. Is it possible? What about n molecules of J?   ▽

Exercise 8.28.   Use *ExPA* to find the extreme paths of the system, making sure that your input file fully represents (8.21). Explain carefully why these are the *only* extreme paths for the system using either the biochemical representation (8.21), your hypergraph representation, or both.   ▽

Table 8.2

r1	I	−1 A	1 B	1 C

Exercise 8.29. Revisit Exercise 8.27, this time using the extreme paths given by the software to determine the proper inputs (if any) necessary for the creation of one molecule of J by the system. ▽

Exercise 8.30. Something interesting happens if you "turn off" an exchange flux at I. Use *ExPA* to find out what happens and then explain in laymen's terms what's going on. ▽

Exercise 8.31. Allow I to be a system output again and use *ExPA* to explore what happens if we also allow G to be a system output. Explain in simple terms what's happening. How do changes in the exchange flux status of G relate to production of J as a system output? ▽

Now that we've had a chance to play around with some "simple" abstract processes, let's take a look at an actual biochemical process.

Exercise 8.32. The file `rbc.expa`, which comes with *ExPA*, is a sample file containing some of the biochemistry of a red blood cell. Use *ExPA* to find the extreme paths. Summarize your findings by creating a table showing only those pieces of the extreme paths relating to exchange fluxes for the system. Afterwards, compare with the graphical representations in [5], and note further discussion there of applications. ▽

Exercise 8.33. The following reaction represents the production of ATP for metabolic energy and NADH for methemoglobin reduction:

$$GLC + 2Pi + 2ADP + 2NAD \longrightarrow 2Pyr + 2ATP + 2NADH + 2H^+ + 2H_2 0.$$

Is this something the red blood cell can do, according to the extreme paths that you found? ▽

Acknowledgments

The author gratefully acknowledges the support of the National Science Foundation under DUE award #0737467. Thanks go to many linear algebra students who experimented with both prototypes and other extended versions of a module, built for that award, from which parts of it became this chapter and the supplementary materials. Many thanks also go Robert J. Kipka for his role in preparing the final section/project, and other contributions as a TEXpert reviewer.

8.4 **SUPPLEMENTARY DATA**

Supplementary files and materials associated with this article can be found from the volumes website http://booksite.elsevier.com/9780124157804

References

[1] Voet D, Voet J. Biochemistry. Wiley and Sons; 2004.

[2] Ackerman, J. How Bacteria in Our Bodies Protect Our Health, Scientific American, June 2012.

[3] <http://commons.wikimedia.org/wiki/File:Glucokinase-1GLK.png?useFormat =mobile> [as of July 2012].

[4] <http://upload.wikimedia.org/wikipedia/commons/archive/a/a0/20091114111 50621Glycolysis.svg [as of July 2012].

[5] Wiback S, Palsson O. Extreme pathway analysis of red blood cell metabolism. Biophys J 2002;83:808–18.

[6] Schilling CH, Palsson BO. The underlying pathway structure of biochemical reaction networks. PNAS 1998;95:4193–8.

[7] Schilling CH, Schuster S, Palsson BO, Heinrich R. Metabolic pathway analysis: basic concepts and scientific applications in the post-genomic era. Biotechnol Prog 1999;15:296–303.

[8] Famili I, Palsson BO. The convex basis of the left null space of the stoichiometric matrix leads to the definition of metabolically meaningful pools. Biophys J 2003;224:16–26.

[9] Famili I, Palsson BO. Systematic metabolic reactions are obtained by singular value decomposition of genome-scale stoichiometric matrices. J Theor Biol 2003;224:87–96.

[10] Lee SY, Hong SH, Moon SY. In silico metabolic pathway analysis and design: succinic acid production by metabolically engineered Escherichia coli as an example. Genome Informatics 2002;13:214–23.

[11] Maertens J, Vanrolleghem PA. Modeling with a view to target identification in metabolic engineering: a critical evaluation of the available tools. Biotechnol Prog 2010;26:313–31.

[12] Llaneras F, Picó J. Which metabolic pathways generate and characterize the flux space? A comparison among elementary modes, extreme pathways and minimal generators. J Biomed Biotech 2010. 13 pages. doi:10.1155/2010/753904 [Article ID 753904].

[13] Schuster S, Dandekar T, Fell DA. Detection of elementary flux modes in biochemical networks: a promising tool for pathway analysis and metabolic engineering. Trends Biotechnol 1999;17:53–60.

[14] Craciun G. Graph theoretic approaches to injectivity in general chemical reaction systems (with Murad Banaji). Adv Appl Math 2010;44:168–84.

[15] Craciun G, Pantea C, Rempala GA. Algebraic methods for inferring biochemical networks: a maximum likelihood approach. Comput Biol Chem 2009;33:361–7.

[16] Shui A, Sturmfels B. Siphons in chemical reaction networks, with Bernd Sturmfels. Bull Math Biol 2010;72:1448–63.

[17] Mithani A, Preston GM, Hein J. A bayesian approach to the evolution of metabolic networks on a phylogeny. PLoS Comput Biol 2010;6:e1000868. doi:10.1371/journal.pcbi.1000868.

[18] Mithani A, Preston GM, Hein J. A stochastic model for the evolution of metabolic networks with neighbor dependence. Bioinformatics 2009;25:1528–35.

Identifying *CpG* Islands: Sliding Window and Hidden Markov Model Approaches

9

Raina Robeva*, Aaron Garrett†, James Kirkwood* and Robin Davies‡

*Department of Mathematical Sciences, Sweet Briar College, Sweet Briar, VA, USA
†Department of Mathematical, Computing, and Information Sciences, Jacksonville State University, Jacksonville, AL, USA
‡Department of Biology, Sweet Briar College, Sweet Briar, VA, USA

9.1 INTRODUCTION

The dinucleotide *CpG* (a cytosine followed by a guanine on a single DNA strand) has a pattern of non-homogeneity along the genome. Regions of relatively low *CpG* frequencies are interrupted by clusters with markedly higher *C* and *G* content, known as *CpG* islands (CGIs). CGIs are often associated with the promoter regions of genes, and methylation of the promoter CGI is associated with the transcriptional silence of the gene. Conversely, promoter-associated CGIs in constitutively expressed housekeeping genes are unmethylated. Appropriate methylation of CGIs is required for normal development, and inappropriate methylation of CGIs in tumor suppressor promoters has been associated with the development of numerous human cancers.

9.1.1 Biochemistry Background

In higher multicellular organisms the genetic composition of an individual is determined by the fusion of sperm and egg nuclei following fertilization. With a few exceptions[1] all cells of a multicellular organism have the same DNA sequence. However, the cells of the multicellular organism have very different patterns of gene expression and thus make very different groups of proteins. During the process of development, cells become differentiated and take on their mature pattern of gene expression. The following question is thus important: Once a tissue is produced during development, how is the tissue-specific pattern of gene expression maintained? Part of the answer lies with the production of specific proteins involved in the transcription of specific genes, part involves the histones which package the DNA, and another part of the answer lies in the chemical alteration of the DNA itself.

In complex organisms a fraction of the cytosine DNA bases may be methylated, with the degree of cytosine methylation varying considerably among fungi, plants, invertebrate and vertebrate animals [1]. Methylation of cytosine occurs on the #5

---

[1] Those include red blood cells, which have no DNA, the lymphocytes of the immune system, in which DNA has been rearranged, and gametes, which have half of the adult's DNA.

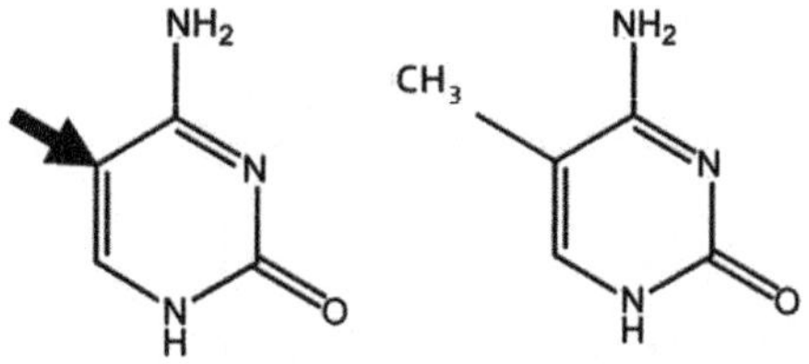

FIGURE 9.1

Comparison of unmethylated (left panel) and methylated (right panel) cytosines. The arrow in the left panel marks the #5 position. (The carbons and nitrogens are numbered, in this case, counter-clockwise beginning with the nitrogen on the bottom.)

position (see Figure 9.1) and the resulting entity is called 5-methyl cytosine. If we were to compare the DNA of a pair of differentiated cells (e.g., liver cells or skin cells), we would observe that they had different patterns of methylation. Patterns of methylation are correlated with patterns of gene expression in an inverse relationship, in which silent (non-expressed) genes are methylated. In a particular differentiated cell type, the pattern of methylation is maintained through successive mitoses by the action of enzymes called maintenance methylases.

In vertebrate animals, methylated cytosines occur in the dinucleotide sequence *CpG*. This dinucleotide is interesting in that its complement on the other strand of DNA is also *CpG*, and if the *C* on one strand is methylated, the *C* on the other strand is too. This state of affairs enables the pattern of DNA methylation to be perpetuated through successive rounds of replication. When a DNA sequence containing methylated *C* in a *CpG* dinucleotide is replicated, the two daughter strands will each have the *C* on the template strand methylated and the *C* on the new strand unmethylated. DNA in this state of half-methylation is the substrate for the maintenance methylase, which will methylate the unmethylated *C* on the new strand and thus restore the methylation pattern of the parent DNA strand. The methylation pattern, and the pattern of gene expression, will be inherited through subsequent mitoses.

Methylation of *CpG* dinucleotides is required for normal embryonic development and patterns of *CpG* methylation must be established following a generalized demethylation that occurs early in embryonic development [2]. The new methylation patterns are established by *de novo* methylases and appear to contribute to lineage restriction during development [3,4]. In other words, when pluripotent stem cells give rise to tissue-specific stem cells with more limited differentiation potential, the promoters of a subset of genes which were formerly active in the pluripotent stem cells become methylated and transcriptionally silent [5].

9.1.2 *CpG* Islands

Over time, both methylated and unmethylated cytosines may undergo random deamination reactions. The impact of these deamination reactions differs depending upon

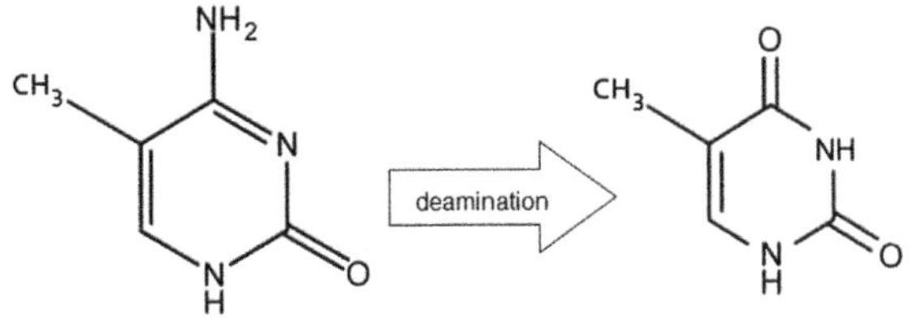

FIGURE 9.2

5-Methyl cytosine is deaminated to thymine.

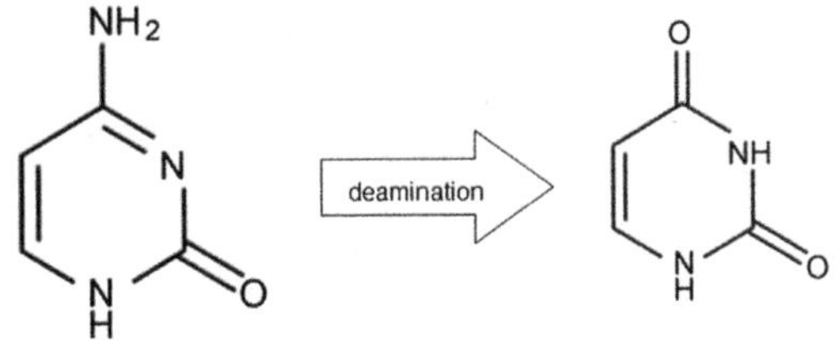

FIGURE 9.3

Cytosine is deaminated to uracil.

the methylation state of the cytosine. Methylated cytosine is deaminated to thymine, while unmethylated cytosine is deaminated to uracil (see Figures 9.2 and 9.3). Uracil bases do not belong in DNA and are recognized and removed by a repair enzyme, which restores the uracil to a C in its position. The T represents a mismatch (as it does not pair with G) and many of these Ts may also be repaired through the action of thymine-DNA glycosylase [6], but many will escape repair and the mutation will be propagated by subsequent rounds of replication. This gives rise to a C to T transition and the loss of a CpG. This means that, on an evolutionary timescale, unmethylated Cs tend to be preserved and methylated Cs tend to be eliminated.

Because of this systemic depletion of methylated cytosines, vertebrates have many fewer CpG dinucleotides than would be predicted by chance. Interestingly, when the DNA of vertebrates is examined, sequence information reveals that many of the CpG dinucleotides that do remain tend to occur in CGIs. The length of these CGIs varies, but they have been reported to range from several hundred to several thousand nucleotides. CGIs are found before genes (i.e., in their promoter regions), in the coding sequences themselves, and after the coding region as well.

The cytosines in CGIs in promoters are normally unmethylated, and many of these un-methylated CGIs are found in the promoters of important housekeeping genes—the genes that must function in order to support life, including the genes for enzymes involved in aerobic respiration, transcription, and translation [7]. Since the Cs in the CpG islands are unmethylated, any deamination events would be detected and repaired and they would not be lost over evolutionary time. Since all cells, even those which give rise to sperm and eggs, must express their housekeeping genes, promoters of housekeeping genes would retain their *CpG* dinucleotides and appear as CGIs [7].

However, CGIs are not confined to the promoters of housekeeping genes. CGIs are found in the promoters and protein coding regions (exons) of about 40% of mammalian genes [8]. Since CGIs seem to be located at the promoter regions of many known genes, identifying CGIs may be a useful method for identifying new genes.

9.1.3 DNA Methylation in Cancer

DNA methylation is a powerful mechanism for gene silencing. Genes which are methylated and repressed in vertebrate organisms are effectively shut off. For example, hemoglobin genes should be methylated (and silent) in skin cells but unmethylated (and actively expressed) in red blood cell precursors.

The generation of cancer requires multiple changes in gene structure and function. A single mutation is not enough to transform a normal cell into a cancerous cell; between three and seven mutations or other genetic insults are required [9, 10]. The affected genes—those which are involved in the progression from normal cell to cancer cell—fall into two different categories: oncogenes and tumor suppressors. Oncogenes must be activated, and tumor suppressors must be inactivated, in order for cancer to develop. Tumor suppressors may be inactivated through mutation or through gene silencing.

Mutations of tumor suppressors were identified first. In hereditary retinoblastoma, a cancer of the retina, affected children have inherited a defective copy of the gene for the tumor suppressor *Rb* from one of their parents. Tumors often arise in both eyes of these children following the loss of the second copy of *Rb* (inherited from the other parent) [11]. Additional tumor suppressor gene mutations were discovered by examining the DNA of many different tumors and comparing that DNA to that of normal tissue. For example, the tumor suppressor *p16*, a cell cycle control protein, was found to be deleted or to have suffered mutations in many different types of cancer [12]. Tumor suppressor genes do not need to be mutated to contribute to the genesis of cancer, however. They merely need to be silenced. If the CGIs in the promoters of *p16* or *Rb* genes are inappropriately methylated and the genes are turned off, then the cell in question will be one step closer to the development of a cancer.

9.2   QUANTITATIVE CHARACTERISTICS OF THE *CpG* ISLAND REGIONS AND SLIDING WINDOWS ALGORITHMS

CGIs were first characterized quantitatively 25 years ago by Gardiner-Garden and Frommer [13] and their definition is still widely in use today. The terms *percent combined C + G content (%C + G)* and *observed over expected CpG ratio (O/E CpG)* introduced by the authors provide a way to identify genomic regions with higher frequencies of *C* and *G* nucleotides and *CpG* dinucleotides. The $\%C + G$

of a sequence is the fraction of the combined number of Cs and Gs in the sequence divided by the total number of nucleotides in the sequence. To define $O/E\ CpG$, we note that if dinucleotides in a DNA sequence were formed by random independent choices of two nucleotides, the expected number of CpG dinucleotides in a sequence of length l would be

$$(\text{number of } Cs \text{ in the sequence}) * (\text{number of } Gs \text{ in the sequence})/l.$$

The observed CpG would be the actual count of CpG dinucleotides found in the sequence of length l. The observed over the expected CpG ratio O/E is defined as the ratio of these two numbers (and, unlike the quantity $\%C + G$, may assume values greater than 1).

    In the original study published in 1987 [13], Gardiner-Garden and Frommer defined CGIs in the vertebrate genome as sequences that have: (1) length of at least 200 bp, (2) $\%C + G \geqslant 50\%$, and (3) $O/E\ CpG \leqslant 0.6$. This definition is still commonly used today but it serves more as a guideline since there is no universal standard for the cutoff values. For instance, Takai and Jones [8] used a more stringent criterion to analyze CGIs in human chromosome 21 and 22: (1) length $\geqslant 500$ bp, (2) $\%C + G \geqslant 0.55$, and (2) $O/E\ CpG \geqslant .65$ motivated by reducing the number of CGIs found within *Alus*.[2]

    Algorithms for extracting CGIs often utilize a sliding windows approach that has been implemented by many web-based software systems including CpGPlot/ CpGReport [14], CpGProd [15], CpGIS [8], and CpGIE [16]. The method calculates the $\%C + G$ and $O/E\ CpG$ for subsequences of fixed length l that differ from one another only by 1 bp (the new subsequence is offset by 1 bp to the right from the previous one). One can visualize the process as sliding a "window" of length l along the genome. If the subsequence in the window meets the specific cutoff values for $\%C + G$ and $O/E\ CpG$, it will be included in a (possibly larger) CpG island region. The details of the specific algorithm implemented by CpGIS are shown in Figure 9.4. An animated version of a sliding windows algorithm is available in the *CpG Educate* suite that has been developed for this chapter and is available at http://inspired.jsu.edu/~agarrett/cpg/.

    The project *Investigating Predicted Genes* available online from the volume's website as part of this chapter utilizes sliding windows software to search for the presence of CGIs in the vicinity of predicted genes. The existence of a CGI in the area of the sequence where the promoter for the predicted gene should be found would be an additional piece of evidence suggesting that the predicted gene may be, in fact, an actual gene.

    Sliding windows algorithms are not based on any specific assumptions of structural mechanisms (mathematical or biological) that can explain the differences in CpG density between the island and non-island regions. As such, they do not utilize any mathematical models, theory, or specialized tools to make the questions of CGI identification more tractable. Sliding windows algorithms often differ from one

---

[2]*Alu* sequences (named for the restriction endonuclease AluI, which cuts in these sequences) are short repetitive sequences with a relatively high $C + G$ content and $O/E\ CpG$ ratio.

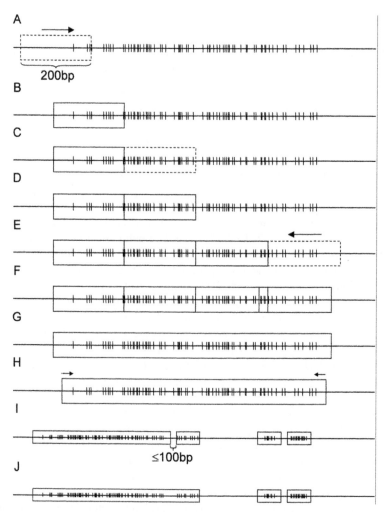

FIGURE 9.4

Schematics for the sliding window algorithms for CGI extraction from human genome sequences from [8]. (A) Set a 200-base window in the beginning of a contig (sequence), compute $C + G$ content and O/E *CpG* ratio. Shift the window 1 bp after evaluation until the window meets the criteria of a CGI. (B) If the window meets the criteria, shift the window 200 bp and then evaluate again. (C and D) Repeat these 200 bp shifts until the window does not meet the criteria. (E) Shift the last window 1 bp toward the 5′ end until it meets the criteria. (G) Evaluate $C + G$ content and O/E *CpG* ratio. (H) If this large CGI does not meet the criteria, trim 1 bp from each side until it meets the criteria. (I) Two individual CGIs were connected if they were separated by less than 100 bp. (J) Values for O/E *CpG* and $C + G$ content were recalculated to remain within the criteria. Reprinted with permission from Takai, D., Jones, P. Comprehensive analysis of CpG islands in human chromosomes 21 and 22. *PNAS* 2002 99 (6) 3740–3745. Copyright (2002) National Academy of Sciences, USA.

another not only in the choice of the threshold values for the length, $\%C + G$, and $O/E\ CpG$ ratio for a sequences to be recognized as a CGI but also in whether islands that are separated by small gaps would be merged if they still meet the cutoff criteria. In addition, some CGI searcher sites use a modified criterion for CGIs, requiring that the $\%C + G$ and $O/E\ CpG$ thresholds are met over an average of several windows (e.g., EMBOSS uses an average of 10 windows). For the same DNA sequences, these differences in the algorithms would generally lead to different regions identified as CGIs. More importantly, it has been shown that the sliding windows method does not guarantee an exhaustive search and that it may fail to identify all regions on the genome that meet the established criteria [17]. The use of alternative methods for CGI identification, such as Hidden Markov Models (HMMs) [18] and clustering methods [19–22] would therefore be preferable.

For the rest of this chapter we consider the HMM approach to locating CGIs. Such models are based on assumptions about the distribution of the nucleotides and di-nucleotide in the genome and provide a convenient mathematical framework within which the question of locating the island regions translates into well-understood mathematical problems. We begin with some examples.

The frequencies in Tables 9.1 and 9.2 present an example of nucleotide frequencies obtained from a sequence of annotated human DNA of about 60,000 nucleotides with known locations and lengths of the islands [23]. There are notable differences in the distributions, in agreement with the expectations that island regions would have elevated $\%C + G$ content and higher frequencies of the CpG dinucleotide. For unannotated sequences those frequencies would be unknown and we would want

Table 9.1 Sample dinucleotide frequencies (from [23]). The first row represents the frequencies of the transitions from A to A, C, T, and G in island and non-island regions and similarly for the other rows. Note that G is a lot more likely to follow C in island regions. The transition frequencies have been computed from annotated DNA as follows. If a_{ij}^{+} stands for the transition frequency (transition probability) from letter i to letter j in the island region, where $i, j \in Q, Q = \{A, C, T, G\}$, then a_{ij}^{+} is computed as the ratio $a_{ij}^{+} = \dfrac{c_{ij}^{+}}{\sum_{k \in Q} c_{ik}}$, where c_{ij}^{+} is the number of times the letter i followed by the letter j in the annotated island regions. The transition probabilities a_{ij}^{-} for the non-island regions are computed in the same way.

	Island ("+") Dinucleotide Frequencies				Non-Island ('−') Dinuceotide Frequencies			
	A	*C*	*T*	*G*	*A*	*C*	*T*	*G*
A	0.180	0.274	0.120	0.426	0.300	0.205	0.210	0.285
C	0.171	0.368	0.188	0.274	0.322	0.298	0.302	0.078
T	0.161	0.339	0.125	0.375	0.248	0.246	0.208	0.298
G	0.079	0.355	0.182	0.384	0.177	0.239	0.292	0.292

Table 9.2 Sample nucleotide frequencies from [23]. The entries are computed by averaging the columns of Table 9.1 for the ("+") and for the ("−") regions.

	A	*C*	*T*	*G*
Island ("+") Frequencies:	0.15	0.33	0.16	0.36
Non-Island ("−") Frequencies:	0.27	0.24	0.26	0.23

to have a way of obtaining some estimates for those from the data. This may initially seem to be an impossible task. After all, how can we estimate those frequencies when the boundaries between the regions are unknown? As we will see below, there is a way to solve this problem, if we view the DNA data as generated by a known underlying mechanism, formally described by a HMM.

The rest of the chapter is organized as follows: Section 3 contains a brief review of Markov chains and HHMs, presenting some examples, highlighting some main concepts, and introducing the notation. Readers without prior experience with Markov chains or HMMs may need to consider a more detailed introduction to these topics such as [25,26]. In Section 4 we present CGI identification methods based on HMMs. The section focuses on the questions of evaluation, decoding, and parameter estimation for HMM. Some final comments and notes on generalizations and ongoing work involving HMMs for CGI identification are gathered in Section 5. A suite of web applications, *CpG Educate*, has been developed by the authors to illustrate the algorithms and is used for many of the examples and exercises in the chapter. *CpG Educate* is freely available at http://inspired.jsu.edu/~agarrett/cpg/.

9.3 DEFINITION AND BASIC PROPERTIES OF MARKOV CHAINS AND HIDDEN MARKOV MODELS

9.3.1 Finite State Markov Chains

A Markov chain is a "process" that at any particular time is in one of a finite number of "states" from a set $Q = \{s_1, \ldots, s_n\}$. We will only consider discrete time, which means that the process can be visualized as running according to a digital timer, possibly changing states at each time step. The *Markov property* assumes the state of the process at the next step depends only on its present state and not on the history of how the process arrived at that state. Thus, the process has no memory beyond the "present" and the only factors that govern the process are (i) its current state, and (ii) the probabilities with which the process moves from its current state to other states.

Assume that $\pi_t \in Q$ denotes the state of the process at time t, where $t = 1, 2, 3, \ldots$ As time goes on, the process transitions from one state to another generating a *path* $\pi = \pi_1 \pi_2 \cdots \pi_l$ (the number l is the *length* of the path). If we have observed the path of the process up to time t, the formal definition of the Markov property is that

$$P\{\pi_{t+1}|\pi_1\pi_2\cdots\pi_t\} = P\{\pi_{t+1}|\pi_t\}, \quad t = 1, 2, 3, \ldots,$$

stating that the probability for the process transitioning to π_{t+1} at time $t + 1$ does not depend on the entire "history" of the process $\pi_1 \pi_2 \cdots \pi_t$ but only on its current state π_t.

For any two states $i, j \in Q$, we consider the transitional probabilitiy $a_{ij} = P\{\pi_{t+1} = j | \pi_t = i\}$, defined as the probability that the process moves from state i to state j when the (discrete) time changes from t to $t + 1$. When $i = j$, a_{ii} is the probability that the process remains in state i at time $t + 1$. The transition probabilities are the same for all values of t, meaning that the transitions depend only on the state of the process and not on when the process visits that state. The transition probabilities are commonly organized in a *transition matrix*

$$
p = \begin{pmatrix} a_{11} & \cdots & a_{1n} \\ \vdots & & \vdots \\ a_{n1} & \cdots & a_{nn} \end{pmatrix}, \quad a_{ij} \geqslant 0,
$$

with $\sum_{j \in Q} a_{ij} = 1$, for all $i \in Q$.[3] The initial state for the process is determined by the *initial distribution* $p_i = P(\pi_1 = i), i \in Q$, where $\sum_{i \in Q} p_i = 1$.

It is often convenient to introduce notation that would allow for the initial distribution and for the ending of the sequence to be treated as transition probabilities and included in the notation, $a_{ij}, i, j \in Q$. To accomplish this, a hypothetical "beginning" state B and an "ending" state E are introduced with the assumption that the process begins at state B at time $t = 0$ with probability 1 and it transitions to E with probability 1 at the end of each path. The probability for transitioning into B after time $t = 0$ is zero, and the probability of transitioning out of E is zero. With these additions, each path $\pi = \pi_1 \pi_2 \cdots \pi_l$ of the Markov chain can be expanded to $\pi = B\pi_1 \pi_2 \cdots \pi_l E = \pi_0 \pi_1 \pi_2 \cdots \pi_l \pi_{l+1}$. We can append a superficial state denoted by 0 to Q, $Q = \{0, s_1, \ldots, s_n\}$ and write $a_{0i} = p_i = P(\pi_1 = i | \pi_0 = B)$ for the initial distribution, for all $i \in Q$, and $a_{i0} = P(\pi_{l+1} = E | \pi_l = i) = 1$, for all $i \in Q$ (at the end of each path $\pi = \pi_0 \pi_1 \pi_2 \cdots \pi_l$, the process moves to E with probability 1). In what follows we will not explicitly append the symbols B and E at the beginning and at the end of all paths but, when it is necessary, the transition probabilities will be interpreted in this generalized sense.

For any observed path $\pi = \pi_0 \pi_1 \pi_2 \cdots \pi_l$, we apply the Markov property to compute its probability as follows:

$$
\begin{aligned}
P(\pi) &= P(\pi_0 \pi_1 \pi_2 \cdots \pi_l) = P\{\pi_l | \pi_{l-1}, \pi_{L-2}, \ldots, \pi_0\} P(\pi_0 \pi_1 \cdots \pi_{l-1}) \\
&= P\{\pi_l | \pi_{l-1}\} P(\pi_0 \pi_1 \pi_2 \cdots \pi_{l-1}) = \ldots =
\end{aligned}
$$

---

[3] This condition simply reflects the fact the process will be in some state from Q at the next time step, unless it terminates.

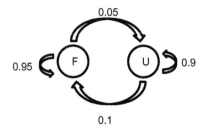

FIGURE 9.5

Directed graph representation of a discrete Markov chain with a state space $Q = \{f,u\}$. The vertices of the digraph correspond to the states of the process and the labels on the directed edges stand for the transition probabilities. The beginning and ending states are not pictured.

$$= P\{\pi_l|\pi_{l-1}\}P\{\pi_{l-1}|\pi_{l-2}\}P\{\pi_{l-2}|\pi_{l-3}\}\cdots P\{\pi_1|\pi_0\}$$

$$= a_{\pi_{l-1}\pi_l}a_{\pi_{l-2}\pi_{l-1}}\cdots a_{\pi_0\pi_1}$$

$$= \prod_{i=1}^{l} a_{\pi_{i-1}\pi_i}.$$

Example 9.1. Assume that a game is played by successively rolling a die. Two dice are available, one fair and one unfair. Either die is equally likely to be chosen for the first game. After that, the process of switching between the dice can be depicted by the transition diagram in Figure 9.5: if the fair die is used in the current game, it will be retained for the next game with probability 0.95 and switched with the unfair die with probability 0.05. If the unfair die is used for the current game, we continue to use it for the next game with probability 0.9 and switch to the fair die with probability 0.1. The process π of switching between the dice can be described by a Markov chain with a state space of two elements $Q = \{F, U\}$, with transition probabilities $a_{FF} = P(\pi_{t+1} = F|\pi_t = F) = 0.95, a_{FU} = P(\pi_{t+1} = F|\pi_t = U) = 0.05, a_{UF} = P(\pi_{t+1} = F|\pi_t = U) = 0.10$, and $a_{UU} = P(\pi_{t+1} = U|\pi_t = U) = 0.90$. The transition matrix is then $P = \begin{pmatrix} a_{FF} & a_{FU} \\ a_{UF} & a_{UU} \end{pmatrix} = \begin{pmatrix} 0.95 & 0.05 \\ 0.10 & 0.9 \end{pmatrix}$ and the initial distribution is $a_{0F} = 0.5$ and $a_{0U} = 0.5$.

We can now easily compute the probability of any path of this process. For example, if $\tilde{\pi} = UFUF$ and $\bar{\tilde{\pi}} = UUFF$, we obtain $P(\tilde{\pi}) = a_{0U}a_{UF}a_{FU}a_{UF} = (0.5) * (0.1) * (0.05) * (0.1) = 0.0025$ and $P(\bar{\tilde{\pi}} = a_{0U}a_{UU}a_{UF}a_{FF} = (0.5) * (0.9) * (0.1) * (0.95) = 0.0428$.

Exercise 9.1. For the Markov chain from Example 9.1, compute the probabilities of the following paths: (a) $\tilde{\pi} = FUUU$; (b) $\bar{\tilde{\pi}} = UUFFUF$; (c) $\tilde{\pi} = FFFFFU$. $\triangledown$

Exercise 9.2. If a Markov chain has a nonzero probability for transitioning from a state to itself, it is possible that a path of the process will feature "runs" of the same symbols, indicating that the process remains in that same state for more than one step. For the Markov chain from Figure 9.5, those will be runs of Us and Fs. Generalizing, assume that the transition matrix in Example 9.1 is $P = \begin{pmatrix} a_{FF} & a_{FU} \\ a_{UF} & a_{UU} \end{pmatrix} = \begin{pmatrix} p & 1-p \\ 1-q & q \end{pmatrix}$, where $0 < p, q < 1$.

Show that the distribution of the lengths of the runs is as follows:

1. For runs of length of at least l that do not start at beginning of the sequence

$$P(\text{a run of } Fs \text{ of length of at least } l) = (1-q)p^{l-1},$$
$$P(\text{a run of } Us \text{ of length of at least } l) = (1-p)q^{l-1}.$$

2. For runs of length at least l that start at beginning of the sequence

$$P(\text{a run of } Fs \text{ of length of at least } l) = a_{0F}\, p^{l-1},$$
$$P(\text{a run of } Us \text{ of length of at least } l) = a_{0U}\, q^{l-1}. \qquad \triangledown$$

Exercise 9.3. In the context of Exercise 9.2, show the following: (1) When p and q are fixed, the probability for runs of length at least l decreases exponentially as l increases. (2) When is fixed, the probability to observe a run of length at least l increases as the probabilities p and q increase. $\qquad \triangledown$

9.3.2 Hidden Markov Models (HMMs)

HMMs generalize Markov chains by assuming that the process described by the Markov chain is not readily observable (it is *hidden*). According to some rules, each hidden state generates (emits) a symbol and only the sequence of emitted symbols is observed. The following example, inspired by the Occasionally Dishonest Casino example by Durbin et al. [23], illustrates the idea.

Example 9.2. The outcome of a game is determined by the roll of a die. If the number of points on the die is 1, 2, 3, or 4, the player wins, and if the number of points is 5 or 6, the player loses. The casino uses a fair or an unfair die for each game. For the fair die, all outcomes are equally likely but the unfair die is heavily biased toward 6 with $p(i) = 0.1$ for $i = 1, 2, 3, 4, 5$ and $p(6) = 0.5$. The process of switching between the dice is a Markov chain with state space $Q = \{F, U\}$, transition matrix $P = \begin{pmatrix} 0.95 & 0.05 \\ 0.10 & 0.9 \end{pmatrix}$, and initial distribution $a_{0F} = 0.5$, $a_{0U} = 0.5$, just as in Example 9.1. The player is unaware which die is being used for each game or when a switch between the fair and unfair dice occurs. Thus the paths π of this Markov process are hidden.

The player records wins and losses from consecutive games in a sequence of Ws and Ls (e.g., $x = WWLLLLLWLL$), where W stands for a win and L stands for a loss. This way, for each sequence of games, the player generates a record

$x = x_1 x_2 x_3 \cdots x_l$ of wins and losses with $x_t \in M = \{W, L\}$, denoting the outcome of game t, $t = 1, 2, \ldots, l$. The chances of winning or losing a game depend on the choice of die used for that game. We can think of each x_t as generated (emitted) by the hidden state π_t with a certain probability. Since a fair die gives the player a win with probability $\frac{2}{3} = 0.67$ and a loss with probability $\frac{1}{3} = 0.33$, we will say that the fair die *emits* a W with probability $e_F(W) = 0.67$ and that it emits an L with probability $e_F(L) = 0.33$. Similarly, the unfair die emits a W with probability $e_U(W) = 0.4$ and an L with probability $e_U(L) = 0.6$. The set $M = \{W, L\}$ is the set of *emitted states* for the hidden process.

The main difference between Markov chains and hidden Markov chains is that the observed sequence x cannot be mapped to a unique path of state-to-state transitions for the hidden process. Multiple hidden paths π can generate x and, as our next example illustrates, they do so with different probabilities.

Example 9.3.   In the context of Example 9.2, assume the following sequence of rolls is generated for four consecutive games: $z = 2561$. The recorded (observed) sequence of wins and losses is then $x = WLLW$. Any hidden sequence $\pi = \pi_1 \pi_2 \pi_3 \pi_4$, $\pi_t \in \{F, U\}$, could have generated this sequence but different sequences will do so with very different probabilities. To illustrate, we will compute the probabilities $P(x, \bar{\pi})$ and $P(x, \bar{\bar{\pi}})$ that x is generated by each of the following hidden sequences: $\bar{\pi} = FUUU$ and $\bar{\bar{\pi}} = FUFU$. For $P(x, \bar{\pi})$, we obtain $P(x, \bar{\pi}) = 0.5 \cdot e_F(W) \cdot a_{FU} \cdot e_U(L) \cdot a_{UU} \cdot e_U(L) \cdot a_{UU} \cdot e_U(W) = 0.000012$. The rationale is as follows. For x and $\bar{\pi}$ to occur, a series of events should take place: the fair die is chosen for the first game, emitting a W; a switch from the fair to the unfair die follows and the unfair die emits an L; the unfair die is retained for the next game, emitting an L, and so on. The probability $P(x, \bar{\pi})$ is then the product of the transition and emission probabilities. Similarly, $P(x, \bar{\bar{\pi}}) = 0.5 \cdot e_F(W) \cdot a_{FU} \cdot e_U(L) \cdot a_{UF} \cdot e_F(L) \cdot a_{FU} \cdot e_U(W) = 0.0000067$, which is different from $P(x, \bar{\pi})$. In principle, if we were to guess which hidden path π generated the observed sequence x, the smart way to do so would be to find the path π^* for which the probability $P(x, \pi)$ is the largest.[4] The probability $P(x)$ of emitting the sequence x, is the sum over all $P(x, \pi)$ probabilities for all hidden sequences π: $P(x) = \sum_{\pi} P(x, \pi)$.

Example 9.4.   (The Occasionally Dishonest Casino example from [23]). Assume the hidden process that switches between the fair and unfair die is as described in Example 9.2 but now the player bets on the specific number of points rolled at each game. It would then make sense to record the sequence $z = z_1 z_2 \cdots z_l$ where $z_t \in \{1, 2, 3, 4, 5, 6\}$ stands for the number of points rolled in game t, $t = 1, 2, 3, \ldots, l$. The hidden states F and U now emit the symbols from the set $M = \{1, 2, 3, 4, 5, 6\}$ with probabilities $e_F(1) = e_F(2) = e_F(3) = e_F(4) = e_F(5) = e_F(6) = \frac{1}{6} = 0.1667$ and $e_U(1) = e_U(2) = e_U(3) = e_U(4) = e_U(5) = 0.1$ and $e_U(6) = 0.5$. We will use this HMM for several of the examples and exercises in Section 4.

---

[4]We will show how this can be done in a computationally efficient way in Section 4.

With this background, we can now ask the following questions: Given an observed sequence $x = x_1x_2x_3 \cdots x_l$, how do we determine the hidden sequence $\pi = \pi_1\pi_2\pi_3 \cdots \pi_l$ that maximizes the probability of observing x? Can we estimate the parameters of the HMM (the transition matrix P and the emission probabilities) from the observed sequence x?

The main reason for considering these gambling examples is that the problem of locating CGIs is in many ways mathematically analogous and may be stated similarly. A DNA sequence x composed of the symbols A, C, T, and G may be viewed as generated by a mechanism switching between two hidden states $Q = \{+, -\}$, analogous to the honest and dishonest dice. One state is that of CGIs (the "+" region), the other is the non-island region (the "−" region). The states A, C, T, and G are the same for both regions, but the transition matrices and/or the emission probabilities will be different.[5] We cannot observe the state-to-state transitions $\pi = \pi_1\pi_2\pi_3 \cdots \pi_l$ of the process directly but we observe the sequences $x = x_1x_2x_3 \cdots x_l$ emitted by the process where each x_t is a nucleotide from the set $M = \{A, C, T, G\}$. The questions above for the casino games are now immediately translated into the following questions: Given a long DNA sequence $x = ACTGTC \cdots TCAC$, how do we determine which hidden path is the most likely to have emitted this sequence? Can we estimate the transition and the emission probabilities of the HMM from the observed sequence? To examine these questions in detail, we first need to introduce the general notation for HMMs.

In general, a HMM consists of a hidden Markov chain with a finite state space Q and a finite set of emission symbols M. The state of the hidden process at time is denoted by π_t. The transition probabilities are $a_{ij} = P(\pi_{t+1} = j | \pi_t = i)$, and the initial distribution is $p_i = a_{0i} = P(\pi_1 = i), i \in Q, \sum_{i \in Q} p_i = 1$. The process emits a symbol $k \in M$ from each of the hidden states $j \in Q$ that it visits. The emission probabilities are denoted by $e_j(k) : e_j(k) = P(x_t = k | \pi_t = j), j \in Q, k \in M$, where (since each of the hidden states emits exactly one symbol) $\sum_{k \in M} e_j(k) = 1$. The set of transition, emission, and initial probabilities forms *the set of parameters* of the HMM.

A convenient way to summarize the parameters of a HMM is to organize them in a table where the transition matrix, emission probabilities, and the initial distribution are given in the columns and where the rows are labeled by the hidden states. Table 9.3 contains the parameters of the HMM from Example 9.3.

Exercise 9.4. Give the table for the parameters of the HMM from Example 9.4.

$\triangledown$

9.3.3 Viewing DNA Sequences As Outputs of a HMM

Tables 9.1 and 9.2 demonstrate two different ways of viewing the quantitative differences between the "+" and the "−" regions in the genome, each one of which can

---

[5] For example, within the "+" and the "−" regions, the transition probabilities could be similar to those in Table 9.1 (with the modification that there should also be small nonzero probabilities of switching between the "+" and the "−" regions).

Table 9.3 The parameter set for the HMM from Example 9.3 in tabular form.

	Transitions		Emissions		Initial Distribution
	F	U	W	L	
F	0.95	0.05	0.67	0.33	0.5
U	0.1	0.9	0.40	0.60	0.5

Table 9.4 A set of probabilities (parameters) of a HMM for a DNA sequence where the model is only concerned with the frequencies of the individual nucleotides. The transition matrix of the HMM is under the "Transitions" heading. Each of the hidden states emits a symbol from the set $M = \{A,C,T,G\}$ with emission probabilities listed under the "Emissions" heading. The hidden process is equally likely to begin in the "+" and "−" state, as stated under the "Initial Distribution" heading.

	Transitions		Emissions				Initial Distribution
	+	−	A	C	T	G	
+	0.90	0.10	0.15	0.33	0.16	0.36	0.5
−	0.05	0.95	0.27	0.24	0.26	0.23	0.5

be used to construct a HMM. When we look only at the nucleotide frequencies as in Table 9.2 we can consider a HMM with a state space $Q = \{+, -\}$, where each of these states can emit a symbol from the set $M = \{A, C, T, G\}$ with emission probabilities as those in Table 9.2. Assuming that hidden process transitions between the "+" and "−" states are as in Figure 9.5 (where in this case, we will identify the state U with "+" and the state F with "−") the parameters for the HMM will be those in Table 9.4.

If we want the model to incorporate information about dinucleotides, as in the case of Table 9.1, the set of emitted symbols is again $M = \{A, C, T, G\}$ but now the emission events at each step are not independent from one another. If, say, the process is in the hidden state "+," the probability for emitting a symbol C will depend upon the symbol emitted by the previous state and whether this symbol was emitted from the "+" or from the "−" hidden state. We can think of it as emitted from one of two hidden states C_+ or C_-. Thus, for each of the emission symbols $k \in M$ we should have states k_+ and k_- in Q, leading to a state space $Q = \{A_+, A_-, C_+, C_-, T_+, T_-, G_+, G_-\}$ for the hidden process. The matrix for the transitions within the subsets of the "+" and "−" states should be close to those in the transition matrices in Table 9.5 but switching between the "+" and "−" subsets $Q_+ = \{A_+, C_+, T_+, G_+\}$ and $Q_- = \{A_-, C_-, T_-, G_-\}$ of Q should also be allowed with some small probability. Table 9.5 presents this scenario.

Exercise 9.5. The HMM from Table 9.4 could be considered to be a special case of the general model from Table 9.5 with state space $Q = \{A_+, A_-, C_+, C_-, T_+, T_-, G_+, G_-\}$. Give a set of HMM parameters for the general HMM from

Table 9.5 The set of parameters of a HMM describing a DNA sequence. The model incorporates information about dinucleotide frequencies. Here the block matrices P^+ and P^- are the "+" and "−" transition matrices from Table 9.1 The process remains in the the "+" and "−" region with probabilities p and q, respectively. When switching between the "+" and "−" regions, this model assumes that all states in the new region are equally likely. For each of the emission symbols $k \in M$, the model assumes that only the states k_+ and k_- from can Q emit k. The initial distribution is uniform.

	Transitions								Emissions				Initial
	A_+	C_+	T_+	G_+	A_-	C_-	T_-	G_-	A	C	T	G	
A_+					$(1-p)/4$	$(1-p)/4$	$(1-p)/4$	$(1-p)/4$	1	0	0	0	0.125
C_+		$(P^+) * p$			$(1-p)/4$	$(1-p)/4$	$(1-p)/4$	$(1-p)/4$	0	1	0	0	0.125
T_+					$(1-p)/4$	$(1-p)/4$	$(1-p)/4$	$(1-p)/4$	0	0	1	0	0.125
G_+					$(1-p)/4$	$(1-p)/4$	$(1-p)/4$	$(1-p)/4$	0	0	0	1	0.125
A_-	$(1-q)/4$	$(1-q)/4$	$(1-q)/4$	$(1-q)/4$					1	0	0	0	0.125
C_-	$(1-q)/4$	$(1-q)/4$	$(1-q)/4$	$(1-q)/4$					0	1	0	0	0.125
T_-	$(1-q)/4$	$(1-q)/4$	$(1-q)/4$	$(1-q)/4$		$(P^-) * q$			0	0	1	0	0.125
G_-	$(1-q)/4$	$(1-q)/4$	$(1-q)/4$	$(1-q)/4$					0	0	0	1	0.125

Table 9.5 to obtain a HMM with transition and emission probabilities such as those in Table 9.4. ▽

The parameters in Table 9.5 present a very special case among all possible HMMs with $Q = \{A_+, A_-, C_+, C_-, T_+, T_-, G_+, G_-\}$ and $M = \{A, C, T, G\}$ and we have included it here since it provides a natural way to include the information from Table 9.1. Ultimately, though, the parameters of the HMM would need to be estimated from the DNA data, and it would be more appropriate to consider the model in its full generality. Thus, any transition matrix P under the Transitions heading, any emission matrix E under the Emissions heading, and any initial distribution could be used as model parameters. In [27], the authors consider such a general model and estimate the parameters from a set of 1000 DNA sequences.

9.4 THREE CANONICAL PROBLEMS FOR HMMS WITH APPLICATIONS TO CGI IDENTIFICATION

We now turn to the mathematical solutions of the questions raised above and describe how they can be solved in a computationally efficient way. After discussing each problem, we provide examples and exercises with connections to CGI identification. We begin with restating the questions in the context of HMMs:

1. *Decoding Problem*: Given an observed path $x = x_1 x_2 x_3 \cdots x_l$ generated by a HMM with known parameters, what is the most likely hidden path $\pi = \pi_1 \pi_2 \pi_3 \cdots \pi_l$ to emit x? In other words, how do we find $\pi_{\max} = \arg\max_\pi P(\pi|x)$?[6] From Example 9.3 we know how to compute $P(x, \pi)$ for any hidden path π and it can be shown (see Exercise 9.6 below) that

$$\pi_{\max} = \arg\max_\pi P(\pi|x) = \arg\max_\pi P(x, \pi).$$

   Since there are finitely many hidden paths, one may be tempted to answer the question by computing $P(x, \pi)$ for all paths π, compare their values, and pick the largest. Such a "brute force" approach is not practical, however, since the number of all hidden paths grows exponentially with the length of the paths l. The number $|Q|^l$ of all such paths is astronomical for large values of l, making the task impossible even for the fastest computers.[7]

2. *Evaluation Problem*: Given an observed path x, what is the probability of this path $P(x)$? Mathematically, this probability can be expressed as

$$P(x) = \sum_\pi P(x, \pi), \tag{9.1}$$

---

[6]The notation is for the argument of the maximum. We need the path(s) π for which $P(x, \pi)$ is maximal.
[7]For an observed sequence of just 90 nucleotides, the number of paths is a bit over 10^{80}, which is the estimated number of atoms in the observable universe. The brute force approach would therefore require years of computing time even if large computing resources are applied. In principle, we would be interested in sequences of tens of thousands of nucleotides.

where the sum is over all possible hidden sequences $\pi = \pi_1\pi_2\pi_3\cdots\pi_l$ with $\pi_t \in Q$ and where

$$P(x, \pi) = a_{0\pi_1} \prod_{i=1}^{l} e_{\pi_i}(x_i)a_{\pi_i\pi_{i+1}}. \tag{9.2}$$

As with the decoding problem, the mathematical solution from Eq. (9.1) has no practical value since the number of such paths grows exponentially as the length of the path l grows.

3. *Training (Learning) Problem:* Given an observed sequence x or a set of observed sequences, what are the HMM parameters that make the sequence x most likely to occur? The answer to this question provides estimates for the parameters of the HMM from a data set of observed sequences.

9.4.1 Decoding: The Viterbi algorithm

The Viterbi algorithm is a recursive and computationally efficient method for computing the most likely state sequence π_{max} for a given observed sequence $x = x_1x_2x_3\cdots x_l$ (see [28] and also [29] for an interesting account of the history by Andrew Viterbi himself).

To understand the recursive step, let's assume that for any state $j \in Q$ we have somehow computed the hidden sequence $\pi_1\pi_2\pi_3\cdots\pi_{t-2}\pi_{t-1}$ of highest probability among those emitting the observations $x_1x_2x_3\cdots x_{t-1}$ up to time $t-1$ with $\pi_{t-1} = j$. That is, we assume that for each $j \in Q$, we have determined the most probable path of length $t-1$ for the data $x_1x_2x_3\cdots x_{t-1}$, ending in state j. Denote the probability of this path by $v_j(t-1)$:

$$v_j(t-1) = \max_{\pi=\pi_1\pi_2\pi_3\cdots\pi_{t-1}} P(\pi_{t-1} = j, x_{t-1}), \quad j \in Q.$$

Next, we can use $v_j(t-1)$ to compute the most probable path of length t ending in each of the states $k \in Q$ and emitting the sequence $x_1x_2x_3\cdots x_t$:

$$v_k(t) = \max_{\pi=\pi_1\pi_2\pi_3\cdots\pi_t} P(\pi_t = k, x_t).$$

The probability $v_k(t)$ of the most likely path of length t ending in k and emitting x_t will be the largest among the probabilities for hidden sequences that get into j at time $t-1$ with a maximal probability, transition from j to k at time t, and emit x_t. Thus

$$v_k(t) = \max_{j\in Q}\{v_j(t-1)a_{jk}e_k(x_t)\} = e_k(x_t)\max_{j\in Q}\{v_j(t-1)a_{jk}\}.$$

Iterating this argument for all $t = 1, 2, \ldots, l$ will allow us to compute the probability of the most likely path $\pi = \pi_1\pi_2\pi_3\cdots\pi_l$ for the data $x = x_1x_2x_3\cdots x_l$. To be able to recover the path $x = x_1x_2x_3\cdots x_l$ itself, for each time t and for each state $k \in Q$, we keep pointers (ptr) to remember the state $r \in Q$ from which the maximal probability path ending in k came. For each $t = 1, 2, \ldots, l$ and $k \in Q$ we record k's predecessor

$\text{ptr}_t(k) = r$, where $v_r(t-1)a_{rk} = \max_{j \in Q}\{v_j(t-1)a_{jk}\}$. The notation $\text{ptr}_t(k) = r$ means that the path with the highest probability ending in state k at time t came from state r at time $t-1$.

Utilizing the agreement that the process always starts in the beginning state B, Viterbi's algorithms can be summarized as follows. To find the most probable path $\pi = \pi_1\pi_2\pi_3\cdots\pi_l$ for the observed data $x = x_1x_2x_3\cdots x_l$, perform the following steps:

- Initialization $t = 0$: $v_B(0) = 1$, $v_j(0) = 0$, $j \in Q$.
- Recursion (repeat for $t = 1, 2, \ldots, l$: $v_k(t) = e_k(x_t)\max_{j \in Q}\{v_j(t-1)a_{jk}\}$; $\text{ptr}_t(k) = r$ where $r = \arg\max_j\{v_j(t-1)a_{jk}\}$.
- Termination: $P(x, \pi^*) = \max_\pi P(x, \pi) = \max_{j \in Q}\{v_j(l)\}$; $\text{ptr}_l(k) = \pi_l^*$.
- The maximal probability path π^* can now be found by backtracking through the recorded pointers.

Our next example illustrates the method.

Example 9.5. Consider the game from Example 9.3 with modified transition probabilities between the hidden states F and U as in Table 9.6. Assume we have observed the sequence $x = x_1x_2x_3 = LWW$. We will apply the Viterbi algorithm to compute the most likely hidden path $\pi = \pi_1\pi_2\pi_3$ for this observed sequence.

$t = 0$:

We begin with probability 1 at the beginning state and thus $v_B(0) = 1$, $v_F(0) = 0$, $v_U(0) = 0$. As explained earlier, $v_B(0) = 1$ is used just for initialization. It stands for the most probable path ending in B at time $t = 0$. Since all paths of the process start from B, $v_B(0) = 1$. Since no path can be in either F or U at time $t = 0$, $v_F(0) = 0$, $v_U(0) = 0$.

$t = 1$:

$$v_F(1) = \max_{\pi_1} P(\pi_1 = F, x_1 = L) = \max\{v_B(0)a_{0F}e_F(L)\}$$
$$= (0.5)(0.33) = 0.1667.$$

The maximum is taken over all paths that start at the beginning state and end at F. There is only one such path. Thus $v_F(1)$ is the probability that the process is at state F at time 1 (which occurs with probability 0.5) and that it emits the symbol $x_1 = L$ from that state (which happens with probability 0.33). Similarly, $v_U(1) = \max_{\pi_1} P(\pi_1 = U, x_1 = L) = \max\{v_B(0)a_{0U}e_U(L)\} = v_B(0)a_{0U}e_U(L) = (0.5)(0.6) = 0.30$.

Table 9.6 HMM parameters for Example 9.5.

	Transitions		Emissions		Initial Distribution
	F	U	W	L	
F	0.7	0.3	0.67	0.33	0.5
U	0.4	0.6	0.40	0.60	0.5

$t = 2$:

$$v_F(2) = \max_{\pi_1 \pi_2} P(\pi_2 = F, x_2 = W)$$
$$= \max\{v_F(1)a_{FF}e_F(W), v_U(1)a_{UF}e_F(W)\}$$
$$= e_F(W) \max\{v_F(1)a_{FF}, v_U(1)a_{UF}\}$$
$$= (0.67) \max\{(0.1667) * (0.7), (0.3) * (0.4)\}$$
$$= (0.67) \max\{0.1167, 0.12\} = (0.67)*(0.12) = 0.0804.$$

Since the max was achieved for $v_U(1)a_{UF}$, the most likely path into F at $t = 2$ came from U and we have $\text{ptr}_2(F) = U$. Similarly,

$$v_U(2) = \max_{\pi_1 \pi_2} P(\pi_2 = U, x_2 = W)$$
$$= \max\{v_F(1)a_{FU}e_U(W), v_U(1)a_{UU}e_U(W)\}$$
$$= e_U(W) \max\{v_F(1)a_{FU}, v_U(1)a_{UU}\}$$
$$= (0.4) \max\{(0.1667)*(0.3), (0.3)*(0.6)\}$$
$$= (0.4) \max\{0.05, 0.18\} = (0.4)*(0.18) = 0.072.$$

Since the max was achieved for $v_U(1)a_{UU}$, the most likely path into U at $t = 2$ came from U and we have $\text{ptr}_2(U) = U$.

$t = 3$:

$$v_F(3) = \max_{\pi_1 \pi_2 \pi_3} P(\pi_3 = F, x_3 = W)$$
$$= \max\{v_F(2)a_{FF}e_F(W), v_U(2)a_{UF}e_F(W)\}$$
$$= e_F(W) \max\{v_F(2)a_{FF}, v_U(2)a_{UF}\}$$
$$= (0.67) \max\{(0.0804) * (0.7), (0.072)*(0.4)\}$$
$$= (0.67) \max\{0.0563, 0.0296\} = (0.67) * (0.0563) = 0.038.$$

Since the max was achieved at $v_F(2)a_{FF}$, the most likely path into F at $t = 3$ came from F and we have $\text{ptr}_3(F) = F$. Finally,

$$v_U(3) = \max_{\pi_1 \pi_2 \pi_3} P(\pi_3 = U, x_3 = W)$$
$$= \max\{v_F(2)a_{FU}e_U(W), v_U(2)a_{UU}e_U(W)\}$$
$$= e_U(W) \max\{v_F(2)a_{FU}, v_U(2)a_{UU}\}$$
$$= (0.4) \max\{(0.0804) * (0.3), (0.072) * (0.6)\}$$
$$= (0.67) \max\{0.0241, 0.0432\}$$
$$= (0.4) * (0.0432) = 0.017.$$

The max was achieved for $v_U(2)a_{UU}$, thus the most likely path into U at $t = 3$ came from U and we have $\text{ptr}_3(U) = U$.

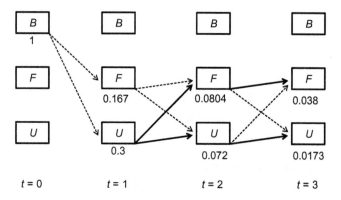

FIGURE 9.6

Trellis diagram for Example 9.5. *B* is the beginning state (the ending state is not pictured). The numbers under each node indicate the maximal probability for all paths ending in this node and emitting sequences in agreement with the data. The dashed arrows indicate possible transitions. For each node, the solid arrow into the node is the pointer indicating where the maximal probability path comes from.

We have reached the end of the sequence $x = x_1 x_2 x_3 = LWW$ and the algorithm terminates. The probability of the most likely path is therefore

$$P(x, \pi^*) = \max_{\pi} P(x, \pi) = \max_{j \in Q}\{v_j(l)\} = \max\{v_F(3), v_U(3)\}$$
$$= \max\{0.038, 0.017\} = 0.038.$$

The maximum was achieved at $v_F(3)$, thus the ending state for the most likely path is $\pi_3^* = F$. Since $\mathrm{ptr}_3(F) = F$, the most likely path into F at time $t = 3$ came from F, and thus $\pi_2^* = F$. Since $\mathrm{ptr}_2(F) = U$, the most likely path into F at time $t = 2$ came from U, and thus $\pi_1^* = U$. Therefore, the hidden sequence of maximal probability 0.038 is $\pi^* = \pi_1^* \pi_2^* \pi_3^* = UFF$.

It is convenient to visualize the Viterbi algorithm by using a *trellis diagram* that makes it easier to follow the process. In a trellis diagram, the columns are labeled with the values of t and the rows correspond to the states of the Markov chain (see Figure 9.6). Starting from the state B and following the arrows can generate all possible paths $\pi = \pi_1 \pi_2 \pi_3$. The number under each node is the maximal probability for the path ending in the node and agreeing with the observed data up to that time (these are the v-values computed by the Viterbi algorithm). The solid arrow into each node is the pointer that keeps track of the origin of the maximal probability path. Locating the largest probability in the last column and backtracking using the solid lines, generates the maximal probability path $\pi^* = \pi_1^* \pi_2^* \pi_3^*$.

It can be shown that the computational complexity of the Viterbi algorithm is no worse than $O(n^2 l)$ in time[8] and $O(nl)$ in space, where $n = |Q|$ is the number of states and l is the length of the sequence. Thus, the complexity of the algorithm is quadratic in time, which, although still computationally demanding, is a huge improvement over the exponential rate of the brute force approach.

Since the Viterbi algorithm involves multiplying many probabilities, the results could get very small, resulting in essentially zero probabilities and causing underflow problems. To avoid this, all calculations are performed on the logarithms of the probabilities $v_k(t), k \in Q, t = 1, 2, \ldots$, using addition instead of multiplication.

Exercise 9.6. Show that the hidden path found by the Viterbi algorithm is exactly the path of maximal probability, given the observed sequence x. That is, show that

$$\pi_{max} = \underset{\pi}{\text{argmax}} \, P(x, \pi) = \underset{\pi}{\text{argmax}} \, P(\pi | x). \qquad \triangledown$$

Obviously, Example 9.5 shows that performing the Viterbi computations by hand is tedious even for very short paths. From now on, we will use the *CpG Educate* suite that implements the Viterbi algorithm for HMMs corresponding to the Dishonest Casino problem from Example 9.4 and for the CGI identification problem.

The applications in the *CpG Educate* suite have the capability to generate an output from a HMM and record the hidden path that produced the emitted sequence. This *simulated data* can then be used to test the "accuracy" of the Viterbi algorithm by comparing the predicted states from the Viterbi decoding with the actual ones from the simulations.

Example 9.6. We used the *Dishonest Casino* application in the *CpG Educate* suite to generate a sequence of length 400 for the Dishonest Casino HMM from Example 9.4. We then applied the Viterbi algorithm to the emitted sequence and compared the output with the known hidden sequence from the simulations. Figure 9.7 depicts the outcome. The gray background identifies rolls made with the unfair die. The outcomes in bold are the predicted unfair die state from the path of maximal probability found by the Viterbi algorithm. Although the overlap is not perfect, Viterbi appears to have recovered the states fairly well. In practice, the Viterbi algorithm will often miss short sequences generated by either die. Since switching between the dice happens with small probabilities, the path of maximal probability would generally tend to "bridge" short gaps and preserve the trends set by longer runs.

When using a HMM for identifying CGIs, the outcome of the Viterbi algorithm will produce the predicted path through the hidden "+" and "−" states. The start of the island will be where the predicted hidden sequence switches from the subset $Q_- = \{A_-, C_-, T_-, G_-\}$ to the subset $Q_+ = \{A_+, C_+, T_+, G_+\}$ and the end of the island will be where the sequence switches back from the "+" to the "−" states.

---

[8] The "big O" notation here means that for very large values of n and l the number of operations required by the Viterbi algorithm has the same order of magnitude as the number $n^2 l$.

663542666266661351334436263331644531243125262121361641256524622336252566556623563

11513244655522446115624245555245436**6666566666624661466**25414242415451263636353636353

15126662454545132125341562114614526525636216362226245343523266**66666412536666624635**

66666526625613615526566363621365646261215414164646441554341346513414316665664121

44233645321243656461555333353216154513131632241546332646132115116**6466664616**2316354

FIGURE 9.7

Viterbi decoding on simulated data of 400 rolls for the Dishonest Casino Example 9.4. The gray background highlights the rolls generated with the unfair die in the simulation. Digits in bold indicate the unfair die states predicted by the Viterbi algorithm.

Exercise 9.7.[*9] Use the *CpG Islands* application in the *CpG Educate* suite to generate simulated sequences of length of 600 bp with HMM parameters from Table 9.5 with $p = 0.9$ and $q = 0.95$. (The file *Exercise_9.7.csv* containing the parameters from Table 9.5 for these values of p and q is provided for you to load into the *CpG Islands* application). Run the Viterbi algorithm on the simulated data and examine how well the decoded sequence matches the states of the HMM from the simulations. Generate several sequences and comment on the outcome. Notice specifically how the maximal probability path generated by the Viterbi algorithm tends to ignore small gaps between the island and non-island regions, as already observed at the end of Example 9.6. ▽

Exercise 9.8. Experiment with applying the Viterbi algorithm to sequences generated by the *Dishonest Casino* application in the *CpG Educate* suite. Consider relatively long sequences of (e.g., 5000 or longer) for different values of the transition probabilities of the HMM (the four probabilities under the Transitions heading). Specifically, consider the following types of transition probabilities: a) transition probabilities are close to uniform (that is, the elements of transition matrix $P = \begin{pmatrix} a_{FF} & a_{FU} \\ a_{UF} & a_{UU} \end{pmatrix}$ are close to 0.5), and b) distributions for which the process retains its current state with a large probability (that is, transition matrices $P = \begin{pmatrix} a_{FF} & a_{FU} \\ a_{UF} & a_{UU} \end{pmatrix} = \begin{pmatrix} p & 1-p \\ 1-q & q \end{pmatrix}$ for which p and q are very close to 1). Consider values as large as 0.9999 for p and q.

For each generated sequence run the Viterbi algorithm and note the general quality of its performance: Is the decoded sequence generally close to the hidden sequence generating the simulated data? If there are misses, are those false negatives (switches to the unfair die in the simulation that have remained undetected by the Viterbi algorithm) or false positives (predicted switches to the unfair die in places

---

[9]Exercises marked with an asterisk indicate that their execution requires downloads from the volume's website.

where the simulations have been generated by using the fair die). Are there any emerging patterns? Be mindful that due to the stochastic nature of the process, you should generate several sequences for each set of parameters. Try to summarize your observations, by answering the following questions:

1. Did you detect any difference in the performance of the Viterbi algorithm for the sets of parameters from part (a) and (b)?
2. If you answered yes in (1), is the performance of the algorithm to decode the hidden sequence better for distributions like those in part a) or for distributions like those in part (b)?
3. Does the performance of the Viterbi algorithm improve when the values of and are very close to 1?
4. What explanations can you provide for your findings for questions (1)? It may be helpful to revisit Exercises 9.1 and 9.2.   ▽

Exercise 9.9. Repeat Exercise 9.8, this time experimenting with the *CpG Islands* application in the *CpG Educate* suite. Consider distributions such as in Table 9.5 and a) values for p and q close to 0.5; b) values for p and q close to 1. Experiment with both short and long sequences and notice that when p or q are very close to 1, the hidden process and/or the maximal probability path may never switch between the "+" and "−" states for relatively short sequence. Use the file *Table_9.5.xlsx* available from the volume's website to generate sets of HMM parameters for different values of p and q. Be sure to save the generated sets of HMM parameters in Comma Separated Values (CSV) format and use the CSV fileto load the parameters into the *CpG Islands* application (refer to the *CpG Educate* tutorial [30] for more details).   ▽

9.4.2 Evaluation and Posterior Decoding: The Forward-Backward Algorithm

In this section we outline a computationally efficient algorithm for computing $P(x)$, the probability for an observed sequence $x = x_1 x_2 \cdots x_l$. We also discuss an alternative decoding method called *posterior decoding*.

The forward algorithm is similar in idea to the Viterbi algorithm. Instead of keeping track of the hidden sequences of maximal probability in state j at time t, the forward algorithm computes the probabilities of being in state j, having observed the first t symbols $x_1 x_2 \cdots x_t$ of the sequence x. Denote these probabilities by $f_j(t) = P(x_1 x_2 \cdots x_t, \pi_t = j), j \in Q$. Recursively, as in the Viterbi algorithm, we can now compute

$$f_k(t+1) = P(x_1 x_2 \cdots x_{t+1}, \pi_{t+1} = k),$$

the probability that the process is in state k at time $t + 1$ with emitted sequence $x_1 x_2 \cdots x_{t+1}$:

$$f_k(t+1) = P(x_1 x_2 \cdots x_{t+1}, \pi_{t+1} = k) = \sum_{j \in Q} f_j(t) a_{jk} e_k(x_{t+1}), \quad k \in Q.$$

The rationale is straightforward. For the process to be in state k at time $t + 1$ with emitted sequence $x_1 x_2 \cdots x_{t+1}$, it must be in one of the states $j \in Q$ at time t with emitted sequence $x_1 x_2 \cdots x_t$ (with probability $f_j(t)$), then transition to state k at time $t + 1$ (with probability a_{jk}) and emit x_{t+1} (with probability $e_k(x_{t+1})$). The probability $f_k(t + 1)$ is then the sum of the product of those probabilities over all states j.

At time $t = 0$, the process is in the beginning state with probability 1 and has emitted no output sequence yet. Thus, we initialize the algorithm by $f_B(0) = 1$, $f_k(0) = 0$, $k \in Q$.

The complete algorithm is:

- Initialization ($t = 0$): $f_B(0) = 1$, $f_j(0) = 0$, $j \in Q$.
- Recursion (repeat for $t = 1, 2, \ldots, l$): $f_k(t) = e_k(x_t) \sum_{j \in Q} f_j(t-1) a_{jk}$, $k \in Q$.
- Termination: $P(x) = \sum_{k \in Q} f_k(l)$.

Example 9.7. Consider again the HMM from Example 9.5. Use the forward algorithm to compute the probability of the sequence $x = LWW$.

Solution:

$t = 0$:

We begin with probability 1 at the beginning state B and, thus, $f_B(0) = 1$, $f_F(0) = 0$, $f_U(0) = 0$.

$t = 1$:

$$f_F(1) = P(x_1 = L, \pi_1 = F) = f_B(0) a_{0F} e_F(L) = (0.5)(0.33) = 0.165.$$
$$f_U(1) = P(x_1 = L, \pi_1 = U) = f_B(0) a_{0U} e_U(L) = (0.5)(0.6) = 0.30.$$

$t = 2$:

$$f_F(2) = P(x_1 = L, x_2 = W, \pi_2 = F) = e_F(W)(f_F(1) a_{FF} + f_U(1) a_{UF})$$
$$= (0.67)((0.165) * (0.7) + (0.3) * (0.4)) = 0.1578.$$
$$f_U(2) = P(x_1 = L, x_2 = W, \pi_2 = U) = e_U(W)(f_F(1) a_{FU} + f_U(1) a_{UU})$$
$$= (0.4)((0.1667) * (0.3) + (0.3) * (0.6)) = 0.0918.$$

$t = 3$:

$$f_F(3) = P(x_1 = L, x_2 = W, x_3 = W, \pi_3 = F)$$
$$= e_F(W)(f_F(2) a_{FF} + f_U(2) a_{UF})$$
$$= (0.67)((0.1578) * (0.7) + (0.092) * (0.4)) = 0.0986.$$
$$f_U(3) = P(x_1 = L, x_2 = W, x_3 = W, \pi_3 = U)$$
$$= e_U(W)(f_F(2) a_{FU} + f_U(2) a_{UU})$$
$$= (0.4)((0.1578) * (0.3) + (0.0918) * (0.6)) = 0.041.$$

Finally, we have $P(x) = P(LWW) = 0.0986 + 0.041 = 0.1396$.

For our main problem of CpG identification, the probability $P(x)$ of a DNA sequence of nucleotides $x = x_1 x_2 \cdots x_l$ is not of much interest by itself but we will need it to compute the *posterior probabilities* $P(\pi_t = k|x), k \in Q, t = 1, 2, \ldots, l$, that symbol x_t in the observed sequence was emitted from state k, which can then be used for decoding.

Since

$$P(\pi_t = k|x) = \frac{P(x, \pi_t = k)}{P(x)}, \qquad (9.3)$$

we need a recursive algorithm for computing $P(x, \pi_t = k)$. The Markov property of the hidden process makes this possible:

$$P(x, \pi_t = k) = P(x_1 x_2 \cdots x_t, \pi_t = k) P(x_{t+1} x_{t+2} \cdots x_l | x_1 x_2 \cdots x_t, \pi_t = k)$$
$$= P(x_1 x_2 \cdots x_t, \pi_t = k) P(x_{t+1} x_{t+2} \cdots x_l | \pi_t = k). \qquad (9.4)$$

The first line is a direct application of the conditional probability formula where the probability that x is generated with $\pi_t = k$ at time t is given as the product of the probabilities of the following events: (1) symbols $x_1 x_2 \cdots x_t$ are emitted up to time t and the process is in state k at time t, and (2) conditioned upon the event (1), the rest of the emitted sequence is $x_{t+1} x_{t+2} \cdots x_l$. The second line follows from the Markov property of the hidden process and restates that the probability to emit the sequence $x_{t+1} x_{t+2} \cdots x_l$ depends only on the state of the process at time t.

Notice that the $P(x_1 x_2 \cdots x_t, \pi_t = k)$ are exactly the probabilities $f_k(t)$, which are computed from the forward algorithm. Denote $b_k(t) = P(x_{t+1} x_{t+2} \cdots x_l | \pi_t = k)$. Equation (9.4) can now be re-written as

$$P(x, \pi_t = k) = f_k(t) b_k(t). \qquad (9.5)$$

The probabilities $b_k(t)$ are computed by the *backward algorithm*. We begin by initializing the algorithm for $t = l$ where, since x_l is the last observed symbol, $\pi_{l+1} = E$ is the end state (that does not emit a symbol). Thus $b_k(l) = P(\pi_{l+1} = E|\pi_l = k) = 1, k \in Q$, since the sequence goes to the end state E with probability 1.

Once we know $b_k(l)$ for all $k \in Q$, for any of the values $t = l - 1, l - 2, \ldots, 1$ we can compute

$$b_j(t) = P(x_{t+1} x_{t+2} \cdots x_l | \pi_t = j)$$
$$= \sum_{k \in Q} P(\pi_{t+1} = k|\pi_t = j) e_k(x_{t+1}) P(x_{t+2} \cdots x_l | \pi_{t+1} = k)$$
$$= \sum_{k \in Q} a_{jk} e_k(x_{t+1}) b_k(t + 1).$$

The justification is as follows: At time $t + 1$ the process can transition from j to any other state $k \in Q$ (this happens with probability a_{jk}), emit x_{t+1} (this happens with probability $e_k(x_{t+1})$), and, being in state k at time $t + 1$, emit the rest of the sequence $x_{t+2} \cdots x_l$ (which happens with probability $b_k(t + 1)$).

Since at time $t = 0$ the process is in the beginning state B with probability 1, $b_0(0) = P(x_1 x_2 \cdots x_l | \pi_0 = B) = P(x_1 x_2 \cdots x_l) = P(x)$. This shows that the probability $P(x)$ can also be computed from the backward algorithm as $b_0(0) = P(x) = \sum_{k \in Q} a_{0k} e_k(x_1) b_k(1)$. If this probability is already computed from the forward algorithm, the backward algorithm will terminate once $b_k(1)$ are computed for all $k \in Q$.

The complete algorithm is:

- Initialization $(t = l)$: $b_k(l) = 1, k \in Q$.
- Recursion (repeat for $t = l-1, l-2, \ldots, 1$): $b_j(t) = \sum_{k \in Q} a_{jk} e_k(x_{t+1}) b_k(t+1)$.
- Termination: $P(x) = \sum_{k \in Q} a_{0k} e_k(x_1) b_k(1)$.

Combining now Eqs. (9.3) and (9.5), we obtain the following equation for the posterior probabilities

$$P(\pi_t = k | x) = \frac{P(x, \pi_t = k)}{P(x)} = \frac{f_k(t) b_k(t)}{P(x)}, \quad k \in Q, \ t = 1, 2, \ldots, l. \quad (9.6)$$

Since we use both the forward and the backward algorithms, we say that the posterior probabilities are computed by the *forward-backward* algorithm.

Example 9.8. Consider again the HMM from Example 9.5. Use the forward-backward algorithm to compute the posterior probabilities of the sequence $x = LWW$.

Solution: $P(x)$ and the probabilities $f_k(t)$ were computed in Example 9.7. We now compute $b_k(t)$ for $t = 3, 2, 1$.

$$t = 3: \quad b_F(3) = 1; b_U(3) = 1.$$

$$t = 2: \quad b_F(2) = a_{FF} e_F(W) b_F(3) + a_{FU} e_U(W) b_U(3)$$
$$= (0.7) * (0.67) + (0.3) * (0.4) = 0.589,$$

$$b_U(2) = a_{UF} e_F(W) b_F(3) + a_{UU} e_U(W) b_U(3)$$
$$= (0.4) * (0.67) + (0.6) * (0.4) = 0.508.$$

$$t = 1: \quad b_F(1) = a_{FF} e_F(W) b_F(2) + a_{FU} e_U(W) b_U(2) = 0.3372,$$

$$b_U(1) = a_{UF} e_F(W) b_F(2) + a_{UU} e_U(W) b_U(2) = 0.2798.$$

The posterior probabilities are now:

$$P(\pi_1 = F | x) = \frac{f_F(1) b_F(1)}{P(x)} = \frac{(0.165) * (0.3372)}{0.1396} = 0.3986,$$

$$P(\pi_1 = U | x) = \frac{f_U(1) b_U(1)}{P(x)} = \frac{(0.3) * (0.2798)}{0.1396} = 0.6014.$$

The process continues similarly until all posterior probabilities are determined (see Exercise 9.10).

Exercise 9.10. Complete the previous example to calculate $P(\pi_2 = F | x)$, $P(\pi_2 = U | x)$, $P(\pi_3 = F | x)$, and $P(\pi_3 = U | x)$.   ▽

Exercise 9.11. Continue the application of the forward algorithm from Example 9.8 to carry out the termination step and compute $P(x)$. Verify that this termination step produces the same value for $P(x)$ as the value computed by the forward algorithm in Example 9.7.   ▽

The posterior probabilities computed here can be used to complement the results from the Viterbi decoding or as a possible alternative decoding method referred to as *posterior decoding*. They could prove useful when there are multiple sequences with probabilities close to the maximal probability sequence(s) generated by the process of Viterbi decoding, in which case it may not be justified to only consider the sequence(s) of maximal probability. The posterior probabilities give the likelihood (based on the entire observed sequence $x = x_1 x_2 \cdots x_l$) that the symbol x_t in position t has been emitted by the hidden state k. Posterior decoding can be quite useful for decoding a hidden process with two states (or two groups of states), which is exactly the case we are concerned with in this chapter. To see this, in the case of the Dishonest Casino example, given the sequence x, we plot the probabilities $P(\pi_t = U|x)$ for each $t = 0, 1, \ldots, l$. The "hills" in the resulting graph would indicate segments in the sequence x that are likely to be emitted from the U state. The remaining segments are more likely to have been emitted by the F state. Analytically, the decoded sequence will be

$$\tilde{\pi}_t = \underset{k}{\operatorname{argmax}}\, P(\pi_t = k|x). \qquad (9.7)$$

Figure 9.8 exemplifies this approach for a simulated sequence of 500 symbols. The gray areas highlight the time steps at which the unfair die was used for the simulations.

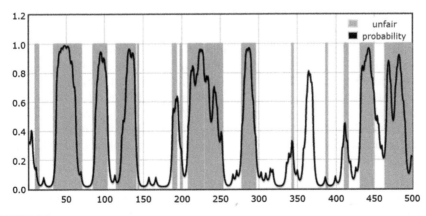

FIGURE 9.8

A plot of the posterior probabilities of being in a state generated by the unfair die for a sequence of 500 simulated runs. The gray highlight identifies the runs obtained from the simulation by using the unfair die. The transition probabilities of the HMM are $a_{UU} = 0.95$; $a_{UF} = 0.05$; $a_{FU} = 0.04$, and $a_{UU} = 0.96$. The default values for the emission probabilities (chosen to match those from Example 9.4) were used.

The condition from Eq. (9.7) in the case of only two hidden states means that if we consider a threshold of 0.5 for Figure 9.8, the hidden states with posterior probabilities plotted above the horizontal line at 0.5 will be predicted as U states. The overlap is not perfect, of course, but, as our next exercise shows, the separation between the states can be even better for hidden processes that only switch between states with very small probability.

Exercise 9.12. Use the *Dishonest Casino* application in the *CpG Educate* suite to experiment with the evaluation algorithms to plot the posterior probabilities of being in a state generated by the unfair die for sequences x of various lengths.[10] Do the same for several sets of transition probabilities, focusing specifically on the two extremes, as in Exercise 9.8: (a) transition probabilities that are close to uniform, and (b) distributions for which the process retains its current state with a large probability. Consider values as large as 0.999 for p and q in this case. Summarize your observations, by answering the following questions: (1) Did you detect any improvement when the sets of parameters are less like those from part (a) and more like those in part (b)? If so, in what sense?; (2) Consider several sets of transition probabilities to illustrate that when the probabilities for switching between the two hidden states get smaller, the performance of posterior decoding improves. ▽

Exercise 9.13. Repeat Exercise 9.12 but, this time, try different values for the emission probabilities. Experiment with posterior decoding to get a sense that its performance improves as the emission distributions for the different states become "more different." If, for instance, both emission distributions are nearly or exactly uniform (e.g., {0.1167, 0.1167, 0.1167, 0.1167, 0.1167, 0.1167} and {0.15, 0.15, 0.2, 0.15, 0.15, 0.2}) the decoding into U and F states will be generally poor. As the emissions distribution for the unfair die becomes more skewed, the performance of the posterior decoding method improves. ▽

Exercise 9.14. Use the *CpG Islands* application in the *CpG Educate* suite to simulate sequences of various lengths and compare the performance of Viterbi decoding vs. Posterior Decoding. Use HMM parameters in the form of Table 9.5 first, then experiment with general sets of HMM parameters. Use the file *Table_9.5.xlsx* from the volume's website to generate sets of HMM parameters in the format of Table 9.5 for different values of p and q. (Do not forget that the file needs to be saved in CSV format before loading the parameters into the *CpG Islands* application. See Exercise 9.9 and [30] for more details.) ▽

Exercise 9.15. Repeat Exercise 9.14 for the *Dishonest Casino* application. Compared to the HMM from Exercise 9.14, the *Dishonest Casino* application has fewer parameters, so you will not need to upload them from a file - just change the parameter values by typing over the default values that are provided. ▽

---

[10] For optimal viewing of the posterior probabilities, use sequences with lengths between 300 and 1200.

9.4.3 Training: The Baum-Welch Algorithm

For all computations until now, we always assumed the parameters of the HMM are known. Those parameters, however, are usually only initial estimates and, in most cases, we may not even have those estimates to begin with. In this section we will discuss how to estimate the HMM parameters from the data. This process is called *training* (or sometimes *learning*). The algorithm presented here was first introduced in [31] (see also [32]). We begin with an observed sequence x or a set of observed sequences $x^1, x^2, \ldots, x^m$ for which we would want to adjust the HMM model parameters to ensure the best possible fit. Once a set of parameters is determined for those sequences, we apply the Viterbi algorithm or use the posterior probabilities to other similar sequences for decoding. The sequences $x^1, x^2, \ldots, x^m$ are called *training sequences* and we say that we will train the model to best fit those sequences.

More formally, this means that given the data $x^1, x^2, \ldots, x^m$ and the HMM, we need to find the values of the model parameters (the transition probabilities a_{jk} and the emission probabilities $e_k(b)$ for $j, k \in Q$ and $b \in M$) that maximize the probability of the data $P(x)$. If we use θ to refer to the whole set of parameters for the model, in this section we will sometimes write $P(x) = P(x|\theta) = P(x|\theta, HMM)$ to emphasize that the likelihood of the data depends on the set of parameters θ and on the model. The goal is to find

$$\theta^* = \underset{\theta}{\mathrm{argmax}}\{P(x|\theta)\}.$$

If the training sequences are annotated and we know the exact paths of the process emitting the training sequences, the estimates for the model parameters will be computed by determining the frequencies of each transition and emission. Assume that A_{jk} is the number of transitions from j to k in the set of training sequences and $E_k(b)$ is the number of times the state $k \in Q$ emitted the symbol $b \in M$. Then the frequencies

$$a_{jk} = \frac{A_{jk}}{\sum_{r \in Q} A_{jr}} \text{ and } e_k(b) = \frac{E_k(b)}{\sum_{c \in M} E_k(c)}, \quad \text{for } j, k \in Q, \ b \in M \qquad (9.8)$$

are maximum likelihood estimates for the HMM [23].

When the paths that generate the training data are unknown, we can no longer determine the counts A_{jk} and $E_k(b)$ but they can be replaced with the expected counts for each transition/emission in the HMM. Obtaining the maximum likelihood estimates in this case can be done by an iterative process known as the Baum-Welch algorithm [31]. This method is no longer guaranteed to find a global maximum for the probability of the data but the iterative steps will converge to a local maximum. The Baum-Welch algorithm is a special case of a more general class of methods known as *Expectation Maximization* algorithms or *EM* algorithms. Below we will outline the computational aspect of the Baum-Welch method but will omit the theoretical proofs of convergence to a maximum. Those proofs together with the general theory of the EM algorithms can be found in [33]. In the description of the Baum-Welch algorithm below we will assume that the HMM is being trained on a single data sequence x. We will comment on the straightforward generalization to multiple sequences $x^1, x^2, \ldots, x^m$ afterwards.

The Baum-Welch algorithm generates a sequence of approximations $\theta_0, \theta_1, \theta_2, \ldots$ for the set of HMM parameters, with each new set of parameters θ improving the value of $P(x|\theta)$ over the previous iteration. The process terminates when two successive iterations produce the same values for $P(x|\theta)$ or values that are closer than any previously chosen tolerance value.

To initialize the Baum-Welch algorithm, choose any set of model parameters, incorporating any prior information that may be available. Absent such information, a uniform or any other arbitrary distribution may be chosen. Denote the set of those initial parameters by θ_0. To obtain improved estimates, we still want to use Eq. (9.8) but this time the counts A_{jk} and $E_k(b)$ will be replaced with the expected counts computed from the data. To compute those expected counts, we will first compute $P(\pi_t = j, \pi_{t+1} = k|x, \theta_0)$—the probability that the hidden process will transition from state j to state k at time t.

The inclusion of the parameter set θ_0 in the notation indicates that these probabilities will be computed based on the initial set of parameters θ_0. We will omit this from the notation from now on for simplicity but all of the expressions below assume that we use this parameter distribution for the computations. Using conditional probabilities, we obtain

$$P(\pi_t = j, \pi_{t+1} = k|x)$$
$$= \frac{P(x_1, x_2, \ldots, x_t, \pi_t = j)P(\pi_{t+1} = k, x_{t+1}, x_{t+2}, \ldots, x_l|x_1, x_2, \ldots, x_t, \pi_t = j)}{P(x)}$$
$$= \frac{f_j(t)P(\pi_{t+1} = k, x_{t+1}, x_{t+2}, \ldots, x_l|\pi_t = j)}{P(x)}, \tag{9.9}$$

the last equation following from the Markov property.[11] We have also used the notation $f_j(t) = P(x_1, x_2, \ldots, x_t, \pi_t = j)$, introduced earlier for the forward algorithm.

The probability $P(\pi_{t+1} = k, x_{t+1}, x_{t+2}, \ldots, x_l|\pi_t = j)$ can further be expressed as

$$P(\pi_{t+1} = k, x_{t+1}, x_{t+2}, \ldots, x_l|\pi_t = j)$$
$$= P(\pi_{t+1} = k|\pi_t = j)P(x_{t+1}|\pi_{t+1} = k)P(x_{t+2}, \ldots, x_l|\pi_{t+1} = k)$$
$$= a_{jk}e_k(x_{t+1})b_k(t+1),$$

where, as in the backward algorithm, we use $b_k(t+1) = P(x_{t+2}, \ldots, x_l|\pi_{t+1} = k)$. The justification is as follows: Once the process is in state j at time t, for the event $\pi_{t+1} = k, x_{t+1}, x_{t+2}, \ldots, x_l$ to take place, the process needs to transition into k, emit the symbol x_{t+1}, and generate the rest of the sequence $x_{t+1}, \ldots, x_l$. Combining this with Eq. (9.9), we obtain

$$P(\pi_t = j, \pi_{t+1} = k|x) = \frac{f_j(t)a_{jk}e_k(x_{t+1})b_k(t+1)}{P(x)}. \tag{9.10}$$

---

[11] The likelihood of the data, given the set of model parameters θ, $P(x) = P(x|\theta)$, is computed by the forward algorithm.

The average number of transitions from j to k in the training sequence x will then be the sum of the probabilities of this transition occurring exactly at position t, over all positions in the sequence x:

$$A_{jk} = \sum_{t=1}^{l} \frac{f_j(t)a_{jk}e_k(x_{t+1})b_k(t+1)}{P(x)}$$

$$= \frac{1}{P(x)} \sum_{t=1}^{l} f_j(t)a_{jk}e_k(x_{t+1})b_k(t+1). \tag{9.11}$$

In a similar way, since the posterior probabilities can be computed from the forward-backward algorithm as $P(\pi_t = k|x, \theta_0) = \frac{f_k(t)b_k(t)}{P(x)}$ (see Eq. (9.6)), the average number of states emitting the symbol $b \in M$ will be

$$E_k(b) = \sum_{\{t=1,\dots,l|x_t=b\}} \frac{f_k(t)b_k(t)}{P(x)} = \frac{1}{P(x)} \sum_{\{t=1,\dots,l|x_t=b\}} f_k(t)b_k(t). \tag{9.12}$$

Next, the expected counts from Eqs. (9.11) and (9.12) are substituted in the Eqs. (9.8), generating improved estimates for the transition and emission probabilities. This is the set of parameters that we have denoted by θ_1. Now, repeating the computations given by Eqs. (9.11) and (9.12) with the new values of the parameters from the parameter set θ_1 and substituting those values in Eqs. (9.8) will generate the set of parameters θ_2 and so on. The process is guaranteed to increase the likelihood of the data, that is, for the sequence of parameter approximations $\{\theta_i\}, i = 0, 1, 2, \dots, P(x|\theta_i) \leqslant P(x|\theta_{i+1})$ [33]. The process terminates when two successive values are either identical or sufficiently close.

As with the other algorithms we have considered in this chapter, to avoid underflows from multiplying small probabilities, the computations are usually carried out in logarithm space. If several sequences $x^1, x^2, \dots, x^m$ are used for training, Eqs. (9.11) and (9.12) should be modified to include the sum of the expected counts from all sequences:

$$A_{jk} = \sum_{i=1}^{m} \frac{1}{P(x^i)} \sum_{t=1}^{l} f_j^i(t)a_{jk}e_k(x_{t+1}^i)b_k^i(t+1),$$

$$E_k(b) = \sum_{i=1}^{m} \frac{1}{P(x^i)} \sum_{\{t=1,\dots,l|x_t=b\}} f_k^i(t)b_k^i(t), \tag{9.13}$$

where the superscripts i indicates that the respective probability is computed for the sequence x^i.

Example 9.9. Once again we will illustrate the method on simulated data from the *Dishonest Casino* application in the *CpG Educate* suite with known parameters. We used the algorithm three times for a simulated sequence of length 500, length

Model Parameters ?

	Transition		Emission						Initial
	Fair	Loaded	1	2	3	4	5	6	
Fair	0.95	0.05	0.1667	0.1667	0.1667	0.1667	0.1667	0.1667	0.5
Loaded	0.1	0.9	0.1	0.1	0.1	0.1	0.1	0.5	0.5

Panel A: Output for a training sequence of length 500

	Transition		Emission						
	Fair	Loaded	1	2	3	4	5	6	Initial
Fair	0.9140	0.0860	0.2429	0.1617	0.1223	0.1641	0.1635	0.1455	1.0000
Loaded	0.0964	0.9036	0.0225	0.1409	0.1773	0.0911	0.1475	0.4206	0.0000

Panel B: Output for a training sequence of length 1000

	Transition		Emission						
	Fair	Loaded	1	2	3	4	5	6	Initial
Fair	0.9499	0.0501	0.1359	0.1620	0.1903	0.1766	0.1528	0.1824	1.0000
Loaded	0.1343	0.8657	0.0991	0.1435	0.0482	0.1151	0.1461	0.4479	0.0000

Panel C: Output for a training sequence of length 100,000

	Transition		Emission						
	Fair	Loaded	1	2	3	4	5	6	Initial
Fair	0.9501	0.0499	0.1692	0.1669	0.1665	0.1662	0.1655	0.1657	0.9788
Loaded	0.0993	0.9007	0.1001	0.1002	0.1006	0.0986	0.0984	0.5021	0.0212

FIGURE 9.9

Output from the Baum-Welch algorithm for simulated data for a sequence of length 500 (Panel A), 1000 (Panel B), and 100,000 (Panel C). The HMM model parameters used for the simulation are presented above Panel A.

1000, and length 10,000, respectively. The results are presented in Figure 9.9. The example illustrates what should be intuitively clear: the longer training sequences yield more accurate estimates for the HMM parameters. Due to the stochastic nature of the process, shorter sequences are more likely to not contain the necessary number of transitions and emissions to accurately estimate the HMM parameters. Notice that since we use a single training sequence in all cases, the estimates for the initial distribution cannot be accurate. Using a set of training sequences vs. a single very long training sequence would be advantageous if obtaining estimates for the initial distribution is important.

Exercise 9.16. Experiment with the Baum-Welch algorithm for the *Dishonest Casino* application in the *CpG Educate* suite. Simulate sequences of various lengths using HMM models with different sets of parameters. How well in your opinion is the Baum-Welch algorithm capable of recovering the HMM parameters for (a) short sequences? (b) long sequences?   ▽

Exercise 9.17. *Repeat Exercise 9.16, this time using the Baum-Welch algorithm for the *CpG Islands* application in *CpG Educate* (a template *CpG_HMM.csv* to save and upload the HMM parameters from a file is available for download from the volume's website).   ▽

9.4.4 Post-Processing

The decoding methods described in this section are purely mathematical and they may not produce completely accurate results when applied to CGIs. Both Viterbi and the posterior decoding methods impose no restrictions on the length of the identified islands or check whether biologically important conditions such as high $\%C + G$ content or high $O/E\ CpG$ ratio (see Section 2) are met. When HMMs are used for CGI identification, the consideration of these properties is done during the *post-processing* stage. At this stage we turn back to the genomic properties of the CGIs that have not been modeled by the HMM. This stage usually includes performing one or more of the following refinements:

- *Combine CGIs separated by short gaps*: Neighboring CGIs that are separated by small gaps of non-island regions are merged into a single larger island. A minimal distance threshold between islands is set in advance and neighboring islands closer than this threshold value are merged. The selection of the threshold values used in the reported literature varies from about 15–20 [27] to up to 100 [8].
- *Check for minimal $\%C + G$ content and $O/E\ CpG$ ratio*: Check to see if the islands identified by the decoding methods meet the biologically relevant thresholds for $\%C + G$ content and $O/E\ CpG$ as described in Section 2. If the identified CGIs do not meet those threshold value requirements, those states will be relabeled as non-islands.
- *Check for minimal length*: As discussed in Section 2, short sequences labeled as CGIs are not of biological interest. Different length-threshold values are used in the literature but those are usually in the range 140–500 bp [8,27]. If the length of a predicted CpG island is less than the threshold value, those states will be relabeled as non-island.

Post-processing is then applied to filter out the regions that do not meet the biological criteria for CGIs with cutoffs.

Example 9.10.   In [27] the authors use the general HMM with states $Q = \{A_+, A_-, C_+, C_-, T_+, T_-, G_+, G_-\}$ and emission symbols $M = \{A, C, T, G\}$ that we considered earlier, with no restrictions on the transition or the emission matrices, training it on a set of 1000 sequences from the database embl173hum of all

Table 9.7 Learned HMM parameters used in the *CpG Discover* System [27]. See the text for details.

	Transitions								Emissions				Initial
	A_+	C_+	T_+	G_+	A_-	C_-	T_-	G_-	A	C	T	G	
A_+	0.2233	0.202	0.1745	0.3106	0.0229	0.0377	0.0196	0.0164	0.8358	0.0607	0.0501	0.0564	0.0742
C_+	0.2667	0.2511	0.3138	0.0746	0.023	0.0465	0.0226	0.0087	0.0382	0.9006	0.0266	0.0376	0.1005
T_+	0.093	0.252	0.245	0.3459	0.0093	0.0184	0.0251	0.0182	0.0624	0.0486	0.8376	0.0544	0.0309
G_+	0.1846	0.2703	0.1815	0.2326	0.0399	0.0272	0.0178	0.0531	0.0344	0.0451	0.0331	0.8904	0.1181
A_-	0.0323	0.0154	0.0154	0.0342	0.2684	0.1679	0.1496	0.3238	0.8699	0.0434	0.0452	0.0445	0.077
C_-	0.0243	0.0392	0.0146	0.0089	0.3052	0.2486	0.3013	0.065	0.0499	0.8827	0.0283	0.0422	0.2492
T_-	0.021	0.0274	0.0209	0.0372	0.0966	0.2415	0.2114	0.3511	0.0729	0.0389	0.8552	0.036	0.0692
G_-	0.048	0.0303	0.0187	0.0287	0.2575	0.207	0.1737	0.2432	0.0382	0.0307	0.0406	0.8935	0.2877

predicted human CpG islands in the EMBL database at the time the work was done. The parameters resulting from the training are provided in Table 9.7.[12]

Post-processing is then applied to filter out the regions that do not meet the biological criteria for CGIs with cutoffs as follows: If the distance between neighboring islands is less than 20 bp, the regions are merged. The newly extended CGIs and all other predicted islands are then tested to be of length $\geqslant$ 140 bp, have $\%C + G \geqslant$ 60%, and $O/E\ CpG \geqslant$ 60%. If those combined conditions are not met, the states previously recognized as CGIs will be relabeled as non-islands. Table 1 in [27] gives the final results for several test sequences used by the authors and compares those with results obtained from using other CGI locating systems.

9.5 CONCLUSIONS AND DISCUSSION

CGIs are of great interest in genomic analysis and are often used as markers for cancer and gene identification, as well as to investigate methylation profiles [34,35]. Following the original definition and algorithm for CGIs identification by Gardiner-Garden and Frommer [13], a number of sliding window algorithms have been developed with the ability to implement different interpretations and cutoff values for $\%C + G$, $O/E\ CpG$, the island length, gaps between the islands, and size of the sliding window [8,14–16]. It has been shown, however, that such methods do not provide an exhaustive search and that they may miss a large percentage of CGIs [17]. In addition, since sliding windows do not make use of any underlying mathematical structure, alternative approaches based on mathematical models are preferable in order to make some of the problems arising in the context of CGI identification more standardized and tractable.

In this chapter we focused mainly on the use of HMM for CGI identification. Viewing DNA data as output from a HMM elucidates the search for CGIs by placing the problems within a well-developed mathematical framework where efficient methods for decoding, evaluation, and training are readily available. In general, HMMs are widely used to analyze sequences and time series of data under the assumption that they have been generated by a process that cannot be directly observed. Instead, information about the process must be inferred from the observed sequences. In addition to CGI identification, this broad-spectrum approach has also been used for speech recognition [36], protein modeling [37], peptide sequencing [38], multiple DNA sequence alignment [39], gene prediction [40], and many others.

The models introduced in this chapter can also be used as a basis for further extensions and improved HMMs for CGI identification. Some work in this direction includes the development of an extensible approach that summarizes the evidence of CpG islands as probability scores [41], a hybrid visualization HMM method [18], a modified HMM approach with Poisson emission probabilities [42], HMM approaches

---

[12]The *CpG Islands* application in the *CpG Educate* suite uses these values as default parameters for the CGI HMM.

to improving the power of pattern detection [43], improved performance Viterbi and EM algorithms [44], and methods for speeding up HMM decoding [45].

The chapter includes a number of exercises of both applied and theoretical nature and an online project *Investigating "Predicted" Genes* available from the volume's website. We encourage the reader to attempt each exercise/project as soon as the relevant material for its execution has been introduced. Most of the applied exercises use the *CpG Educate* suite of web applications developed for this chapter and available at http://inspired.jsu.edu/~agarrett/cpg/[30]. *CpG Educate* uses the General Hidden Markov Model library (http://ghmm.org/) for the implementation of all HMM algorithms. A key feature of *CpG Educate* is that it provides the option to simulate data for most of the HMMs used in the chapter and compare the results from decoding and training with those from the simulations. After completing the chapter exercises, we invite the reader to undertake further experimentations with the decoding and learning methods for various simulated and actual data.

Acknowledgments

Authors Davies, Kirkwood, and Robeva gratefully acknowledge the support of the National Science Foundation under DEU award #0737467.

9.6 SUPPLEMENTARY MATERIALS

A supplementary project and additional files and data associated with this article can be found, in the online version, at http://dx.doi.org/10.1016/B978-0-12-415780-4.00019-3 and from the volume's website http://booksite.elsevier.com/9780124157804.

References

[1] Bird A. DNA methylation patterns and epigenetic memory. Genes Dev 2002;16:6–21.

[2] Klose RJ, Bird AP. Genomic DNA methylation: the mark and its mediators. Trends Biochem Sci 2006;31:89–97.

[3] Sorensen AL, Timoskainen S, West FD, Vekterud K, Boquest AC, Ahrlund-Richter L, et al. Lineage-specific promoter DNA methylation patterns segregate adult progenitor cell types. Stem Cells Dev 2010;19:1257–66.

[4] Isagawa T, Nagae G, Shiraki N, Fujita T, Sato N, Ishikawa S, et al. DNA methylation profiling of embryonic stem cell differentiation into the three germ layers. PLoS One 2011;6:e26052.

[5] Collas P. Programming differentiation potential in mesenchymal stem cells. Epigenetics 2010;5:476–82.

[6] Neddermann P, Jiricny J. The purification of a mismatch-specific thymine-DNA glycosylase from HeLa cells. Journal Biol Chem 1993;268:21218–24.

[7] Straussman R, Nejman D, Roberts D, Steinfeld I, Blum B, Benvenisty N, et al. Developmental programming of CpG island methylation profiles in the human genome. Nature Struct Mol Biol 2009;16:564–71.

[8] Takai D, Jones PA. Comprehensive analysis of CpG islands in human chromosomes 21 and 22. Proc Natl Acad Sci USA 2002;99:3740–5.

[9] Ashley DJB. The two hit and multiple hit theories of carcinogenesis. Br J Cancer 1969;23:313–28.

[10] Renan MJ. How many mutations are required for tumorigenesis? Implications from human cancer data. Mol Carcinog 1993;7:139–46.

[11] Schappert-Kimmijser J, Hemmes JGD, Nijland R. The heredity of retinoblastoma. Ophthalmologica 1966;151:197–213.

[12] Noburi T, Miura K, Wu DJ, Lois A, Takabayashi K, Carson D. Deletions of the cyclin dependent kinase-4 inhibitor gene in multiple human cancers. Nature 1994;368:753–6.

[13] Gardiner-Garden M, Frommer M. CpG Islands in Veribrate Genome. J Mol Biol 1987;196:261–82.

[14] Rice P, Longden I, Bleasby A. EMBOSS: The European Molecular Biology Open Software Suite. TIG 2000;16:276–7.

[15] Ponger L, Mouchiroud D. CpGProD: identifying CpG islands associated with transcription start sites in large genomic mammalian sequences. Bioinformatics 2002;18:631–3.

[16] Wang Y, Leung FCC. An evaluation of new criteria for CpG islands in the human genome as gene markers. Bioinformatics 2004;20:1170–7.

[17] Hsieh F, Chen SC, Pollard K. A nearly exhaustive search for CpG islands on whole chromosomes. Int J Biostatistics 2009;5(1), Art. 14.

[18] Rambally G, Rambally R. A hybrid visualization Hidden Markov Model approach to identifying CG-islands in DNA sequences, In Southeastcon, 2008. IEEE; 3–6 April 2008. p. 1–6.

[19] Hackenberg M, Previti C, Luque-Escamilla P, Carpena P, Martinez-Aroza J, Oliver J. CpGcluster: a distance-ased algorithm for CpG-island detection. BMC Bioinform 2006;7:446.

[20] Hackenberg M, Barturen G, Carpena P, Luque-Escamilla PL, Previti C, Oliver JL. Prediction of CpG-island function: CpG clustering vs. sliding-window methods. BMC Genom 2010;26:327.

[21] Sujuan Y, Asaithambi A, Liu Y. CpGIF: an algorithm for the identification of CpG islands. Bioinformation 2008;2:335–8.

[22] Chuang LY, Huang HC, Lin MC, Yang CH. Particle swarm optimization with reinforcement learning for the prediction of CpG islands in the human genome. PLoS One 2011;6:e21036.

[23] Durbin R, Eddy S, Krogh A, Mitchison G. Biological sequence analysis. Probabilistic models of proteins and nucleic acids. Cambridge, UK, Cambridge University Press; 1998.

[24] Pahter L, Sturmfels B. Algebraic statistics for computational biology. Cambridge, UK: Cambridge University Press; 2005.

[25] Norris JR. Markov chains. Cambridge: Cambridge University Press; 1997.

[26] Elliot JR, Aggoun L, Moore JB. Hidden markov models: estimation and control (corrected 3rd printing). New York: Springer; 2008.

[27] Lan M, Xu Y, Li L, Wang F, Zuo Y, Tan CL, et al. CpG-Discover: a machine learning approach for CpG island identification from human DNA sequence. In: Proceedings of international joint conference on neural networks, Atlanta, Georgia, USA; June 14–19, 2009. p. 1702–7.

[28] Viterbi, A. Error bounds for convolutional codes and an asymptotically optimum decoding algorithm. IEEE Trans Inform Theory 1967;IT-13:260–9.

[29] Viterbi A. A personal history of the viterbi algorithm. IEEE Signal Process Mag 2006;23:120–42.

[30] Garrett, A. CpG EducateSoftware tutorial; 2012, <http://inspired.jsu.edu/~agarrett/cpg/CpGEducate.pdf>.

[31] Baum LE, Petrie T, Soules G, Weiss NA. Maximization technique occurring in the statistical analysis of probabilistic functions of Markov chains. Ann Math Stat 1970;41:164–71.

[32] Welch LR. The Shannon lecture: hidden Markov models and the Baum-Welch algorithm. IEEE Inform Soc Newslett 2003;53(4), December 2003.

[33] McLachlan G, Krishnan T. The EM algorithms and extensions. 2nd ed. Hoboken, NJ: Wiley; 2008.

[34] Illingworth PR, Bird AP. CpG islands – a rough guide. FEBS Lett 2009;583: 1713–20.

[35] Bobbie PO, Reams R, Suther S, Brown CP. Finding molecular signature of prostate cancer: an algorithmic approach. In: Proceedings of the 2006 international conference on bioinformatics & computational biology, BIOCOMP'06, Las Vegas, Nevada, USA, June 26–29, 2006, p. 265–9.

[36] Rabiner LR. A tutorial on hidden Markov models and selected applications in speech recognition. Proc IEEE 1989;77:257–85.

[37] Krogh A, Brown M, Mian IS, Sjlander K, Haussler D. Hidden Markov models in computational biology. Application to protein modeling. J Mol Biol 1994;235:1501–31.

[38] Fischer B, Roth V, Roos F, Grossmann J, Baginsky S, Widmayer P, et al. NovoHMM: a hidden Markov model for de novo peptide sequencing. Anal Chem 2005;77:7265–73.

[39] Do CB, Mahabhashyam MS, Brudno M, Batzoglou S. ProbCons: probabilistic consistency-based multiple sequence alignment. Genome Res 2005;15:330–40.

[40] Bernal A, Crammer K, Hatzigeorgiou A, Pereira F. Global discriminative learning for higher-accuracy computational gene prediction. PLoS Comput Biol 2007;16:e54.

[41] Wu H, Caffo B, Jaffee HA, Irizarry RA, Feinberg AP. Redefining CpG islands using hidden Markov models. Biostatistics 2010;11:499–514.

[42] Irizarry RA, Wu H, Feinberg AP. A species-generalized probabilistic model-based definition of CpG islands. Mamm Genome 2009;20:674–80.

[43] Zhai Z, Ku SY, Luan Y, Reinert G, Waterman MS, Sun F. The power of detecting enriched patterns: an HMM approach. J Comput Biol 2010;17:581–92.

[44] Lam T, Mayer A. Efficient algorithms for training the parameters of hidden Markov models using stochastic expectation maximization (EM) training and Viterbi training. Alg Mol Biol 2010;5:38.

[45] Lifshits Y, Mozes S, Weimann O, Ziv-Ukelson M. Speeding up HMM decoding and training by exploiting sequence repetitions. Algorithmica 2009;54:379–99.

Phylogenetic Tree Reconstruction: Geometric Approaches

10

David Haws[*], Terrell L. Hodge[†] and Ruriko Yoshida[‡]

[*]*IBM T. J. Watson Research Center, Yorktown Heights, NY, USA*
[†]*Department of Mathematics and College of Arts and Sciences Dean's Office, Western Michigan University, Kalamazoo, MI, USA*
[‡]*Department of Statistics, University of Kentucky, Lexington, KY, USA*

10.1 INTRODUCTION

In this section, we give a brief, high-level overview that encompasses the main topics to be covered in this chapter, and the relationships between them. As will be seen, phylogenetic tree reconstruction is a multi-layered process, and covering all aspects of it would take far more than a single chapter. In the next section, we begin in earnest our investigations into a subcollection of the concepts raised in the Introduction that we intend to treat in more detail (and from scratch, with definitions). For comprehensive mathematical treatments of terms, methods, and details that we do not cover, see, e.g., [1,2] for graph-theoretical/combinatorial approaches; see also [3] for an algebraic geometry and statistics perspective on many of these topics. For a less mathematical, more biologically oriented text, see, e.g., [4].

Through the ages, the diversity of species on earth and the sources of that diversity have always featured prominently in human thought, as evidenced by Aurignacian cave paintings, religious scriptures, and modern science. A prevailing biological concept is that the diversification of life forms is related to the separation of gene pools. If geographical or other barriers separate gene pools, the process of gene flow can be insufficient to counteract genetic drift, and genetic or behavioral barriers emerge against future gene flow (even after the removal of physical barriers) [5]. Of course, genetic isolation alone is insufficient to explain diversity, which further requires the raw material of genetic mutation, inevitably acted upon by natural selection.

A *phylogenetic tree* (or *phylogeny*) is a diagram/graph which represents relations of evolutionary descent of different species, organisms, or genes from a common ancestor. A phylogenetic tree is a useful tool to understand and organize information on biological diversity, structuring classifications, and providing insight into events that occurred during evolution. An excellent preliminary guide to the biological perspective, upon which we build a more mathematical approach in this chapter, is [6]; see `http://www.nature.com/scitable/topicpage/reading-a-phylogenetic-tree-the-meaning-of-41956`.

Mathematical Concepts and Methods in Modern Biology. http://dx.doi.org/10.1016/B978-0-12-415780-4.00010-7

One may reconstruct a phylogenetic tree among distinct groups or species from morphological data obtained by measuring and quantifying the phenotypic properties of representative organisms via, for example, parsimony. However, recent phylogenetic analysis uses nucleotide sequences encoding genes or amino acid sequences encoding proteins as the basis for classification. Therefore in this chapter we focus on phylogenetic tree reconstruction methods based on sequenced genes or genomic data sets. Through evolutionary history, a character (e.g., a nucleotide) of the sequence might be changed to another one, deleted or inserted. Thus, before reconstructing a phylogenetic tree, we have to align an input sequence data set, e.g., identify sequences of characters (DNA bases, amino acids, etc.) which are thought to be representing the same ("homologous") regions of genes (from different species or different gene families, etc.) and then line up the sequences so that nucleotides in differing sequences can be compared site-by-site, where one site may vary from another by mutation, i.e., insertions, deletions, or substitutions of characters. Such a line-up of two sequences is an *alignment*; if we have multiple sequences then the result is properly called a *multiple alignment*, although we will often abuse terminology and use "alignment" for both. Aligning multiple sequences is generally known to be an NP-hard problem [7,8], and there are scores of approaches to creating alignments heuristically. Here we assume that we have a perfectly aligned sequence data set and focus on the reconstruction of phylogenetic trees from a multiple alignment.

There are several methods to infer a phylogenetic tree from a given alignment, including the maximum-likelihood (ML) method, distance-based methods, parsimony-based methods, Bayesian inference methods, and so on (see [1] for more details). In this chapter we will focus on distance-based methods.

Distance-based methods build phylogenetic trees from aligned sequences by making use of pairwise "distances" between the sequences. These distances arise from a model of sequence evolution, or *evolutionary model*, which encodes certain hypotheses about how sequences evolve, often including the probabilities that one character at a given site will transition to another, as well as assumptions about how sequences as a whole then transform, via evolution, into one another. Models of sequence evolution are needed to address the problem that observed sequences may have experienced much more change over time that their elements alone might show; by way of analogy, just as an observation that a light switch is currently off does not alone completely specify the number of times previously that it had been flipped on and off again. Many common models of sequence evolution are, in the lingo of mathematicians, given by a continuous-time Markov model with a substitution rate matrix whose entries are the probabilities of characters changing to one another.

Each such evolutionary model given by continuous-time Markov chains corresponds to a phylogenetic tree for which each node is a sequence, and the (directed) edges link sequences which evolve from one to another by data given by substitution rate matrices. In this way, the phylogenetic tree summarizes the relationships between the species (or other organisms represented by the sequences) in terms of common ancestors (nodes) and evolutionary changes via edge lengths (e.g., times since divergence, number of substitution events, etc.).

Under the assumption that a phylogenetic tree describes evolution via a continuous-time Markov process, the main problem of phylogenetic tree reconstruction is to find a tree when only the character sequences, such as DNA sequences or protein sequences, etc., are given. In the lingo of trees and Markov processes, one assumes that the only sequence data that observed is at the tips, i.e., in the *leaves*, while other information on the phylogenetic tree, the particulars of the substitution events, and edge lengths are missing. In a distance-based approach, distances between sequences (i.e., what should be sums of edge lengths on the path in the phylogenetic tree between the nodes for the sequences) are derived directly from the sequences by using the evolutionary model to compute the most likely distance between each pair of sequences. This is formalized in mathematical and statistical terms by so-called maximum likelihood estimates.[1] Computing these estimates will not be our focus; using them will be.

Pairwise distances, along with a distance-based method for reconstructing phylogenetic trees from the set of all pairwise distances, can be used to reconstruct a particular phylogenetic tree that relates the sequences. Usually the set of all pairwise distances computed from an alignment, collectively often called a *distance matrix* (or an example of a *dissimilarity map*), does not give a *tree metric*, which is a distance matrix realizing a phylogenetic tree. Thus a distance-based method tries to find a tree metric closest to the given set of pairwise distances computed from the alignment under some criteria.

To date, of all the tree reconstruction methods, distance-based methods for phylogeny reconstruction have been seen to be the best hope for accurately building phylogenies on very large sets of taxa such as the data sets for tree of life for the insects Hymenoptera [2,9]. More precisely, distance-based methods have been shown to be statistically consistent in all settings (such as the long branch attraction problem; see [1] for details), in contrast with other methods, such as parsimony methods, e.g., [10–12]. Distance-based methods also have a huge speed advantage over parsimony and likelihood methods, and hence enable the reconstruction of trees on greater numbers of taxa. However, a distance-based method is not a perfect method for reconstructing a phylogenetic tree from a given sequence data set. This is because, in computing pairwise distances, one ignores both the interior nodes and the overall tree topology and so there is a concern that one loses some information from the input data sets. Therefore, it is important to understand how a distance-based method works and how robust it is with noisy data sets. (However, it is noteworthy that from an information-theoretic point of view, a recent article [13] argues that at least some distance-based methods can be proved to preserve more information than may be obtainable from maximum likelihood methods for tree reconstruction, contrary to commonly expressed concerns in the mathematical and biological literature.)

A distance-based method is related to geometry and combinatorics. In fact, one can describe the space of phylogenetic trees, i.e., a set of all tree metrics over the set of all distance matrices, as points in a high-dimensional space which form a union of so-called polyhedral cones. In Section 10.3, we will take an elementary approach to

---

[1] Separate from the ML approach to tree reconstruction.

realizing trees as points, and pave the way for further study of the space of phylogenetic trees from the view of polyhedral geometry.

There has been much work to understand distance-based methods, such as the balanced minimum evolution method and neighbor-joining method. In 2002, Desper and Gascuel introduced a balanced minimum evolution (BME) principle, based on a branch length estimation scheme of Pauplin [14]. The guiding principle of minimum evolution tree reconstruction methods is to return a tree whose total length (sum of branch lengths) is minimal, given an input dissimilarity map. The BME method is a special case of these distance-based methods wherein branch lengths are estimated by a weighted least-squares method (in terms of the input dissimilarity map and the tree in question) that puts more emphasis on shorter distances than longer ones. Each labeled tree topology gives rise to a vector, called herein *the BME vector*, which is obtained from Pauplin's formula.

Implementing, exploring, and better understanding the BME method have been focal points of several recent works. The software FastME, developed by Desper and Gascuel, heuristically optimizes the BME principle using nearest-neighbor interchanges (NNI) [15]. In simulations, FastME gives superior trees compared to other distance-based methods, including one of biologists' most popular distance-based methods, the Neighbor-Joining (NJ) Algorithm, developed by Saitou and Nei [16]. In 2000, Pauplin showed that the BME method is equivalent to optimizing a linear function, the dissimilarity map, over the BME representations of binary trees, given by the BME vectors [14]. Eickmeyer et. al. defined the *nth BME polytope* as the convex hull of the BME vectors for all binary trees on a fixed number n of taxa. Hence the BME method is equivalent to optimizing the input dissimilarity map (a linear function), over a BME polytope. In 2010, Cueto and Matsen [17] studied how the BME method works when the addition of an extra taxon to a data set alters the structure of the optimal phylogenetic tree. They characterized the behavior of the BME phylogenetics on such data sets, using the BME polytopes and the *BME cones*, i.e., the normal cones of the BME polytope. We will discuss some details of BME method and NJ method from the view of polyhedral geometry in Section 10.4, again opening the doors to further reading and review.

10.2 BASICS ON TREES AND PHYLOGENETIC TREES

Trees are ubiquitous objects in any theory in which branching processes occur. Most generally, as a graph, a *tree* $T = (V, E)$ is a directed, connected graph consisting of a nonempty set of *vertices* (or *nodes*), V, and set of *edges*, $E \subseteq V \times V$, in which there are no cycles or loops. We will assume all our trees are finite, that is, V is a finite set. Each edge $e \in E$ identifies with a single ordered pair (a, b) $(a \neq b)$ wherein, by definition, a is the *initial vertex* of e and b is the *terminal vertex* of e, so that e is drawn as the arrow $a \rightarrow b$, and we also often say e starts at a and ends at b. Given a vertex $v \in V$, the number of edges ending in v is the *indegree indeg*(v) of v, while the number of edges starting in v is the *outdegree outdeg*(v) of v. The *leaves* of a tree T are those vertices

$v \in V$ for which the outdegree of v is 0 (also sometimes called terminal vertices). A tree may be *rooted* or *unrooted*; by definition, if T is rooted, then there is a vertex v_o which has $indeg(v_o) = 0$, and otherwise T is unrooted. An *interior vertex* is one which is not a node, although if T is rooted and $|V| > 2$, the root is usually not regarded as an interior vertex or as a leaf. A pair of two leaves $\{i, j\}$ is a *cherry*, if i and j have a common parent node,[2] say, a, i.e., there are edges starting at a and ending at i, respectfully, j. More generally, any leaves sharing a common parent node are called *sisters*.

Alternatively, when working with unrooted trees, as we will most often do below for reconstruction methods, one can dispense with directionality (orientation) and start by defining trees simply as connected graphs (on nonempty vertex sets) with no cycles or loops. In this case, edges are defined by pairs $\{a, b\}$ of (distinct) elements in V. It is common to represent an edge $e = \{a, b\} \in E$ simply as a line segment $a - b$ (or $b - a$). One simply defines the degree $deg(v)$ of $v \in V$ to be the number of edges incident to v. Leaves are vertices v with $deg(v) = 1$, and $v \in V$ is an *interior vertex* if $deg(v) \geqslant 2$. Binary trees are then trees for which $deg(v) = 3$ for all interior vertices. Cherries are then pairs of leaves joined by two edges that meet in an interior vertex v, with $deg(v) = 3$ (and v is still the *parent*, or *ancestor*). Sisters are then defined analogously. An unrooted tree can be rooted by picking a leaf and adding an orientation directing every edge away from it; more commonly for biological purposes, a node is added along some interior edge, the two new edges split from it, as well as the remaining edges are, again, directed so that the node has indegree zero. If T is a directed tree in the earlier sense, then upon "forgetting" the directions of the edges, one gets an (undirected) tree in this new sense, and for any vertex v of T, $deg(v) = indeg(v) + outdeg(v)$. For unrooted trees, it is convenient to dispense with orientation, but helpful to view unrooted trees as graphs that could be directed and rooted as needed. We will assume unrooted trees are just graphs, rather than directed graphs, and rooted trees are directed graphs. We may use the term "directed tree" where there might be need or confusion.

There are many ways of representing trees, and many types of software available for drawing and viewing trees. In the following exercise, we explore the Newick format for representing rooted phylogenetic trees, along with some of the tree-drawing programs using Newick-style input to output rooted and unrooted trees. We do so through free access at the website, Phylogeny.fr at `http://www.phylogeny.fr`. This website is specifically geared toward providing phylogenetic tools for nonspecialists. First, we note that a compact representation for a rooted tree is to group sisters by parentheses, e.g., `(A, (B,C))`; represents the tree with cherry $\{B, C\}$ and then a root joining this cherry to A, as in Example 10.1. Such a format is the *Newick format* or Newick representation of the tree. (Note that the end semicolon in `(A, (B,C))`; is part of the standard Newick format, and not meant to be punctuation in regular English.)

Example 10.1.   The rooted tree given by `(A, (B,C))`; see Figure 10.1.

---

[2] We use this terminology although in a biological interpretation, interior nodes may not be viewed as ancestors, but only as demarcation points for speciation.

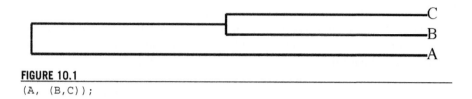

FIGURE 10.1

```
(A, (B,C));
```

Note that neither the root nor the parent node of the cherry $\{B, C\}$ is shown explicitly; instead, vertical lines are used to indicate that a branching event occurs. Shrinking these vertical edges to points would produce the nodes that would be commonly shown in a mathematical graphical representation of a tree. This is a common difference between rooted trees as often represented in biology, and trees in mathematical graph theory. Also, if a tree is unrooted, it is common to represent it in Newick format by arbitrarily picking a root. In this case, still, drawing programs may neglect to display interior vertices; they are to be inferred at a branching point. (This will be seen in Exercise 10.1 below.)

Exercise 10.1.

1. Obtain the graph in Example 10.1 yourself, as follows:

   a. Go to the link "TreeDyn" in the list "Tree Viewers" under the heading "Online Phylogeny Programs" at the bottom of the homepage of Phylogeny.fr at `http://www.phylogeny.fr/`, or similarly from the pop-up menu for "TreeDyn" that appears under "Tree Viewers" when mousing over the heading "Online Programs" in the bar across the top of the Phylogeny.fr homepage.

   b. In the available textbox at the top of the "TreeDyn" page, type `(A, (B,C));` click the "Submit" button, and await the output.

   c. Once a picture of a tree is displayed, scroll down the page to "Tree Style," and change from the default option "Phylogram" to the option "Cladogram." The program will automatically redraw the graph.

2. Compare the result of part (1) with the original drawing (produced by "Phylogram"), and with the figure in Example 10.1 provided above. How do they differ?

3. Replace `(A, (B,C));` by `(A, (C,B));` in parts (1)–(2) with the "Cladogram" setting as above. Describe how the output differs from Example 10.1. Note that you can restart "TreeDyn," or more easily, after completing any drawing, you can click on the "Data and Settings" tab at the top of the page to return to the data you originally input into the textbox, and edit it.

4. Draw a rooted tree on three leaves A, B, C with cherry $\{A, B\}$.

5. Draw an unrooted tree by inputing `(A, (B,C));` but selecting "Radial (by Drawtree)" under the "Tree Style" option. What, if anything, changes when the letters A, B, and C are permuted in the Newick input?

6. Use the Newick input (A, (B, (C,D))); and the "Cladogram" and "Radial (by Drawtree)" tree style options as in parts (1)–(5) above to create graphs of rooted and unrooted trees. Examine also the outcome of permuting the leaves in the Newick format.

7. Repeat part (6) of this exercise, but with a Newick representation of a tree with two pairs of cherries, {A, B} and {C, D}, with the rooted version having these two pairs join in a common parent node.

8. Apply the setting "Radial (by Drawtree)" (or just select the "Drawtree" program directly) with the Newick input (, (,)); and compare with part (5) of this exercise.   ▽

Exercise 10.2. Create other examples of a tree that is rooted and another that is unrooted. List the indegrees, outdegrees, and degrees for the vertices in your trees, as appropriate.   ▽

Exercise 10.3. Given an arbitrary tree T, show that if a root of T exists, it is unique. (If there is a root, there is only one root.)   ▽

An important means for comparing trees is that of an isomorphism. Directed trees $T = (V, E)$ and $T' = (V', E')$ are *isomorphic* if they are isomorphic as graphs, meaning that there is a bijection $\varphi : V \to V'$ which extends to a bijection on the edges by setting $\varphi(e) = (\varphi(a), \varphi(b)) \in E'$, given $e = (a, b) \in E$. Without loss of generality, we write $\varphi : T \to T'$ for this isomorphism. Likewise, one may define isomorphisms of undirected trees $T = (V, E)$ and $T = (V', E')$ as bijections for $\varphi : V \to V'$ with $\varphi(\{a, b\}) = \{\varphi(a), \varphi(b)\} \in E'$, for $e = \{a, b\} \in E$. Informally, the notion of an isomorphism captures the idea that trees are like mobiles connecting a set of objects, with edges that are springs. No matter how they are suspended by a string from the ceiling, and no matter how they are stretched or spun about, if they can be turned and their edges can be elongated or compressed so that the only difference between mobiles is the particular choices of the objects dangling from the ends (but not how many objects there are or the ways in which those objects are hooked together), then the trees the mobiles represent are isomorphic. Accordingly, isomorphisms of directed trees preserve the indegree and outdegree of each vertex, so if T and T' are isomorphic as directed trees, then T is rooted if and only if T' is.

As we will focus on applications and representations of trees in phylogenetics, we will be particularly interested in relationships between trees and their sets of leaves. In this context, for a finite set X, a *phylogenetic X-tree* (rooted or unrooted) is a tree $T = (V, E)$ for which $X \subset V$ is the set of all leaves of T, and, by definition, its leaves are labeled by X. That is, when pictured, the leaves are visibly labeled by the elements of X, while interior nodes are left unlabeled. An alternative, more formal mathematical formulation is to say that a phylogenetic X-tree is a pair (T, ψ) consisting of a tree $T = (V, E)$ and a leaf-labeling ψ, which, by definition, is an injection $\psi : X \to V$, but the informal idea will suffice. Although the interior vertices of a phylogenetic X-tree are, by definition, unlabeled, in what follows, it will be helpful in several scenarios to add labelings to interior nodes so as to describe or compare phylogenetic X-trees that underlie these more fully labeled trees.

Example 10.2.

1. The trees in parts (1)–(5) of Exercise 10.1 were phylogenetic X-trees with the leaves labeled by $X = \{A, B, C\}$, whereas the interior vertices were not labeled. The tree in part (8) of Exercise 10.1 had no labeled leaves and hence is not a phylogenetic X-tree, but does represent the "underlying tree" or "tree topology" of an unrooted phylogenetic X-tree on three leaves, as will be defined a little later.

2. An approximation to the evolutionary history of all organisms alive today would be a phylogenetic X-tree wherein the leaves X consist of all species currently in existence; in this case, $|X| > 1,800,000$. The Tree of Life project `http://tolweb.org/tree/`, while noting that a tree is an imperfect model, uses phylogenetic X-trees as an organizational framework, and the use there is representative of hundreds of more specialized scientific articles employing phylogenetic trees. The reader is encouraged to explore this website for current versions and many visualizations of phylogenetic trees on multiple organizational levels of species and other biological taxa.

A tree may simply be called *phylogenetic* if it is a phylogenetic X-tree for some (finite) set X. Without loss of generality, it is possible to use $[n] = \{1, 2, \ldots, n\}$ in place of X for $|X| = n$. A binary tree that is a phylogenetic tree is called a *binary phylogenetic tree*. Binary phylogenetic X-trees form the most commonly used group of trees in biology, since branching events that arise from evolutionary events, like mutations or speciation, correspond to a splitting of one lineage into two, rather than more than two simultaneously. However, non-binary trees do arise when ancestry is uncertain (e.g., the order of speciation or other branching events is uncertain), and in this case, there may be sisters that are not just cherries. An extreme example is a *star tree* on n leaves, which has $n + 1$ nodes and n edges, with the single interior vertex joined to each and every leaf.

Exercise 10.4.

1. Show any rooted binary tree on $n \geqslant 3$ leaves always has a cherry, and hence conclude any unrooted binary tree on $n \geqslant 3$ leaves does, too.

2. Explain why the total number of vertices of any unrooted binary tree on $n \geqslant 3$ leaves is $2n - 2$.

3. Argue that the number of edges of any unrooted binary tree on $n \geqslant 3$ leaves is $2n - 3$.

4. Suppose $n \geqslant 3$. Using the previous part of this exercise, argue that the number of unrooted binary phylogenetic X-trees for a set X of n leaves is the double factorial $(2n-5)!!$. As a hint, consider how to build an unrooted binary phylogenetic X- tree on n leaves from any one on $n - 1$ leaves by adding a new edge ending in the new leaf to an edge on the tree with one less leaf. (By definition of the double factorial, $1!! = 1$, $3!! = 3 \cdot 1 = 3$, $5!! = 5 \cdot 3!! = 5 \cdot 3 \cdot 1 = 15$, $7!! = 7 \cdot 5!! = 7 \cdot 5 \cdot 3 \cdot 1 = 105$, and, generally, $(2n-5)!! = (2n-5) \cdot (2(n-1)-5)!! = (2n-5) \cdot (2n-7) \cdots 5 \cdot 3 \cdot 1$. (Alternatively, one can show that $(2n - 5)!! = \frac{(2n-4)!}{(n-2)!2^{n-2}}$, for $n \geqslant 2$.) ▽

Suppose $T = (V, E)$ is a phylogenetic X-tree and $T' = (V', E')$ is a phylogenetic X'-tree. Since T and T' are trees with additional properties, and since isomorphisms of directed trees preserve indegrees and outdegrees, respectfully, preserves degrees for undirected trees, a function $\psi : T \to T'$ can be an isomorphism of the phylogenetic trees X and X' only if ψ restricts to a bijection $\psi : X \to X'$ on the leaf sets. Necessarily, then, $|X| = |X'|$. A tree isomorphism $\varphi : T \to T'$ for which the restriction $\varphi : X \to X'$ of $\varphi : V \to V'$ is in fact an identity map (so $X = X'$ and $\varphi(v) = v$ for all $v \in X$), is, by definition, an isomorphism of phylogenetic X-trees. In our biological context, the notion of isomorphism makes explicit those ways in which pictures of phylogenetic trees might differ, but still represent the same evolutionary relationships among the leaves.

Example 10.3. For $X = \{A, B, C, D\}$, let T_1 be the unrooted binary phylogenetic X-tree $((A,B),(C,D))$; where the cherry $\{A, B\}$ has ancestor u and the cherry $\{C, D\}$ has ancestor v. Let T_2 be the unrooted binary phylogenetic tree $((A,C),(B,D))$; where cherry $\{A, C\}$ has ancestor s and cherry $\{B, D\}$ has ancestor t. Then setting $\varphi : T_1 \to T_2$ given by $\varphi(A) = C, \varphi(B) = D, \varphi(u) = s, \varphi(v) = t, \varphi(C) = A$, and $\varphi(D) = B$ creates an isomorphism of T_1 with T_2 as phylogenetic X-trees. (Note that the lengths of edges play no part here, only the connectivity relationships.)

Exercise 10.5.

1. For $X = \{A, B, C, D\}$, let T_1 be the unrooted binary phylogenetic X-tree $((A,B),(C,D))$; and let T_2 be the unrooted binary phylogenetic tree $((A,C),(B,D))$; and let $\varphi : T_1 \to T_2$ be as in as in Example 10.3.

    a. Explain why although $\varphi : T_1 \to T_2$ is an isomorphism of trees, it is not an isomorphism of phylogenetic X-trees.

    b. Find another distinct isomorphism of T_1 and T_2 just as trees.

2. Explain why trees $(A,(B,C))$; and $(A,(C,B))$; are isomorphic as phylogenetic $X = \{A, B, C\}$ trees, but $(A,(B,C))$; and $(B,(A,C))$; are not.

3. Explain why if $(A,(B,C))$; and $(2,(1,3))$; are isomorphic as phylogenetic trees, then $(A,(B,C))$; and $(1,(2,3))$; are not.   ▽

Formally, then, the *unlabeled* or *underlying* tree associated to a phylogenetic X-tree T is an equivalence class of all trees T' isomorphic to T. Informally, it is represented by any tree T' isomorphic to T, and T' is said to have the same *topology* or *tree topology* as T.

Example 10.4. There is only one tree topology for unrooted phylogenetic X-trees on a set X of four leaves; it is given in Newick form by $((,),(,))$; (see Figure 10.2).

If there is an isomorphism $\varphi : T \to T'$ of rooted phylogenetic X-trees, then T and T' are said to be *equivalent trees*, and likewise for two unrooted phylogenetic X-trees.

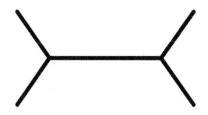

FIGURE 10.2

Quartet ((,),(,));

Exercise 10.6.

1. (Quartets) There are three distinct unrooted binary phylogenetic X-trees on a set of four leaves X. That is, up to isomorphisms of binary phylogenetic X-trees, there are only three distinct such non-equivalent trees—every other unrooted binary phylogenetic X-tree on four leaves $X = \{1, 2, 3, 4\}$ will be an unrooted binary phylogenetic X-tree equivalent to one of these. Draw (representatives of) these three trees.

2. Up to equivalence, how many distinct rooted binary phylogenetic X-trees on a fixed set X of three leaves are there? Draw representatives of each.

3. For $X = \{1, 2, 3\}$, how many distinct tree topologies are there underlying the rooted phylogenetic X-trees? What if these X-trees are unrooted?   ▽

   More formally, there is an equivalence relation on the set of all rooted phylogenetic trees (resp., unrooted phylogenetic trees) given by setting two such trees to be related if they are equivalent phylogenetic X-trees. Just as with equivalence relations in general, one often picks a representative of each equivalence class and identifies the whole class with any representative (e.g., just as a fraction in lowest common terms represents all fractions which reduce to it). Consequently, from now on, for any fixed X, when we speak of the set of "all rooted phylogenetic X-trees" or "all unrooted phylogenetic X-trees" we understand that any particular such tree can be represented by one on $X = [n]$ with the same underlying tree topology. For (rooted) phylogenetic X-trees, this formalizes the notion in biology of a "cladogram," so when drawn in the plane, neither horizontal axis nor vertical axis has any particular meaning, e.g., as in [4, p. 15].

   From now on, for any set X of n elements, *we will let $\mathcal{T}_n$ denote the set of all unrooted binary phylogenetic X-trees T*. For example, by Exercise 10.4, $\mathcal{T}_4$ consists of three distinct such trees (again, up to equivalence), and by Exercise 10.5, all these leaf-labeled trees in $\mathcal{T}_4$ have the same underlying tree topology.

Exercise 10.7.

a. Identify the subsets of trees below which are isomorphic to one another simply as trees (see Figure 10.3).

b. For the same collection of trees, identify which are isomorphic as phylogenetic X-trees, for $X = [6]$.

c. How many distinct tree topologies are there for unrooted binary trees with
$n = 3, 4, 5$, and 6 leaves?

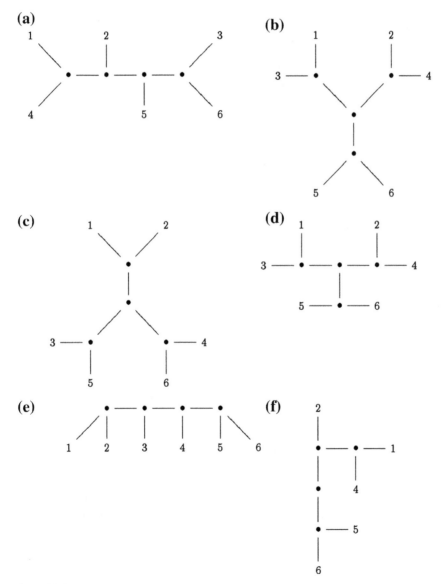

FIGURE 10.3
Figure for Exercise 10.7

d. How many distinct (unrooted) binary phylogenetic X-trees are there in $\mathcal{T}_n$, if $X = [n]$, for $n = 3, 4, 5, 6$? ▽

10.3   TREE SPACE

10.3.1   Trees as Points

In the last section, we discussed trees, phylogenetic X-trees (and more so, binary ones) in particular, as graphs, and developed a bit of intuition supporting their interpretation in some biological situations. In this section, we move beyond the visual appeal of trees, as graphical models, in order to interpret them in a geometric framework that better sets the stage for implementing algorithmic and computational methods to reconstruct the phylogeny of a set of species (or genes, or other organisms) and for describing geometrically the advantages and pitfalls of those methods.

By definition, phylogenetic X-trees have leaves that are labeled, but for phylogenetic tree reconstruction, one may also want to label the edges. Thinking of the vertices V, of a phylogenetic X-tree $T = (V, E)$, as species for the moment, then labels on the edges can, depending on the biological context, be regarded as providing some attribute of the sequences, often a measure of evolutionary change from one species to the next. In graph-theoretic terms, a labeling of the edges E of T is an *edge-weighting*, a function $\omega : E \to \mathbb{R}$ that assigns a real number $\omega(e)$, say, to every edge $e \in E$. In many cases, such edge-weightings $\omega : E \to \mathbb{R}$ are assumed to take nonnegative values, but for phylogenetic tree reconstruction algorithms, it is useful to allow for more general edge-weightings. In phylogenetics, the graphical notion of an edge-weighting corresponds to an *evolutionary distance map*. Determination of evolutionary distances is a process anchored in the choice of a so-called *model of evolution*, which describes how (and how frequently or with what probabilities) sequences change, e.g., via substitutions, insertions, or deletions in their character strings. This is a very large and significant area of research for which there are a number of nice treatments in both biology and biomathematics texts; see the references in the Introduction.

We also do not discuss in any detail here the many ways by which evolutionary distances are defined and determined, we do give two examples in Exercise 10.8 below.

Trees $T \in \mathcal{T}_n$, together with weightings ω, will be the outcomes of the distance-based reconstruction methods we examine. *We let $\mathfrak{T}_n = \{(T, \omega) \mid T = (V, E) \in \mathcal{T}_n, \omega : E \to \mathbb{R}^+\}$ denote the set of all ordered pairs of unrooted binary phylogenetic X-trees T together with positive edge weightings ω on T.* Some distance-based reconstruction methods may produce edge weightings with negative values or zero, so it may be useful to extend $\mathfrak{T}_n$ to include these weightings.

Exercise 10.8.

1. The *Hamming distance*, $dH(x, y)$, is defined to be the number of characters at which two sequences x and y differ. Suppose we have two sequences x, y over

$\{A, G, C, T\}$; for example:

$$x = GATTACATTC,$$
$$y = GCCATACTTC.$$

Compute the Hamming distance $dH(x, y)$ between x, y.

2. The Jukes-Cantor correction dJC to the Hamming distance is defined as

$$dJC(x, y) = -\frac{3}{4} \log \left(1 - \frac{4}{3} f \right),$$

where f is the frequency of the different sites between two sequences. For example, suppose we have two sequences x, y of length 10, for which the Hamming distance between x, y is $dH(x, y) = 6$. Then $f = 6/10 = 0.6$.

   a. Compute the Jukes-Cantor correction for the sequences x, y as in part (1) of this exercise.
   b. Observe that for any two sequences x, y of the same length, $dJC(x, y) = dJC(y, x)$ (and the same is true for dH). What can you say about $dJC(x, y)$ (respectfully, $dH(x, y)$) if x and y are the same?   ▽

The Hamming distance represents an easy, but rather crude measure of difference between sequences. For example, it fails to take into account the possibility that sequences could have characters change over time, and then change back. Also, there is no accounting for well-known biochemical phenomenon such as the fact that the probability that one DNA character might change to another is not generally uniform, but likely differs on the particular DNA bases themselves, or how they are arranged or grouped along the sequence. So-called "evolutionary models" describe special sets of additional assumptions that are made to account for these kinds of issues, and methods for determining the related evolutionary distances between any two given leaves x, y, that are represented by two aligned sequences (of DNA, RNA, proteins, etc.) s_x, s_y, generally depend on the choice of particular models of evolution. For example, the model of evolution that eventually gives rise to the Jukes-Cantor correction in Exercise 10.8 above is obtained from the Jukes-Cantor model of evolution; see e.g., [4] or [1], for more details on this model and [3] a derivation of this distance from the model by an algebraic geometry approach.

The evolutionary model for Hamming distance is very simple; it assumes all bases have equal probability of changing into one another, all sites in the string of DNA are independent, and that the only change that has occurred over time is that which is observed. As with the Hamming distance or the Jukes-Cantor model and Jukes-Cantor correction, in general, evolutionary distances incorporate information differences or distinctions between sequences s_x and s_y, and hence give rise to a so-called *dissimilarity map*, to be defined precisely further below. Sequences which are the same provide no new information; it's the sense in which they are dissimilar that shows change and hence evolution.

10.3.2 Fitting Trees: A Distance-Based Approach

With the previous comments in mind, one approach (a "distance-based approach") to solving the central problem of phylogenetic tree reconstruction is to

1. start with the set of (aligned) sequences s_x, $x \in X$,
2. use an evolutionary model to derive a measure of evolutionary distance $d(x, y)$ between x and y on the basis of the information encoded in s_x and s_y, and then
3. use the numerical data $d(x, y)$, $x, y \in X$, to try to reconstruct a phylogenetic tree T on X which is consistent with this data.

Each of the three steps above is highly significant; for this chapter, we will be focusing on the last step. In particular, since real sequence data contains "noise," it is not usually possible to find a tree which "fits" the data exactly, so in the last step above it becomes necessary to look for a tree T on X which "best fits" the data $d(x, y)$, $x, y \in X$, and we will be exploring some ways to do this in much greater detail. As preparation, we want first to return to some mathematical formulations that will be instrumental in what follows. We start with the definition of a dissimilarity measure (also called a dissimilarity map).

Mathematically, for any finite set X, a *dissimilarity map* $D = [d(x, y)]_{x,y \in X}$ on X is simply a real-valued ($d(x, y) \in \mathbb{R}$ for all $x, y \in X$) symmetric ($d(x, y) = d(y, x)$ for all $x, y \in X$) square matrix with diagonal entries zero ($d(x, x) = 0$ for all $x \in X$).

Exercise 10.9. Assign values to the as-yet unspecified entries $d(i, j)$ of the matrix $D = [d(i, j)]_{i,j \in [4]}$ below so as to make D a dissimilarity map on $X = [4]$.

$$D = \begin{bmatrix} d(1, 1) & 13 & 3 & -13 \\ 13 & 0 & d(2, 3) & 2 \\ d(3, 1) & \pi & d(3, 3) & d(3, 4) \\ -13 & 2 & d(4, 3) & 0 \end{bmatrix} \qquad \triangledown$$

Exercise 10.10. (Just for fun; the result will not be used or needed in the rest of this chapter.) For those familiar with the linear algebraic interpretation of a real-valued n by n square matrix A as a linear transformation $T_A : \mathbb{R}^n \to \mathbb{R}^n$, via the matrix product $T_A(v) = Av$, for $v \in \mathbb{R}^n$ represented as an n by 1 column vector, what is the effect of applying T_D for D a dissimilarity map, to the standard bases vectors $\mathbf{e}_1, \ldots, \mathbf{e}_n \in \mathbb{R}^n$? $\qquad \triangledown$

As matrices, dissimilarity maps show up frequently in mathematics. For example, by the polar decomposition theorem, every real-valued square matrix M can be expressed as a sum of a symmetric and a skew-symmetric matrix via $M = \frac{1}{2}(M + M^T) + \frac{1}{2}(M - M^T)$, so if M begins as a zero-diagonal matrix, then the symmetric component, $\frac{1}{2}(M + M^T)$, will be a dissimilarity map. As another example, a map $\delta : \mathbb{R} \times \mathbb{R} \to \mathbb{R}$ is called a *metric* if $\delta(x, x) = 0$, $\delta(x, y) = \delta(y, x)$, and $\delta(x, z) \leqslant \delta(x, y) + \delta(y, z)$ for all $x, y, z \in \mathbb{R}$. The last inequality is the well-known *triangle inequality*, familiar from Euclidean geometry; think of points x, y, and z as lying on the vertices of a triangle. Recall that, for $x \in \mathbb{R}$, $|x|$ is the distance from x to 0 on the real line, and $|x - y| = \sqrt{(x - y)^2}$ is the distance between any two real numbers on the number line. Likewise, for two points $x = (x_1, x_2)$, $y = (y_1, y_2)$ in the plane $\mathbb{R}^2$, setting $d(x, y) = \sqrt{(x_1 - y_1)^2 + (x_2 - y_2)^2}$ gives the usual distance between the two points (length of line segment joining x and y). For two points $x = (x_1, x_2, x_3)$ and $y = (y_1, y_2, y_3)$ in three-space $\mathbb{R}^3$, by reducing to a picture with two triangles, each in their own plane, it is not hard to generalize the Pythagorean theorem and show that setting $d(x, y) = \sqrt{(x_1 - y_1)^2 + (x_2 - y_2)^2 + (x_3 - y_3)^2}$ gives the standard distance between the two points (length of line segment joining x and y) in $\mathbb{R}^3$. Distances can take on other forms, but going back to general metrics, we have, by definition, that they are dissimilarity maps with the triangle inequality as an additional property. Metrics are prominent in the study of mathematical real analysis, as they generalize distance measures like the absolute value, as we have just seen.

Exercise 10.11.

a. Give an argument to show that the standard distance map $d(x, y) = \sqrt{(x_1 - y_1)^2 + (x_2 - y_2)^2}$ for $x, y \in \mathbb{R}^2$ truly is a metric.

b. Given the comments above, generalize the distance $D = [d(x, y)]$ given by the Pythagorean theorem for $\mathbb{R}^2$ and its analog in $\mathbb{R}^3$ to points x, y with four coordinates, and then five. It is a very important point that the notion of the Cartesian coordinate plane $\mathbb{R}^2$ generalizes to spaces $\mathbb{R}^4$, $\mathbb{R}^5$, and so on for $\mathbb{R}^n$, for any integer $n \geqslant 1$, where these spaces exist as collections of all the ordered n-tuples with n entries drawn from $\mathbb{R}$.

c. If you have not (or not recently) seen a proof of the Pythagorean theorem for points in $\mathbb{R}^3$, look for one in any linear algebra book, multivariable calculus text, or find an online source. Generalizing to $\mathbb{R}^n$ for any positive integer $n \geqslant 1$ is a common exercise in courses on linear algebra and advanced calculus. For a suggestive picture, see http://en.wikipedia.org/wiki/File:Cube_diagonals.svg.   ▽

Turning back to biology, starting with a labeled X-tree $T = (V, E)$, $X \subset V$ (T rooted or unrooted) with edge-weighting $\omega : E \to \mathbb{R}$, a dissimilarity map $D_{T,\omega}$ arises naturally. As demonstrated in the example below, simply sum up the edge-weightings $\omega(e)$, over all distinct edges e on the path $\mathcal{P}_{x,y}$ in T joining any two leaves x, y in X, to get each value $d_{T,\omega}(x, y) := \sum_{e \in \mathcal{P}_{x,y}} \omega(e)$.

Example 10.5. For the quartet tree T with leaves $X = \{a, b, c, d\}$ and the edge-weighting ω indicated above (see Figure 10.4),

$$D_{T,w} = \begin{bmatrix} d_{T,w}(a, a) & d_{T,w}(a, b) & d_{T,w}(a, c) & d_{T,w}(a, d) \\ d_{T,w}(b, a) & d_{T,w}(b, b) & d_{T,w}(b, c) & d_{T,w}(b, d) \\ d_{T,w}(c, a) & d_{T,w}(c, b) & d_{T,w}(c, c) & d_{T,w}(c, d) \\ d_{T,w}(d, a) & d_{T,w}(d, b) & d_{T,w}(d, c) & d_{T,w}(d, d) \end{bmatrix}$$

$$= \begin{bmatrix} 0 & 2.3 & 1.6 & 3.6 \\ 2.3 & 0 & 1.5 & 2.9 \\ 1.6 & 1.5 & 0 & 2.8 \\ 3.6 & 2.9 & 2.8 & 0 \end{bmatrix}. \tag{10.1}$$

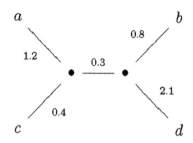

FIGURE 10.4

Edge-weighted quartet tree T on $X = \{a; b; c; d\}$; for Example 10.5

Exercise 10.12.

a. For the tree T and edge-weighting ω as in Example 10.5, for each set of three elements $i, j, k \in X$, calculate $d_{T,\omega}(i, k) + d_{T,\omega}(k, j)$ and compare it to $d_{T,\omega}(i, j)$. For example, for $i = a$, $j = b$, and $k = d$, $d_{T,\omega}(i, k) + d_{T,\omega}(k, j) = 2.5 + 2.5 = 5$, while $d_{T,\omega}(i, j) = 3$. How do these compare? What happens for the other combinations, including when two or more of i, j, k are equal? Interpret your observations graphically.

b. Starting again with the notation of Example 10.5, for the fixed choices $i = a$, $j = b$, $k = c$, $l = d$, calculate the quantities $d_{T,\omega}(i, j) + d_{T,\omega}(k, l)$, $d_{T,\omega}(i, k) + d_{T,\omega}(j, l)$, $d_{T,\omega}(i, l) + d_{T,\omega}(j, k)$. What do you notice? Does your conclusion change if you take other choices of $i, j, k, l \in X$?

c. For any quartet tree (that is, any unrooted tree with four leaves) for which i and j are sisters (cherries, here), use a graphical interpretation of relevant paths along the tree to argue that

$$d_{T,\omega}(i, j) + d_{T,\omega}(k, l) < d_{T,\omega}(i, k) + d_{T,\omega}(j, l),$$

and

$$d_{T,\omega}(i, k) + d_{T,\omega}(j, l) = d_{T,\omega}(i, l) + d_{T,\omega}(j, k),$$

for any positive edge-weighting ω on T.   ▽

The *four-point condition* for a dissimilarity map D on a set X is satisfied if

$$d(u, v) + d(x, y) \leqslant max\{d(u, x) + d(v, y), d(u, y) + d(v, x)\},$$

for any $u, v, x, y \in X$, where by definition the right-hand side of the inequality is the largest ("max") of the two sums indicated.

Exercise 10.13.

1. Show that any quartet tree satisfies the four-point condition.

2. If T is a phylogenetic X-tree, for any set X with $|X| \geqslant 4$, and with any positive edge-weighting ω and induced dissimilarity map $D_{T,\omega}$, show that the four-point condition holds for T.   ▽

Returning to an arbitrary labeled X-tree $T = (V, E)$, $X \subset V$ with edge-weighting $\omega : E \to \mathbb{R}$, we have, by assumption, that there's no edge between a leaf and itself, so $d_{T,\omega}(x, x) = 0$ for any $x \in X$. This is consistent with notions of evolutionary distance: there's zero dissimilarity between any sequence s_x and itself. It is also presumed that the information encoded by evolutionary distance measures on pairs of sequences s_x, s_y is independent of the ordering of the two (i.e., it does not matter which is "first" or "second"). From Exercise 10.8, one can see explicitly that the evolutionary distances dJ and dH are dissimilarity maps.

We have seen that if one begins with an edge-weighted X-tree T, using the tree and the weighting, one can use the edge labels to find a value $d_{T,\omega}(x, y)$ for each pair of leaves $x, y \in X$, that has an interpretation as a "distance," namely, the path length between x and y along the tree T, if the "length" of an edge e is taken to be its edge weight $\omega(e)$. The resulting matrix of all values $D_{T,\omega} = [d_{T,\omega}(x, y)]_{x,y \in X}$ is a dissimilarity map. Again, the fundamental biological problem of phylogenetics is to start the other way around—from a relevant set X (species, genes, etc. or sequences standing in for the species, genes, and so on) and a collection $\{d(x, y)_{x,y \in X}\}$ of relevant "distances," find a tree T and an edge-weighting ω for which the natural dissimilarity map $D_{T,\omega}$ fits the data $D = [d(x, y)]_{x,y \in X}$ "well." This is what is meant by "reconstructing a phylogenetic tree" from the given data, using a distance-based approach. In the most ideal case of "fitting well," $D_{T,\omega}$ fits D exactly, that is, they agree as functions: $d_{T,\omega}(x, y) = d(x, y)$ for all $x, y \in X$. If one begins with a phylogenetic X-tree T, and a nonnegative edge-weighting ω for T, and takes as data D by setting $D = D_{T,\omega}$, then trivially $D_{T,\omega}$ fits D exactly. This raises the issue of *consistency* in tree reconstruction methods, that is, whether a tree reconstruction method applied to data D that is derived from a tree T (perhaps weighted, with weight ω) actually outputs this tree T (and the corresponding weights ω). One can also speak more generally of *statistical consistency* of a tree reconstruction method, namely, the probability that the method outputs the correct tree given sufficient data about the input.

Exercise 10.14.

1. For the quartet tree T with cherry $\{x, w\}$ having parent node u and cherry $\{y, z\}$ having parent node v, and for the representative sequences on the leaves $X = \{x, y, w, z\}$ below, if $D = dH$, can you find an edge-weighting ω for T so that $D_{T,\omega} = D$?

$$s_w = GATTTCCTTC, s_x = GACATACTTC,$$
$$s_Y = GATTACATTC, s_z = GATTAAACTTC.$$

2. In the basic parsimony method of tree reconstruction, cherries are selected by linking nodes for which dH is minimal. For the same quartet tree T with cherry $\{x, w\}$ having parent node u and cherry $\{y, z\}$ having parent node v, suppose $d_H(u, v) = 1, d_H(u, w) = 1 = d_H(v, z)$, and $d_H(x, u) = 5 = d_H(v, y)$, with corresponding edge weight ω. (That is, if $e = (i, j), i, j \in V = \{x, y, w, z, u, v\}$, is an edge, we let $\omega(e) = d_H(i, j)$.)

a. Would the parsimony method applied to T with ω reconstruct this tree? Explain your answer.

b. At least how many characters long would the sequences $s_i, i \in X$, have to be to create a tree T with the given edge-weightings?

3. If D is a dissimilarity map on X for $|X| = 4$, is D necessarily equal to $D_{T,\omega}$ for some choice of quartet T and some edge-weighting ω?   ▽

10.3.3 Tree Metrics and Tree Space

Dissimilarity maps D for which there is *some* labeled tree T and *some* nonnegative edge-weighting ω for T such that $D = D_{T,\omega}$ are called *tree metrics*.

Exercise 10.15.

a. Show that a tree metric is necessarily a metric.

b. Show that every tree metric satisfies the four-point condition.   ▽

These results lead, in part, to a fundamental characterization theorem of tree metrics:

Theorem 10.1.

1. *A dissimilarity map D on a set X is a tree metric if and only if D satisfies the four-point condition.*

2. *If a dissimilarity map D on a set X is a tree metric for some ω edge-weighted tree $T = (V, E)$ on X, and likewise for an X-tree $T' = (V', E')$ and edge-weighting ω' on T, then T and T' are isomorphic phylogenetic trees. Moreover, for $\varphi : T \to T'$ an isomorphism, for any edge $e \in E$, if $e' = \varphi(e)$, then $\omega'(e') = \omega(e)$.*

Part (2) of Exercise 10.15 gives one half of part (1) of Theorem 10.1, showing how to recognize tree metrics out of all possible dissimilarity maps; for a proof, see, e.g., [2, Theorem 7.2.6]. Part (2) of Theorem 10.1 shows that any dissimilarity map which is a tree metric for a phylogenetic X-tree not only comes from such a tree, but corresponds to essentially one and only one pair (T, ω) of a phylogenetic X-tree and edge-weighting ω on T; see [2, Theorem 7.1.8] and its proof. Part (2) of Theorem 10.1 implies that tree metrics on X correspond exactly to edge-weighted phylogenetic trees on X.

If a given dissimilarity map D on a set X is in fact a tree metric, then by definition, there is a tree and an edge weighting that fits D exactly. As previously noted, in general, evolutionary distance data D from real sequence data for species, genes, etc., is noisy, and in our new language, although D will be a dissimilarity map, D will usually not already be a tree metric. Consequently, one looks for ways to find a phylogenetic X-tree T with edge-weighting ω for which $D_{T,\omega}$ and D will be "close" in some specified way, and in this sense, say a tree with a "best fit" has been found. To discuss "closeness," it would be helpful to have a notion of distance that could be used to compare dissimilarity maps D coming from sequence data with induced tree metrics $D_{T,\omega}$. Having discussed previously how to find distances between two points on a line $\mathbb{R}$, in the plane $\mathbb{R}^2$, or in three space $\mathbb{R}^3$, and more generally in $\mathbb{R}^n$ for $n \geqslant 1$ by means of a standard distance (via Pythagorean theorem analog), we might be motivated to seek a way to regard both D and $D_{T,\omega}$ as points, and then seek a distance between those points that we would then try to minimize in order to find an edge-weighted tree T that "best fits" the original data. Taking that step is our next focal point.

Since any matrix $M = [m_{i,j}]$ is really just a nicely ordered list of elements (i.e., listed by row and cross-listed by column), there is already an immediate way to regard matrices as points (vectors): just start writing out the entries of the matrix across row 1; when you get to the end, start adding the elements of row 2, and so on, until all elements are listed. For example, if M is a 4×4 matrix, then there is a correspondence of M and a vector with $16 = 4 \cdot 4$ entries:

$$M = [m_{i,j}] \quad \longleftrightarrow \quad (m_{1,1}, m_{1,2}, m_{1,3}, m_{1,4}, m_{2,1}, m_{2,2}, m_{2,3},$$
$$m_{2,4}, m_{3,1}, \ldots, m_{3,4}, m_{4,1}, \ldots, m_{4,4}). \qquad (10.2)$$

Other orderings of the elements of M to create a 16-tuple could be taken, of course, but as long as we are consistent this has no effect on our definition of distance.

Given the symmetry and existence of all zero-diagonal elements of any dissimilarity map $D = [d(x, y)]_{x,y} \in X$, there is a lot of redundancy in the entries; in fact, if $|X| = n$, then for the non-zero entries of D, there are at most $m = \frac{1}{2}(n^2 - n)$ distinct entries $d(i, j)$ in the n by n matrix D. Starting with D, it is straightforward to create a streamlined vector v_D in $\mathbb{R}^m$ carrying the same information as D, by choosing an ordering for reading off, and then listing in that order, the elements of D above the diagonal. This is illustrated for one choice of such an ordering below, for D as in Eq. (10.1):

Associated vector $v_D = (2.3, 1.6, 3.6, 1.5, 2.9, 2.8)$
$$\longleftrightarrow (d_{T,w}(a, b), d_{T,w}(a, c), d_{T,w}(a, d), d_{T,w}(b, c), d_{T,w}(b, d), d_{T,w}(c, d)).$$

Exercise 10.16. We have just seen how to take any n by n dissimilarity map D and, having fixed an ordering of elements in D, turn it into a point (vector) $v_D \in \mathbb{R}^m$, for $m = \frac{1}{2}(n^2 - n)$. Starting from any vector v in $\mathbb{R}^m$, explain how to reverse this process to obtain an n by n dissimilarity map D_v. Apply this to the particular case $v = (1.1, 0, 3.3, 2.5, 1.4, 5)$. Is D_v a tree metric? $\triangledown$

With this adjustment, it is now possible to represent any two n by n dissimilarity maps D, D' by vectors $v_D, v_{D'} \in \mathbb{R}^m$, where $m = \binom{n}{2} = \frac{1}{2}(n^2 - n)$. One can use the

standard Euclidean distance measure $\delta(D, D') = \sqrt{\Sigma_{1 \leqslant i < j \leqslant n}(d(i, j) - d'(i, j))^2}$ to measure a distance between D and D' in $\mathbb{R}^m$. In particular, suppose D is a dissimilarity map coming from evolutionary distances between elements in a set X (or their representative sequences). Suppose $D' = D_{T,\omega}$ for an ω edge-weighted phylogenetic X-tree T (with $|X| = n$). Then one can phrase the problem of finding a (n edge-weighted) phylogenetic tree T that "best fits" the observed data D on X as one of finding a pair $(T, \omega) \in \mathcal{I}_n$ so that $\delta(D, D_{T,\omega})$ is as small as possible. Equivalently, one may take a *least-squares approach*, since minimizing $\delta(D, T_{T,\omega})$ is the same as minimizing $\Sigma_{1 \leqslant i < j \leqslant n}(d(i, j) - d_{T,\omega}(i, j))^2$. When there is an exact fit, $\delta = 0$.

To summarize:

- For a set X, $|X| = n$, each positive edge-weighted phylogenetic X-tree (T, ω) gives rise to a tree metric $D_{T,\omega}$.
- For a fixed ordering of reading off upper diagonal elements in the matrix $D_{T,\omega}$ one obtains a vector $v_{D,T,\omega} \in \mathbb{R}^m$ $m = \binom{n}{2}$.
- There are one-to-one correspondences between

   - pairs $(T, \omega) \in \mathcal{I}_n$ of trees $T \in \mathcal{T}_n$ and positive edge-weightings,
   - the n by n dissimilarity maps D that are tree metrics $D_{T,\omega}$, and
   - the vectors $v_{D,T,\omega} \in \mathbb{R}^m$.

- In this way, edge-weighted phylogenetic X-trees become points in $\mathbb{R}^m$, and matching such an edge-weighted tree to a dissimilarity map D arising from evolutionary distances on sequences corresponding to a set X of species (or genes, etc.) becomes a problem of minimizing distances of the vectors $v_{D,T,\omega}$ and v_D in $\mathbb{R}^m$.
- Regarding positively (or nonnegatively) edge-weighted trees (T, ω) with $T \in \mathcal{T}_n$ as a subset of points $D_{T,\omega} \in \mathbb{R}^m$, the goal is, when presented with a point $D \in \mathbb{R}^m$, to seek one of these points $D_{T,\omega}$ as close to D in $\mathbb{R}^m$ as is possible.

In this way, a fundamental problem of biology is turned into a geometric problem. In particular, one wants to understand better the set $\mathcal{I}_n = \{(T, \omega) \mid T = (V, E) \in \mathcal{T}_n, \omega : E \to \mathbb{R}^+\}$ regarded as a set of points $\mathcal{I}_n \subset \mathbb{R}^m$. The set $\mathcal{I}_n$ is often called "tree space," though there is one for each n.

Example 10.6. If $n = 3$, then there is only one tree topology, and there is only one (unrooted binary) phylogenetic tree $T \in \mathcal{T}_3$, but infinitely many edge-weightings ω for any choice of T. More precisely, for each such T, there are three edges, and for each edge e, $\omega(e)$ can take all positive real values $\mathbb{R}^+$, independent of the other edges. The edge-weighting ω on T is completely described by the ordered triple $(\omega(e), \omega(g), \omega(h))$. In this way, geometrically, the significance of a choice of edge-weighting ω for T is that ω describes a point in the positive orthant $\mathbb{R}^{+3}$ of $\mathbb{R}^3$ associated to T. Taking all possible positive weightings on ω on T corresponds precisely to the full positive orthant $\mathbb{R}^{+3}$ for T. For $n = 4$, for any $T \in \mathcal{T}_4$, there are $(2n - 3) = (8 - 3) = 5$ edges and $(2n - 5)!! = (8 - 5)!! = 3$ phylogenetic X-trees $T \in \mathcal{T}_4$. The pairs (T, ω) correspond to points in three copies of $\mathbb{R}^{+5}$. (For those not so familiar with high-dimensional geometry, this is analogous to the way in which, for every point in time, there is a three-dimensional copy of space at that point in time, only here the finitely many trees $T \in \mathcal{T}_n$ play the role of selecting out just finitely many points in time to consider, and the $\mathbb{R}^{+5}$ are subsets of -dimensional space, rather than 3-dimensional space.) In more mathematical terms, $\mathfrak{T}_4$ lives nicely in a product vector space $\mathcal{T}_4 \times \mathbb{R}^5$, and looks like three copies of (the positive orthants of) $\mathbb{R}^5$. More generally, for $n \geqslant 3$, $\mathfrak{T}_n$ is the product set $\mathcal{T}_n \times \mathbb{R}^{+s}$, of $(2n-5)!!$ many copies of $\mathbb{R}^s$, where $|\mathcal{T}_n| = (2n-5)!!$ and $s = |E| = 2n-3$, both as per Exercise 10.4.

The treatment of the tree space $\mathfrak{T}_n$ is handled somewhat differently by different authors, to take advantage of simplifications in representing trees (and to use rooted trees in some cases, unrooted in others). For instance, if one extends to include nonnegative edge-weightings (that is, some values $\omega(e)$ can be zero), then one may include positively weighted, (unrooted) nonbinary trees on n leaves as part of tree space, and then compare how edge-weighted binary n-trees are related to one another through the collapsing of one or more edges. For an original definition of tree space and helpful pictures of this sort in the rooted tree case, as well as a discussion of how the positive orthants are "glued" together along spaces corresponding to (nonbinary) trees with collapsed edges, see [18]; for another analogous but different perspective on the unrooted case, also with nice pictures, see [19]. Using another variation on tree space, there has been a lot of work done to examine distances, particularly geodesics, in tree space; see [18,20,21]. Regardless of the precise formulation of tree space, as the example and brief treatment above may geometrically suggest, it can be useful to group points (T, ω) by their phylogenetic trees T or even by their underlying topologies. This is, again, the perspective that for each $T \in \mathcal{T}_n$ there is the "same" copy of $\mathbb{R}^s$ attached to it, so to determine a point in $\mathfrak{T}_n$, it is often useful to focus on determining just the tree $T \in \mathcal{T}_n$ alone, and then think of the associated weighting ω, even though the two are linked. (Again, this is analogous to thinking of a point in our everyday four-dimensional space-time lives first by its time coordinate, then as a point in space at that time.)

In any case, for practical purposes, however, the notion of standard distance in $\mathbb{R}^m$ may not be the best distance measure for comparing trees and fitting trees to dissimilarity maps. In fact, it is shown in [22] that taking a least-squares approach to picking an edge-weighted tree T with weighting ω and induced metric $D_{T,\omega} = [d_{T,\omega(i,j)}]$ to best fit a given dissimilarity map $D = [d(i, j)]$ is an NP-complete problem. The same holds true for minimizing $\Sigma_{1 \leqslant i < j \leqslant n}|d(i, j) - d_{T,\omega}(i, j)|$ (the f-statistic) rather than the corresponding sum of squares.

In the search for heuristics to carry out tree reconstruction, many fitting algorithms proceed instead by wandering among the trees (points) in tree space, and rather than use a least-squares or f-statistic approach, use metrics that record points $v_{T,\omega}, v_{T',\omega'}$ as being "close" when their trees $T, T' \in \mathcal{T}_n$ (edge-weighted by ω, resp., ω') can be transformed into one another explicitly by means of a small number of iterations of a few well-known operations on trees as graphs. In the case of Robinson-Foulds (RF) topological distance, the two relevant operations are a form of merging nodes and then splitting nodes on the underlying tree topologies (ignoring weightings), as illustrated in the picture below for two trees of RF-distance apart (see Figure 10.5).

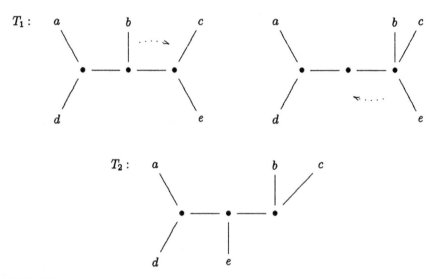

FIGURE 10.5

Trees T_1 and T_2 of RF distance one apart

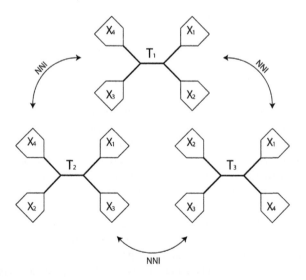

FIGURE 10.6

Generic NNI moves; each tree is related to the other by a single NNI move

Other well-known operations in tree space $\mathcal{T}_n$ include the nearest-neighbor interchange (NNI move) the subtree-pruning-and-regrafting move (SPR move), and the tree bisection and reconnection move (TBR move). A generic NNI move is illustrated in Figure 10.6 above, where each triangle represents a *clade*, that is, a subtree arising from a node and all its descendants and edges connecting them; here it could be a single leaf. For clades A, B, C, D, an NNI move is essentially a move on a quartet, exchanging the neighbor of one cherry for a neighbor of the other cherry (see Figure 10.6). The SPR moves and TBR moves are more general but can be similarly pictured; see, e.g., [2, Section 2.6] for definitions and representative pictures, or [23, Section 3] for a definition and picture of an SPR move relevant to our later discussion.

10.4 NEIGHBOR-JOINING AND BME

10.4.1 Neighbor-Joining

The Neighbor-Joining (NJ) Algorithm [16] takes as input a dissimilarity map $\mathbf{c} \in \mathbb{R}^{\binom{n}{2}}$ and builds an edge-weighted unrooted binary phylogenetic X-tree $(T, \omega) \in \mathfrak{T}_n$, by iteratively picking cherries out of X. There are three essential steps in the NJ Algorithm:

1. Pick a cherry $\{i, j\}$, for leaves $i, j \in X$.
2. Create a node $a_{i,j}$ joining leaves i and j.
3. Compute the distances from other nodes (including i, j and all other leaves in X) to the new node $a_{i,j}$ (and so produce a new dissimilarity map).

If $|X| > 3$, one then continues the procedure by eliminating $\{i, j\}$ from the set of leaves X (i.e., replace $\{i, j\}$ by $a_{i,j}$ in the set X) and seeking the next cherry in the remaining set of leaves to join, until the total number of remaining leaves n is 3. The result is an edge-weighted unrooted binary phylogenetic X-tree. Visually, one can proceed by picturing a star tree on $|X|$-leaves and seeing the NJ Algorithm as a set of instructions for successively splitting off leaves onto their own branches (see Figure 10.7).

The key component of the NJ Algorithm is the method for picking cherries, accomplished by the Q-criterion, described in Theorem 10.2 below. The formulation of the cherry-picking step of the NJ Algorithm, as a problem of minimizing the Q-values, comes from a reworking of [16] by [24].

Theorem 10.2. (Cherry-picking criterion (also known as Q-criterion) [16,24])

Let $\mathbf{c} \in \mathbb{R}^{\binom{n}{2}}$ be a tree metric for a tree $T \in \mathcal{T}_n$ with leaf set X, and define the $n \times n$-matrix $Q_{\mathbf{c}}$ with entries:

$$Q_{\mathbf{c}}(i, j) = (n-2)c_{i,j} - \sum_{k=1}^{n} c_{i,k} - \sum_{k=1}^{n} c_{k,j} = (n-4)c_{i,j} - \sum_{k \neq j} c_{i,k} - \sum_{k \neq i} c_{k,j}. \quad (10.3)$$

Then any pair of leaves $\{i^*, j^*\} \subset X$, for which $Q_{\mathbf{c}}(i^*, j^*)$ is minimal, is a cherry in the tree T.

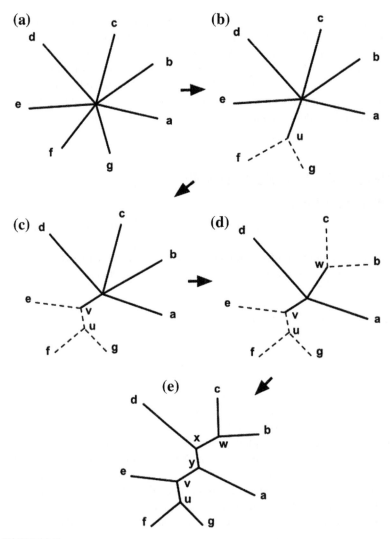

FIGURE 10.7

Pictorial representation of successive steps in the Neighbor-Joining Algorithm. From http://en.wikipedia.org/wiki/File:Neighbor-joining_7_taxa_ start_to_finish.png.

The NJ Algorithm is predicated upon the Q-criterion in the sense that *if* one starts with a tree metric, then the minimum value of Q gives a cherry, so if one is looking to build a tree from an arbitrary dissimilarity map, one might likewise try calculating the Q-values, and hope it works. As it turns out, this is practically and mathematically effective, yielding an algorithm with running time $O(n^3)$ [24], but in its current formulation, the rationale for the Q-criterion is nonobvious. The next exercise helps shed some light on the cherry-picking criterion, when $|X| = 4$.

Exercise 10.17. For the quartet $T = ((\texttt{A},\texttt{B}),(\texttt{C},\texttt{D}))$; as in Example 10.3, there are five edges (u, A), (u, B), (u, v), (v, C), (v, D). No matter what values an edge-weighting ω actually assigns to these five edges, if we use the tree metric $D_{T,\omega}$ in place of $\mathbf{c}$ in the cherry-picking criteria, then, using the first expression for Q in Theorem 10.2, the summands in the expression for the Q-values are basically just sums of path lengths along T. For example, the first term in $Q(A, B)$ is $(4 - 2)(\omega(u, A) + \omega(u, B))$, or twice the length of the path from A to B. The next summand subtracted off is the sum of three terms: the length of the path from A to B, the length of the path from A to C, and the length of the path from A to D. The last summand subtracted off is the sum of three terms just like the last, but starting from B and going to A, to C, and then to D. Thus the path lengths from A to B cancel out completely, and in what remains, each length from a leaf to its parent node (i.e., $\omega(A, u)$, $\omega(B, u)$, $\omega(C, v)$, and $\omega(D, v)$) appears twice, while the length $\omega(u, v)$ from u to v appears four times, and all are then multiplied by -1 to give $Q(A, B) = -2(\omega(A, u) + \omega(B, u) + \omega(C, v) + \omega(D, v)) - 4(\omega(u, v))$. (By tracing over the graph, one can even just graphically represent all the path pieces described above and visually see the cancellations/reduction to this expression for $Q(A, B)$.) To complete this exercise:

1. Do a similar analysis for $Q(A, C)$ and explain why $Q(A, C) > Q(A, B)$.
2. Explain why likewise $Q(A, D) > Q(A, B)$.
3. Observe by symmetric arguments that only two possible things can happen: $Q(A, B) < Q(i, j)$ for all other pairs $\{i, j\}$ in X OR $Q(A, B) = Q(C, D)$, and otherwise both are less than $Q(i, j)$ when $\{i, j\} \neq \{A, B\}$ or $\{C, D\}$.   ▽

Exercise 10.17 shows that the cherry-picking criteria at least makes sense for unrooted binary quartet trees, since there is only one topology to consider. There is a much stronger connection between quartets and the NJ Algorithm, as is given by the next theorem, shown, e.g., in [25].

Theorem 10.3. *If $|X| = 4$ and D is a dissimilarity map on X, then the NJ Algorithm returns a tree that satisfies the four-point condition.*

As suggested by the last part of Exercise 10.17, in general, while the NJ Algorithm will pick out the pair of cherries with minimal distance between them first, if there are two or more pairs of cherries with the same distance between them, any of these may be chosen first. Thus, more generally, the order of choosing cherries in the NJ Algorithm may not be unique (i.e., when multiple pairs of leaves have the same Q-values).

If the NJ Algorithm selects taxa $\{k, l\}$ as a cherry, and $a_{k,l}$ is the new node joining $\{k, l\}$, then the new dissimilarity map $\mathbf{c}' \in \mathbb{R}^{\binom{n-1}{2}}$ is defined to be

$$\text{if } i \neq a_{k,l} \neq j \qquad c'_{i,j} = c_{i,j}$$

$$\text{else} \qquad c'_{i,a_{k,l}} = \frac{1}{2}\left(c_{i,k} + c_{i,l} - c_{k,l}\right).$$

Exercise 10.18.

1. Apply the NJ Algorithm to the dissimilarity map D specified by the information given below. (Recall that any element marked $*$ can be inferred from the data given.) Take note of the order in which cherries are picked, and when more than one choice of cherry could be made.[3]

$$D = \begin{bmatrix} * & 5 & 4 & 7 & 6 & 8 \\ * & * & 7 & 10 & 9 & 11 \\ * & * & * & 7 & 6 & 8 \\ * & * & * & * & 5 & 9 \\ * & * & * & * & * & 8 \\ * & * & * & * & * & * \end{bmatrix}$$

2. The software package BIONJ is an improved version of the NJ Algorithm that generally produces the same trees when the number of leaves is small. Access BIONJ at http://www.phylogeny.fr, e.g., from the homepage under "Phylogeny" pop-up, located under the heading "Online Programs" in the bar near the top of the page.

   a. First, click on the option to "load an example of a distance matrix." Scroll down and click the "Submit" button to see the output. Click on the "Data and Settings" tab above the text box to return to the input dissimilarity map.
   b. Based on that experience, now replace the example data with the corresponding distance matrix data for D in part (1) of this problem, adjusted accordingly, and apply BIONJ. Use the leaf names "A, B, C, D, E, F," in that order, with A corresponding to the first row/column of the dissimilarity map D, and so on. If Java is supported, the button "Visualize your tree with ATV" can also be used for another view of the tree.
   c. For more practice with real species:
      i. Go to http://www.atgc-montpellier.fr/fastme/ fastme_ex.txt Three dissimilarity maps appear there. Try loading the first one into BIONJ. (Note: For implementation at http://www. phylogeny.fr you will need to put the species name and the corresponding numerical data for its row together on the same line.)
      ii. Predict, from the output, the first cherry that would be picked by the NJ Algorithm, and then verify your guess by computing the Q-criterion for the dissimilarity map (i.e., carry out the first step in the NJ Algorithm).
      iii. What species are being related, here? You may wish to make use of the search feature at http://tolweb.org/tree/. ▽

---

[3] Though tedious, carrying out these computations will provide the experience necessary to make sense of the algorithm. A tutorial for the first step, using an equivalent arithmetic formulation of the Q-criterion presented here, is given at http://www.icp.ucl.ac.be/~opperd/private/neighbor html.

Recall our geometric perspective that, for a given a fixed $n = |X|$, and a dissimilarity map (distance matrix) D, to "fit" a tree to D is to find a pair T, ω so that $D_{T,\omega}$ and D are "close." The NJ method is at least consistent: If a tree metric $D_{T,\omega}$ is given as the input D, then $D_{T,\omega}$ will be returned as the output.

Exercise 10.19. Suppose D is an $n \times n$ dissimilarity map ($n \geqslant 3$) and $D' = \lambda D$, for some positive constant λ, that is, $d'(i, j) = \lambda d(i, j)$ for all $1 \leqslant i, j \leqslant n$.

1. Explain why, in the first iteration of the Q-criterion for D and D', there will be the same choices of cherries to pick, i.e., if $\{i, j\}$ is a cherry for D then it is also one for D', and vice versa.
2. Continuing, suppose that if there is more than one choice of cherry to pick, then one picks the same cherry for both D and D'. Now explain why the distance from the new node joining this cherry for D' is λ times that for D.
3. If one always continues to pick the same cherries (whenever a choice is offered), explain why the trees returned by applying the NJ Algorithm to D and to D' will have the same underlying leaf-labeled tree topology. $\triangledown$

The NJ Algorithm constructs a weighted (binary) phylogenetic X-tree T, ω from a dissimilarity map D on X. As a tree, T has vertices V, with $X \subset V$, and the NJ Algorithm proceeds by constructing the set V, edges E, and edge-weights ω from the data D on X. An *order σ of picking cherries* in the NJ Algorithm when outputting the pair T, ω from D could be defined as a sequence of pairs $\sigma = (\{i_1, j_1\}, \{i_2, j_2\}, \{i_3, j_3\}, \ldots, \{i_r, j_r\})$ of elements $i_a, j_b \in V$, for which $\{i_k, j_k\}$ is the *k*th cherry to be picked. The first cherry $\{i_1, j_1\}$ is a subset of X, but the remaining pairs may involve interior vertices. For $|X| = n$ fixed, no matter what phylogenetic X-tree T is constructed, by Exercise 10.4, the number of vertices $|V|$ is always the same. Without loss of generality (that is, by relabeling the elements of V), it makes sense to speak of an *ordering of cherries* or *cherry-picking order* as a sequence of pairs of elements of $V = [2n - 2]$ (with, say, the elements of $X \subset V$ corresponding to $[n]$), for which the NJ Algorithm produces some binary phylogenetic X-tree T from some input dissimilarity map $D \in \mathbb{R}^m$, by picking cherries according to σ.

For a cherry-picking order σ on V, with $X \subset V$, $|X| = n$, and $|V| = 2n - 2$, for $m = \binom{n}{2}$, take C_σ to be the set of all dissimilarity maps D in $\mathbb{R}^m$ for which the NJ Algorithm applied to D proceeds using σ. (The NJ Algorithm also outputs a weighting ω on T, but that's not important for the definition of C_σ.) Applying Exercise 10.19, if $D \in C_\sigma$ then $\lambda D \in C_\sigma$ for any positive multiple λ. Thus, C_σ is a cone in $\mathbb{R}^m$. This cone is called a *neighbor-joining (NJ) cone*. The entire space $\mathbb{R}^m$ is the union of NJ cones, running over all cherry-picking orderings σ. Take C_T to be the set of all dissimilarity maps D in $\mathbb{R}^m$ for which the NJ Algorithm outputs the unrooted binary phylogenetic X-tree $T = (V, E) \in \mathcal{T}_n$ with some weighting ω. For each (binary) phylogenetic X-tree $T \in \mathcal{T}_n$, one has $C_T = \bigcup_\sigma C_\sigma$ for cherry-picking orders σ that are compatible with T. In some cases, a tree $T \in \mathcal{T}_n$ can arise from more than one ordering, and in that case, the boundaries of the corresponding NJ cones intersect. Neighbor-joining cones describe the NJ Algorithm geometrically—the output of the NJ on a point $D \in \mathbb{R}^m$ is a weighted tree with underlying topology T if and only if D is in the union of NJ cones C_T.

Example 10.7. For an unrooted binary phylogenetic X-tree $T = (V, E)$ on a set $X = \{i, j, k, \ell, b\}$ of five leaves, one has $|V| = 8$, and the orders of picking cherries in the NJ Algorithm applied to any dissimilarity map $D \in \mathbb{R}^{\binom{5}{2}}$ will consist of sequences of pairs from V. Necessarily, T has two pairs of cherries, say $\{i, j\}$ and $\{k, \ell\}$, one other leaf b, and three interior nodes $u, v, w \in V$. In the NJ Algorithm, for any order of picking cherries σ that yields T (with some weighting) the first cherry picked will be either $\{i, j\}$ or $\{k, \ell\}$. After this, if, say, u is the created ancestor node for $\{i, j\}$, then either $\{k, \ell\}$ is a cherry to be picked, or $\{u, b\}$ a cherry to be picked. Once either cherry is chosen, the other is completely determined for the next step, since there is only one (binary, unrooted) tree topology for $n = 5$. Thus, for $T \in \mathcal{T}_5$ and cherries $\{i, j\}$ and $\{k, \ell\}$, the NJ cones C_σ, and $C_{\sigma'}$ are the same, for σ starting with the subsequence $\{i, j\}, \{u, b\}, \{k, l\}$ and for σ' starting with $\{i, j\}, \{k, \ell\}, \{u, b\}$ or $\{i, j\}, \{k, \ell\}, \{v, b\}$, where v is the ancestor of $\{k, \ell\}$. Switching the order of $\{i, j\}$ and $\{k, \ell\}$ in σ and σ' does yield different orders and cones. Thus, for any binary phylogenetic X tree T on $|X| = 5$ leaves, there are two essentially distinct NJ cones which encompass all dissimilarity maps D that output T; in [26] these are labelled $C_{ij,b}$, and $C_{k\ell,b}$. Consequently, there are $30 = 15 \cdot 2$ total NJ cones for $n = 5$. For more details, see [26].

Neighbor-joining cones are defined and studied in detail for small trees ($n \leqslant 6$) in [26]. By studying and comparing the volumes of the NJ cones, [26] can geometrically formulate results about the behavior of the NJ Algorithm. These include how likely the NJ Algorithm is to return a particular tree topology given a random input vector (dissimilarity map), and the robustness of the NJ Algorithm. More complete information about these NJ cones, including information about "dimensions" of the cones, rays of the cones, faces, and other results, also appears in [27]. In both [27,23], geometric information is used to gain further insight into a comparison between the NJ Algorithm as a phylogenetic tree reconstruction method and another distance-based method, the balanced minimum evolution (BME) method, a topic we now explore.

10.4.2 Balanced Minimum Evolution

Recall that for any set $|X|$ with $|X| = n$, and any edge-weighted phylogenetic X-tree $(T, \omega) \in \mathcal{T}_n$, with $T = (V, E)$, there is a naturally associated tree metric $D_{T,\omega}$. One can set the *total length of the tree* to be the total sum of all the edge weights: $\omega(T) = \sum_{e \in E} \omega(e)$. The *minimum evolution principle* for tree reconstruction uses the idea of minimizing tree length in order to find the best-fitting tree $(T, \omega) \in \mathcal{T}_n$ to a given dissimilarity map $D \in \mathbb{R}^{\binom{n}{2}}$. Specifically, given a dissimilarity map $D \in \mathbb{R}^{\binom{n}{2}}$ as input, a minimum evolution method seeks to locate as output an edge-weighted tree (T, ω) so that $D_{T,\omega}(i, j) = d(i, j)$ for all $i, j \in X$, and $\omega(T)$ is as small as possible. Biologically, the minimum evolution principle is driven by the idea that evolution is efficient. Since branch lengths represent amount of evolutionary change leading from one, say, species to the next, the total tree length $\omega(T)$ is a measure of that efficiency across the whole tree, and hence should be minimal under a minimum evolution perspective.

There are many different minimum evolution methods; for all, it is necessary to start from D and come up with not only with the phylogenetic tree T, but also the edge weighting (branch lengths) ω. Since there are finitely many possible phylogenetic X-trees, one approach is, starting with any dissimilarity map $D \in \mathbb{R}^{\binom{n}{2}}$ and starting with any tree $T \in \mathcal{T}_n$, to come up with a branch length estimation scheme ω for T based on the data D. One then computes the resulting tree length $\omega(T)$ and, looking over all possible T, seeks T so that $\omega(T)$ in minimized.

One approach for producing ω from D, the Balanced Minimum Evolution (BME) method, uses an approach of Pauplin's [14]. Pauplin's scheme for branch length estimation from D, applied to the case of just a quartet, should now be very familiar to us from our previous exercises on the four-point condition and the NJ Algorithm. Specifically, if $e \in E$ is the interior edge joining u and v in the (unrooted) quartet tree $((A,B), (C,D))$; as in Example 10.3, then $\omega(e)$ is given by setting $\omega(e) = \frac{1}{4}(d(A,C) + d(A,D) + d(B,C) + d(B,D)) - \frac{1}{2}(d(A,B) + d(C,D))$. This idea is extended in [14] to the case when A, B, C, D are replaced by entire subtrees in an arbitrary tree $T \in \mathcal{T}_n$. The result gives an iterative formula for $\omega(e)$ for any interior edge in T that accounts for the relative size (say, number of leaves) in each subtree. (This is the "balance" of the BME method.) The (easier) case of edges that are external is also handled there. Rather than say more about these formulas, however, we instead make use of a very nice result by which the total tree length $\omega(T)$ can be calculated more easily, without first finding all the branch lengths $\omega(e)$.

Starting from $T \in \mathcal{T}_n$, define $w^T(i,j)$ to be one divided by the product of deg$(x) - 1$ for every interior node x encountered on the path $\mathcal{P}_{i,j}$ in T from i to j. Equivalently, set $y^T(i,j)$ to be the number of edges between leaves i and j on the path $\mathcal{P}_{i,j}$. In this case, we can see that $w^T(i,j) := 2^{1-y^T(i,j)}$. Now set

$$w^T := (w^T(1,2), w^T(1,3), \ldots, w^T(1,n), w^T(2,3), w^T(2,4), \ldots,$$
$$w^T(2,n), \ldots, w^T(n-1,n)) \in \mathbb{R}^{\binom{n}{2}}.$$

Notice that w^T only depends on the topology of T, so we may call it the *BME vector* of T.

Example 10.8. For $T \in \mathcal{T}_5$ shown in Figure 10.8, below, one obtains the BME vector

$$w^T = \left(\frac{1}{4}, \frac{1}{8}, \frac{1}{8}, \frac{1}{2}, \frac{1}{4}, \frac{1}{4}, \frac{1}{4}, \frac{1}{8}, \frac{1}{8}\right).$$

Exercise 10.20. Find the BME vectors w^T for each $T \in \mathcal{T}_4$.   ▽

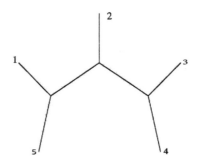

FIGURE 10.8

An example of $T \in \mathcal{T}_5$; see Example 10.8

Using the BME vectors, *Pauplin's [14] formula* for the balanced tree length esti-mation (i.e., estimated BME length) $\omega(T)$, starting with the dissimilarity map D, is given by

$$\omega(T) = \sum_{i,j:i<j} w^T(i, j)d(i, j). \tag{10.4}$$

Equivalently, in the standard language of innner products of vectors, upon rewriting D as a vector d using the ordering from before, one gets $\omega(T) = w^T \cdot d$. The BME method proceeds by seeking $T \in \mathcal{T}_n$ so that $\omega(T)$ in Eq. 10.4 is minimal.

Exercise 10.21. Using the results of Exercise 10.20 and the dissimilarity map from Example 10.5, use the BME method to find a tree $T \in \mathcal{T}_4$ that best fits D under Pauplin's branch length estimation scheme. Also find $\omega(e)$ for the interior edge e of T in this case as well. ▽

In [28], the authors in fact defined terms $w^T(i, j)$ for any phylogenetic X-tree T (not necessarily binary) in terms of certain cyclic permutations of ("circular orderings") of X that respect the structure of T. In the case of edge-weighted binary X-trees, one recovers the expression for $w^T(i, j)$ in the BME vector and Pauplin's formula. In [28] this perspective was used to establish the consistency of the balanced tree length estimation. The consistency (and statistical consistency) of the BME method as a whole was established in [29].

Like all methods that would require a complete search over all trees $T \in \mathcal{T}_n$, combinatorial explosion is a problem. The large number of trees to consider quickly overwhelms computer capacity, so that in practice, one must come up with a heuristic. In [29, 15], the theoretical underpinnings for the BME method were further examined, and a heuristic proposed and implemented for the BME method via the program FastME. This program essentially uses only nearest-neighbor interchange (NNI) moves among trees to seek out a tree T/BME vector w^T that is a good candidate for having minimal $\omega(T)$. For more details, see [29, 15].

Exercise 10.22. To access FastME, go to http://www.atgc-montpellier. fr/fastme/

1. Use the example file (under input data) to explore the option "Balanced GME." Note that one can click on "file" in "example file" to see the data, some of which will be familiar from previous exercises. Results will be e-mailed back to you in Newick format.
2. Review [15] for a discussion of the success of this implementation versus the NJ Algorithm.   ▽

Options in `FastME` also include implementations of the NJ Algorithm and its improved version, the BioNJ Algorithm. There is a special reason why. Although the NJ Algorithm was long a favorite among molecular and other biologists for its effectiveness in practice, after many years of investigation, reservations about the NJ Algorithm remained, due in part to an ongoing sense of uncertainty about just what the algorithm was doing (and hence why it could be successful), from a more theoretical perspective. Illumination came in the form of a result by [29] linking the BME method and NJ Algorithm by showing that the NJ Algorithm is a so-called greedy algorithm for implementing the BME principle. Since the BME method is well grounded in standard mathematical approaches (minimization problems), this link shed light on the "true" nature of the NJ Algorithm. As a greedy algorithm, the NJ Algorithm acts locally, however, and like any greedy algorithm, does not necessarily actually produce a tree T which is optimal for the BME principle (i.e., has total length $\omega(T)$ minimal).

The BME method for phylogenetic tree reconstruction can be reformulated still further in a neat geometric way. By definition, a convex set B (say $B \subset \mathbb{R}^m$) is one for which any point $z \in \mathbb{R}^m$ that is on any line ℓ joining any two points x, $y \in B$ is also in B, i.e., z on ℓ implies $z \in B$. It is an easy exercise to see that the intersection of any set of convex sets is also convex. Let $\mathcal{B}_n$ be the convex hull in $\mathbb{R}^{\binom{n}{2}}$ of all the BME vectors w^T for $T \in \mathcal{T}_n$; by definition, the convex hull of a set of points is the intersection of all convex sets containing these points. This produces a polytope (analog of a polygon in higher dimensional space), which we call the *nth BME polytope* $\mathcal{B}_n$, inside $\mathbb{R}^{\binom{n}{2}}$. As a consequence of the definition of the BME vectors and the BME method, the vertices of the BME polytope are actually the BME vectors w^T, $T \in \mathcal{T}_n$. Using the generalized notion of BME vectors as in [2], the BME vector of the star phylogeny lies in the interior of the BME polytope $\mathcal{B}_n$, and all other BME vectors lie on the boundary of the BME polytope.

Exercise 10.23.   Explain why $\mathcal{B}_4$ is just a triangle in $\mathbb{R}^6$.   ▽

By Exercise 10.23, the BME polytope for $n = 4$ is a familiar two-dimensional object. More generally, the dimension of $\mathcal{B}_n = \binom{n}{2} - n$. If $n \leqslant 6$, then thinking not of the full polytope in $\mathbb{R}^{\binom{n}{2}}$, but considering only vertices and edges of $\mathcal{B}_n$ alone, there is a graph isomorphism between $\mathcal{B}_n$ and the complete graph on n vertices.

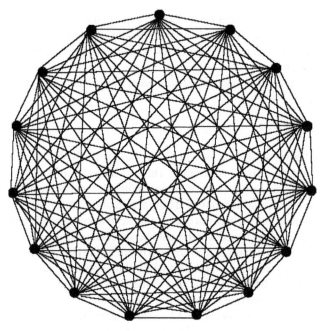

FIGURE 10.9

A representation of the BME polytope $\mathcal{B}_5$

Example 10.9.   For $n = 5$, $|\mathcal{T}_5| = 15$, so $\mathcal{B}_5$ is graph isomorphic to the complete graph on 15 vertices, shown below (see Figure 10.9).

These and many other facts about BME polytopes appear in [27,23]. For $n = 7$, [27] observed that in the actual polytope $\mathcal{B}_7 \subset \mathbb{R}^{\binom{7}{2}}$, the graph isomorphism with the complete graph on $n = 7$ vertices no longer holds: there are pairs of trees $T, T' \in \mathcal{T}_7$ for which the BME vectors are not connected by an edge of the polytope. That is, the line joining w^T and $w^{T'}$ is not itself an edge of the polytope, but must sit somewhere inside $\mathcal{B}_7$, since $\mathcal{B}_7$ is convex. See [p. 5][27] for a picture and description of these pairs, and other information about faces, facets, and more for small n. For $n > 7$, computational complexity prevents the hands-on geometric methods explored in [27] from being exploited for higher n.

Describing more fully the edges and other features of the the BME polytope could have practical implications for phylogenetic tree reconstruction. In the language of the well-known mathematical methods of linear optimization, the BME method becomes a linear programming problem: using the correspondence between $T \in \mathcal{T}_n$ and $w^T \in \mathbb{R}^{\binom{n}{2}}$, carrying out the BME method for D is the same as minimizing the linear objective d over the (convex) BME polytope $\mathcal{B}_n$. This framing of the problem motivated further theoretical studies of the BME polytope in [23], where a new version of a cherry-picking algorithm, this time for the BME method, was developed and used to show that both NNI moves and SPR (subtree-pruning-and-regrafting) moves between trees $T \in \mathcal{T}_n$ give rise to edges in the BME polytope. Since FastME uses NNI moves to heuristically implement the BME method, Haws et al. [23] implies that FastME is an edge-walk along the BME polytope. With better knowledge of the BME polytope, it might be possible to find an even faster heuristic. Results in [23] show that clades (subtrees obtained by taking all children of a parent node and the edges that join them) correspond to "faces" in the BME polytope and identify themselves with BME polytopes $\mathcal{B}_k$ values $k \leqslant n$ inside $\mathcal{B}_n$.

Finally, as it was for NJ cones, it is possible to define BME cones. For any $T \in \mathcal{T}_n$, a BME cone is simply the set of all $D \in \mathbb{R}^{\binom{n}{2}}$ for which $\omega(T)$ as in Eq. 10.4 is minimal. Again, geometrically, the BME cone describes the BME method, for the cone for T consists of all D for which the BME method applied to D produces T (with weighting ω). As before, $\mathbb{R}^{\binom{n}{2}}$ is partitioned into the BME cones, running over all $T \in \mathcal{T}_n$, although a given D may lie in more than one BME cone. A consequence in [23] of this analysis, in combination with previous work on the NJ cones, is a further refinement of the relationship between the relationship of NJ as a greedy algorithm for the BME principle. In mathematical lingo, for any cherry-picking order σ, the intersection of the NJ cone $C_{T,\sigma}$ and the BME cone for T has positive measure. In particular, this shows that for *any* order of joining nodes to form a tree, there is a dissimilarity map so that the NJ Algorithm returns the BME tree, a useful phylogenetic result. The performance of the NJ Algorithm and BME method can be further analyzed through studying these cones, and the implications of their differing geometries are explored further in [27,23]. A complete characterization of edges of the BME polytope in terms of tree operations remains an open question, however.

Also, in terms of further work, [17] studied how the BME method works when the addition of an extra taxon (e.g., species) to a data set alters the structure of the optimal phylogenetic tree. They characterized the behavior of the BME phylogenetics on such data sets, using the BME polytopes and the BME cones which are the normal cones of the BME polytope, as noted in [27]. A related study for another tree reconstruction method, UPGMA, was carried out in [30].

10.5 SUMMARY

Many open questions in the treatment given in this chapter of some distance-based methods for phylogenetic tree reconstruction, via geometric and computational perspectives, remain. For example, giving a complete characterization of edges of the BME polytope in terms of tree operations has yet to be done. Other opportunities for further research include expanding from trees to networks, since it is known that trees do not capture well several aspects of molecular biology, including hybridization. Currently, distance-based methods have been superseded (at least in common use by many biologists) by other approaches. These other approaches include maximum likelihood methods and Bayesian approaches, the latter of which produce not just a single tree, but a distribution of possible trees.

Still, there remain good reasons for using distance-based methods. We have, quite misleadingly used trees with small examples of leaves in this chapter in order to make the underlying ideas accessible, and because much of the underlying mathematical theory can be reduced to small trees, like quartets. However, in the "real world" of biology, the most obvious reason for continuing interest in distance-based methods is due to their effectiveness via good polynomial time algorithms in the face of combinatorial explosion and computational complexity when the number of leaves is large. Even $n > 20$ can overwhelm many other methods, but we have (also rather scandalously), with the exclusion of consistency, not gone into details in this chapter about comparing the success, via additional factors such as efficiency, robustness, power, and falsifiability, of distance-based algorithms and their competitors. (See [4, Section 6.1.4] for a brief but comprehensive treatment, up to its publication date.)

The link between the NJ Algorithm and the BME method described above, but not as well known as it might be, also addresses some issues which have perhaps mistakenly led biologists to pull back from using the NJ Algorithm or its variants. A very recent paper [13] also provides some thought-provoking arguments against another commonly voiced concern among biologists, that in providing only information between leaves, distance-based methods may lose (too much) information (e.g., see [4, Section 6.2.4] for a sample of such concerns in a biology-friendly text). Looking at covariances in the dissimilarity maps reveals more information that has commonly disregarded, given calculations like that for the Q-criterion which explicitly use only the values $d(i, j)$ for an input dissimilarity map D. Still, we hope this chapter's treatment will open the doors to these questions and issues, and to further readings in the articles and books referenced herein.

References

[1] Felsenstein J. Inferring phylogenies. Sinauer Associates, Inc.; 2004.
[2] Semple C, Steel M. Phylogenetics. Oxford lecture series in mathematics and its applications, vol. 24. Oxford: Oxford University Press; 2003.
[3] Pachter L, Sturmfels, B, editors. Algebraic statistics for computational biology. Cambridge University Press; 2005.
[4] Page RDM, Holmes EC. Molecular evolution: a phylogenetic approach. Blackwell Science Ltd; 1998.
[5] Turelli M, Barton NH, Coyne JA. Theory and speciation. Trends Ecol Evol 2001;16:330–43.
[6] Baum D. Reading a phylogenetic tree: the meaning of monophyletic groups. Nature Educ 2008;1.

[7] Wang L, Jiang T. On the complexity of multiple sequence alignment. J Comput Biol 1994;1:337–48.

[8] Elias I. Settling the intractability of multiple alignment. J Comput Biol 2006;13:1323–39.

[9] Desper R, Gascuel O. Theoretical foundation of the balanced minimum evolution method of phylogenetic inference and its relationship to weighted least-squares tree fitting. Mol Biol Evo 2004;2:587–98.

[10] Felsenstein J. Cases in which parsimony and compatibility methods will be positively misleading. Syst Zool 1978;27:401–10.

[11] DeBry RW. The consistency of several phylogeny-inference methods under varying evolutionary rates. Mol Biol Evol 1992;9:537–51.

[12] Denis F, Gascuel O. On the consistency of the minimum evolution principle of phylogenetic inference. Discrete Appl Math 2003;127:63–77.

[13] Roch S. Toward extracting all phylogenetic information from matrices of evolutionary distances. Science 2010;327:1376–9.

[14] Pauplin Y. Direct calculation of a tree length using a distance matrix. J Mol Evol 2000;51:41–7.

[15] Desper R, Gascuel O. Fast and accurate phylogeny reconstruction algorithms based on the minimum-evolution principle. J Comp Bio 2002;9:687–705.

[16] Saitou N, Nei M. The neighbor joining method: a new method for reconstructing phylogenetic trees. Mol Biol Evol 1987;4:406–25.

[17] Cueto MA, Matsen FA. Polyhedral geometry of phylogenetic rogue taxa; 2010. Preprint <arXiv:1001.5241>.

[18] Billera LP, Holmes SJ, Vogtman K. The geometry of space of phylogenetic trees. Adv Appl Math 2001;27:733–67.

[19] Nye T. Principle component analysis in the space of phylogenetic trees. Ann Stat 2011;39:2716–39.

[20] Owen M. Computing geodesic distances in tree space. SIAM J Discrete Math 2011;25:1506–29.

[21] Owen M, Provan JS. A fast algorithm for computing geodesics in tree space. IEEE/ACM Trans Comput Biol Bioinform 2011;8:2–13.

[22] Day W. Computational complexity of inferring phylogenies from dissimilarity matrices. Bull Math Biol 1987;49:461–7.

[23] Haws D, Hodge T, Yoshida R. Optimality of the neighbor joining algorithm and faces of the balanced minimum evolution polytope. Bull Math Biol 2011. doi:10.1007/s11538-011-9640-x.

[24] Studier JA, Keppler KJ. A note on the neighbor-joining method of Saitou and Nei. Mol Biol Evol 1988;5:729–31.

[25] Mihaescu R, Levy D, Pachter L. Why neighbor-joining works. Algorithmica 2009;54:1–24.

[26] Eickmeyer K, Yoshida R. The geometry of the neighbor-joining algorithm for small trees. Lecture notes in computer science. Springer-Verlag, vol. 1547; 2008. p. 81–95.

[27] Eickmeyer K, Huggins P, Pachter L, Yoshida R. On the optimality of the neighbor-joining algorithm. Algorithms Mol Biol 2008;3. <http://www.almob.org/content/3/1/5>.

[28] Steel M, Semple C. Cyclic permutations and evolutionary trees. Adv Appl Math 2004;32:669–80.

[29] Gascuel O, Steel M. Neighbor-joining revealed. Mol Biol Evol 2006;23: 1997–2000.

[30] Davidson R, Sullivant S. Polyhedral combinatorics of UPGMA cones; 2012. Preprint, <http://arxiv.org/abs/1206.1621>.

Index

Printed and bound by CPI Group (UK) Ltd, Croydon, CR0 4YY

03/10/2024

01040323-0002